Causal Analysis with Event History Data Using Stata

This third edition of *Causal Analysis with Event History Data Using Stata* provides an updated introduction to event history modeling along with many instructive Stata examples. Using the latest Stata software, each of these practical examples develops a research question, points to useful contextual background information, gives a brief account of the underlying statistical concepts, describes the organization of input data and the application of Stata statistical procedures, and assists the reader in interpreting the content of the results obtained.

Emphasizing the strengths and limitations of continuous-time event history analysis in different fields of social science applications, this book demonstrates that event history models provide a useful approach to uncover causal relationships or to map a system of causal relationships. In particular, this book demonstrates how long-term processes can be studied, how different forms of duration dependencies can be estimated using nonparametric, parametric and semiparametric models, and how parallel and interdependent dynamic systems can be analyzed from a causal-analytical point of view using the method of episode splitting. The book also shows how changing contextual information at the micro, meso and macro levels can be easily integrated into a dynamic analysis of longitudinal data. Finally, the book addresses the problem of unobserved heterogeneity of time-constant and time-dependent omitted variables and makes suggestions for dealing with these sometimes difficult methodological problems.

Causal Analysis with Event History Data Using Stata is an invaluable resource for both novice and experienced researchers from a variety of fields (e.g. sociology, economics, political science, education, psychology, demography, epidemiology, management research and organizational studies, as well as medicine and clinical applications) who need an introductory textbook on continuous-time event history analysis and who are looking for a practical handbook for their longitudinal research.

Hans-Peter Blossfeld, Prof., Dr. rer. pol. Dr. h. c., is Emeritus of Excellence at the Graduate Centre Trimberg Research Academy (TRAc) at the University of Bamberg in Germany since April 2020. He held the Chair of Sociology I at the Faculty of Social Sciences, Economics and Business Administration at the University of Bamberg and was Professor of Sociology at the European University Institute in Florence, Italy.

Götz Rohwer was Professor Emeritus of Methods of Social Research and Statistics at Ruhr-University Bochum in Germany. He passed away in March 2021.

Gwendolin J. Blossfeld is a Postdoc at the Faculty of Social Sciences, Economics and Business Administration at the University of Bamberg in Germany.

Causal Analysis with Event History Data Using Stata

Third Edition

Hans-Peter Blossfeld, Götz Rohwer and Gwendolin J. Blossfeld

With a Chapter on Sequence Analysis by Brendan Halpin

Routledge
Taylor & Francis Group
NEW YORK AND LONDON

Designed cover image: Getty Images

Third edition published 2026
by Routledge
605 Third Avenue, New York, NY 10158

and by Routledge
4 Park Square, Milton Park, Abingdon, Oxon, OX14 4RN

Routledge is an imprint of the Taylor & Francis Group, an informa business

First edition published by Routledge 2007
Second edition published by Routledge 2019

ISBN: 9781032708096 (hbk)
ISBN: 9781032657783 (pbk)
ISBN: 9781032708102 (ebk)

DOI: 10.4324/9781032708102

Typeset in Computer Modern Roman

Access the Support Material: www.routledge.com/9781032657783

Publisher's Note
This book has been prepared from camera-ready copy provided by the authors.

Contents

Preface

Event history analysis today can draw on a number of established statistical tools to describe and causally analyze continuous-time event history data. Since we want to emphasize that we discuss causal inference with event history models, we have changed the title of this third edition of our book to 'Causal Inference with Event History Data Using Stata'. This third edition provides an updated introduction to event history modeling along with many instructive Stata examples. Each of these practical examples develops a research question, points to useful contextual background information, gives a brief exposition of the underlying statistical concepts, describes the organization of input data and the application of Stata's statistical procedures, and assists the reader in interpreting the content of the results obtained. The book also demonstrates the power of event history analysis using many research applications from the economic and social science literature. The book is therefore aimed both at beginners who need an introductory textbook to continuous-time event history analysis as well as experienced researchers (from sociology, economics, political science, education, psychology, demography, epidemiology, medicine, and clinical research) who are looking for a practical handbook for their longitudinal research. With this large number of examples and computer exercises, our book should be particularly useful for undergraduate classes and graduate teaching.

We use the latest Stata software for our examples. Stata is now a widely used software package in the economic and social sciences that offers many useful tools for data analysis, data management and graphical visualization. For more information, we refer the reader to the Stata homepage.[1] The data used in our examples come from the German Life History Study (GLHS), but are anonymized for data protection reasons. The GLHS study was directed by Karl Ulrich Mayer at the Max Planck Institute for Human Development and Education in Berlin. The original data collection was funded by the German Research Foundation (DFG). We would like to thank Professor Mayer for his kind permission to use a sample of 600 job episodes from the GLHS as the basis for our practical examples. This third edition of our book also offers many exercises with data from the National Educational Panel Study (NEPS) in the appendix. The NEPS is a research infrastructure project that provides data on educational processes and trajectories in Germany with several starting cohorts. One cohort focusses on adults and has collected detailed retrospective information in the tradition of the GLHS. From 2008 to 2013, the collection of NEPS data was funded by the Federal Ministry of Education and Research (BMBF). The NEPS survey has been conducted since 2014 by the Leibniz Institute for Educational Trajectories (LIfBi) in Bamberg in cooperation with a Germany-wide network of educational experts (https://www.neps-data.de/tabid/2665/language/en-US). The NEPS data used in our book come from a non-representative sub-sample limited to western Germany, which was also completely anonymized by LIfBi using advanced strategies. We would like to thank the LIfBi Executive Board and Daniel Fuß, the head of the Research Data Center at the LIfBi for access to this subsample and especially Tobias Koberg from the LIfBi team for data preparation and anonymization. The GLHS and NEPS data files used throughout the book, the set of files containing the Stata do-files for the examples in the book and the appendix can be downloaded from `https://doi.org/10.5157/NEPS:Campus:1.0.0`. This gives the reader the unique opportunity to run and modify all the application examples in the book on a computer. We recommend that beginners in event history analysis work through the application examples in the book step by step on themselves. Based on our own teaching experience from many workshops and introductory courses, this seems to be the most efficient and easiest way to familiarize yourself with the sometimes complex analysis techniques.

Our book shows that event history models provide a useful approach to uncovering causal relationships or mapping a system of causal relationships. Event history models are naturally linked to a causal understanding of social processes, as they allow us to model a theoretically assumed change in future outcomes of a process in relation to its history and the (time-constant or changing) conditions of other related processes. In particular, this book demonstrates how long-term processes can be studied, how different forms of duration dependencies can be estimated using nonparametric, parametric and semiparametric models, and how parallel and interdependent dynamic systems can be analyzed from a causal-analytical point of view using the method of episode splitting. The book also shows how changing contextual information at the micro, meso and macro levels can be easily integrated into a dynamic analysis of longitudinal data. Finally, the book addresses the problem of unobserved heterogeneity of time-constant and time-dependent omitted variables and makes suggestions for dealing with these sometimes difficult methodological problems.

Depending on theoretical considerations regarding the process time axis and the degree of measurement accuracy of event times, a distinction is made in the literature between discrete-time and continuous-time event history models (see e.g. Tuma and Hannan 1984; Blossfeld, Hamerle and Mayer 1989; Mills 2011;

Blossfeld 2025). Discrete-time event history methods are either (1) based on a discrete process time axis (i.e., that events can only occur at fixed time intervals) or (2) assume a continuous process time axis (i.e., that events can occur at any point in time, but these event times can only be measured as a sequence of rather coarse time intervals). For example, in some national panel studies, events such as job changes are only recorded at annual intervals. Various publications on discrete-time models can be found in the literature (see, e.g., Allison 1982, 2014; Yamaguchi 1991; Singer and Willet 1993; Vermunt 1997; Rohwer and Pötter 2007; Steele 2008; Blossfeld 2025). In a companion book to this volume, Gwendolin J. Blossfeld has made a detailed comparison of continuous-time and discrete-time event history analysis using the examples in this book. This gives the reader the unique opportunity to compare the continuous-time event history models of this book with the corresponding discrete-time models based on different observation intervals in order to analyse how increasingly longer observation intervals affect the results of the estimated models. Continuous-time event history models, which are the focus of this book, assume that the timing of events is measured precisely. How precisely the timing of an event actually needs to be captured is largely a question of content. For long careers, for example, monthly measurements of job mobility seem to be sufficient. Of course, all temporal observations are ultimately discrete, but for continuous models they should be captured more accurately. More precise measurements of the temporal sequence of events lead to fewer tied observations in the data set and facilitate the reconstruction of the causal sequence of events from different interrelated or interdependent processes. In retrospective life history studies, for example, data is often collected at the level of one month. At this point, it should be noted that although discrete-time and continuous-time models use different statistical approaches, the results of these models are very similar in many specific applications, especially when the observation intervals are quite small (see Blossfeld 2025).

As in the second edition, this book also includes a chapter on sequence analysis (see Chapter 11), written by Brendan Halpin. Sequence analysis is an approach that treats the entire sequence of events as a unit and defines measures of similarity and difference between pairs of sequences. Sequence analysis can be used for several purposes in the social sciences, including (1) the data-driven classification of event sequences, (2) mapping sequences to empirical or theoretical ideal types, (3) describing how the diversity of sequences differs across groups, and (4) examining the determinants of dyadic similarity (e.g., the event sequences of marital or non-marital partners).

Acknowledgments. We have received support for our work from various sources and organisations. In particular, we received support for the second edition of this volume from the European Research Council (ERC) funded project 'Education as a Lifelong Process – Comparing Educational Trajectories in Modern Societies' (eduLIFE), which was carried out at the University of Bamberg, Germany, and the European University Institute in Florence, Italy. For the second edition of this book, we received further support from the Faculty of Social Sciences at the University of Bochum, Germany, and the Institute of Sociology at the University of Leipzig, Germany. Finally, we would like to thank Katrin Golsch, the third author of the first edition of our book, and Thorsten Schneider, the third author of the second edition of our book, for their expertise and support in applying Stata to event history analysis. In the third edition, Gwendolin J. Blossfeld, an expert in (continuous-time and discrete-time) event history modeling and the Stata software, has joined the authors. As the third edition builds on the second edition of this book, we would also like to thank the following people for their contributions to the second edition: Ulrich Pötter for helpful suggestions, Jonathan Harrow for thorough proofreading and linguistic revision of the earlier manuscript, and Thorsten Schneider, Till Hovestadt, Mirjam Odile Nanko and Fabienne Wöhner, all from the University of Leipzig, for the final formatting of the second edition. Finally, we would like to point out that we have once again used Donald Knuth's typesetting program TeX in conjunction with Leslie Lamport's LaTeX and Tomas Rokicki's DVIPS PostScript driver to create the artwork for our book.

Hans-Peter Blossfeld, University of Bamberg
Götz Rohwer, Ruhr University of Bochum
Gwendolin J. Blossfeld, University of Bamberg

Notes

[1] https://www.stata.com/

Chapter 1

Introduction

Over the last three decades, social scientists have increasingly collected and analyzed event history data. This is not a random trend, nor does it reflect a prevailing fashion in survey research or statistical analysis. Rather, it points to the growing realization among social scientists that event history data is often the most appropriate empirical information that can be obtained about the substantive process under study.

James Coleman (1981, p. 6) characterized this kind of substantive process as follows: (1) There is a collection of units (which may be individuals, organizations, societies, or whatever), each of which moves between a finite (usually small) number of states; (2) these changes (or events) can occur at any point in time (i.e., they are not restricted to predetermined points in time); and (3) there are time-constant and/or time-dependent factors influencing the events.

Illustrative examples of this type of substantive process can be given for a variety of research fields: In *labor market studies*, workers move between unemployment and employment,[1] full-time and part-time work,[2] or between different types of jobs;[3] in *social inequality studies*, people become a home-owner over the life course;[4] . in *demographic analyses*, men and women enter into consensual unions, marriages, or into father-/motherhood, or divorce;[5] in *sociological mobility studies*, workers move between different types of jobs, social classes, or industries over the course of their lives;[6] in *studies of organizational ecology and entrepreneurship*, companies, unions, or organizations are founded or closed down;[7] in *political science research*, governments break down, voluntary organizations are founded, or countries undergo a transition from one political regime to another;[8] in *migration studies*, people move between different regions or countries;[9] in *marketing applications*, consumers switch from one brand to another or purchase the same brand again; in *criminological studies*, prisoners are released and commit another criminal act after some time; in *communication analyses*, interaction processes such as interpersonal and small group processes are studied;[10] in *educational studies*, students drop out of school before completing their degrees, enter into a specific educational track, or start a program of further education later in life;[11] in *analyses of ethnic conflict*, racial and ethnic confrontation, protest, riot, and attack are studied;[12] in socialpsychological studies, aggression behaviors are analyzed;[13] in *psychological studies*, human development processes are studied;[14] in *psychiatric analyses*, people may show signs of psychoses or neuroses at a certain age;[15] in *social policy studies*, entry to and exit from poverty, the transitions into retirement, or the changes in living conditions in old age are analyzed;[16] in *medical and epidemiological applications*, patients switch between the states "healthy" and "diseased" or go through different phases of an addiction career;[17] and so on.

Technically, in all these different examples, the units of analysis occupy a discrete state in a theoretically meaningful state space, and transitions between these states can occur at virtually any time.[18] Given an event history data set, the typical problem of the social scientist is to apply appropriate statistical methods to describe the process of change, discover the causal relationships among events, and assess their importance.

This book was written to help the applied social scientist achieve these goals. In this introductory chapter, we first discuss different observation plans and their consequences for causal modeling. We also discuss several concepts of causal inference and summarize the basic concepts of event history analysis and show that the change in the transition rate is a natural way to represent the causal effect in a statistical model. The remaining chapters are organized as follows:

- Chapter 2 describes event history data sets and their organization. It also shows how to use such data sets with Stata.
- Chapter 3 discusses basic nonparametric methods for describing event history data, in particular the life table and the Kaplan-Meier (product-limit) estimation methods as well as cumulative incidence functions, and finally a suitable nonparametric method for processes with competing risks.
- Chapter 4 deals with the basic exponential transition rate model. Although this very simple model is almost never appropriate in practical applications, it serves as an important starting point for all other transition rate models.
- Chapter 5 describes a simple generalization of the basic exponential model, the so-called *piecewise constant exponential model*. In our view, this is one of the most useful models for empirical research, and we devote a full chapter to discussing it.

- Chapter 6 discusses time-dependent covariates. The examples are limited to exponential and piecewise exponential models, but the topic—and part of the discussion—is much more general. In particular, we introduce the problem of how to model parallel and interdependent processes.
- Chapter 7 introduces a variety of models with a parametrically specified duration-dependent transition rate, in particular Gompertz–Makeham, Weibull, log-logistic, and log-normal models.
- Chapter 8 discusses the issue of goodness-of-fit testing for parametric transition rate models. In particular, the chapter describes simple graphical checks based on transformed survivor functions and generalized residuals.
- Chapter 9 presents basic and advanced semiparametric transition rate models based on an estimation approach proposed by D. R. Cox (1972). The advanced section on Cox models describes fixed effects models for multiepisode data.
- Chapter 10 discusses problems of model specification and, in particular, transition rate models with unobserved heterogeneity. The discussion of the specification of more than one error term at the same level is mainly critical, and examples are limited to the use of a gamma mixing distribution. This chapter also provides a brief introduction to transition rate models with random coefficients in a multi-level framework.
- Chapter 11 by Brendan Halpin sequence analysis is presented as a complementary alternative to event history analysis that offers exploratory and descriptive advantages. Sequence analysis focuses on longitudinal data such as life course trajectories as a whole and calculates distances between sequences. These distances can be used to create data-based typologies using cluster analysis, to calculate distance to theoretically or empirically defined reference sequences, to compare trajectory diversity across variables such as cohort, to compare the similarity of dyadic pairs of sequences such as couples' time use, and so on.
- The Appendix contains many exercises designed to help the reader gain a deeper understanding of the (basic) concepts, become more familiar with the use of event history data and estimation commands in Stata, and gain greater skills in estimating results.

1.1 Causal Modeling and Observation Plans

When modeling event histories, design issues related to the type of substantive process are crucial. It is assumed that the methods of data analysis (e.g., estimation and testing techniques) cannot depend solely on the type of data alone (e.g., cross-sectional data, panel data, etc.) as is the case when applying more traditional statistical methodologies. Rather, the characteristics of the specific kind of social process itself must "guide" both the *design of data collection* and *the way these data are analyzed and interpreted* (Coleman 1973, 1981, 1990).

Various observational designs have been used to collect data generated by a continuous-time, discrete-state substantive process (Coleman 1981; Tuma and Hannan 1984; Jenkins 2005). With regard to the level of detail on the process of change, one can distinguish between cross-sectional data, panel data, event count data, event sequence data, and (continuous-time and discrete-time) event history data.

In this book, we do not treat *event count data* (see, e.g., Hannan and Freeman 1989; Olzak 1992; Andersen et al. 1993; Barron 1993; Olzak and Shanahan 1996; Minkoff 1997; Olzak and Olivier 1998a, 1998b). These simply record the number of different types of events for each unit (e.g., the number of upward, downward, or lateral moves in the employment career during a period of 10 years). Event sequence Data (see, e.g., Rajulton 1992; Abbott 1995; Halpin and Chan 1998) which document sequences of states occupied by each unit are only used in chapter 11. In this section, we focus on cross-sectional and panel data as the main standard sociological data types (Tuma and Hannan 1984) and compare these with continuous-time event history data. For example, as shown in Figure 1.1.1, an individual's family career is observed in a cross-sectional survey, a panel survey, and a continuous-time event-oriented survey. A detailed comparison between continuous-time and discrete-time event history data and their estimation procedures can be found in Blossfeld (2026).

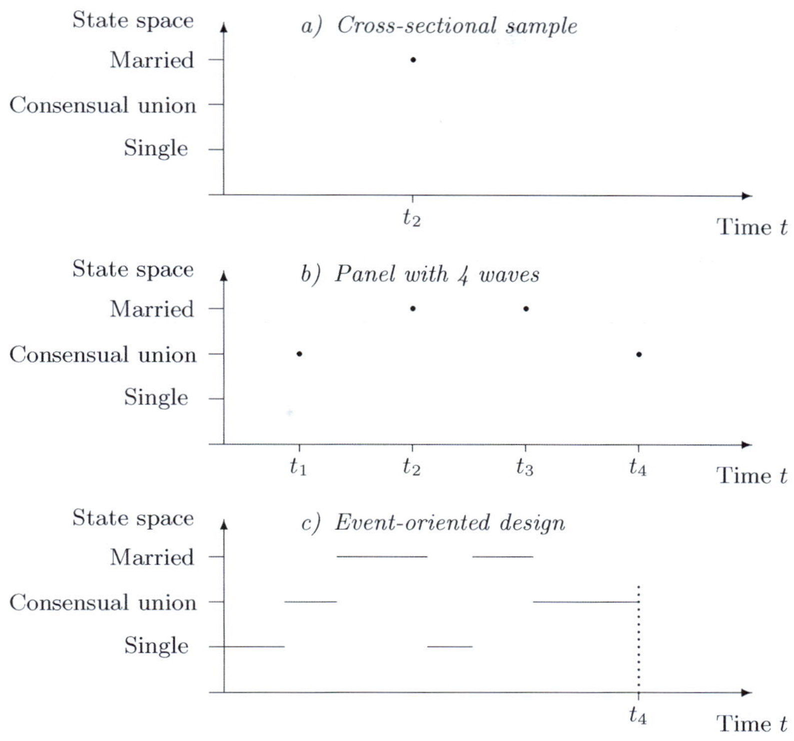

Figure 1.1.1 Observation of an individual's family career on the basis of a cross-sectional survey, a panel study, and an event-history-oriented design.

1.1.1 Cross-Sectional Data

Let us first discuss the *cross-sectional observation*. In the social sciences, this is the most common form of data used to assess sociological hypotheses. The family history of the individual in Figure 1.1.1 is represented in a cross-sectional study by one single point in time: her or his marital state at the time of the interview. A cross-sectional sample is therefore only a "snapshot" of the substantive process being studied. Normally, the point in time when researchers take that "picture" is not determined by hypotheses about the dynamics of the substantive process itself, but by external considerations such as acquiring research funds, finding an appropriate institute to conduct the survey, and so on.

Coleman (1981) has demonstrated that one must be careful when drawing inferences about the effects of explanatory variables in logit models on the basis of cross-sectional data because, implicitly or explicitly, social researchers have to assume that the substantive process under study is in some kind of *statistical equilibrium*. Statistical equilibrium, steady-state, or *stability of the process* mean that although individuals (or any other unit of analysis) may change their states over time, the state probabilities are fairly trendless or stable. Therefore, an equilibrium of the process requires that the inflows to and the outflows from each of the discrete states are broadly the same over time. Only under such stationary conditions it is possible to interpret the estimates of *logit* and/or *log-linear analyses*, as shown by Coleman (1981).

Even if the assumption of a steady state is justified in a particular application, the effect of a causal variable in a logit and/or log-linear model should not be taken as evidence that it has a particular effect on the substantive process (Coleman 1981; Tuma and Hannan 1984). This effect can be interpreted ambiguously for the process under study, because causal variables often influence the inflows to and the outflows from each of the discrete states in different ways. For example, it is known that people with higher education have a lower probability of becoming poor (e.g., receive social assistance); at the same time, however, educational attainment does not have a significant effect on the probability of escaping poverty (see, e.g., Leisering and Leibfried 1998; Leisering and Walker 1998; Zwick 1998). This means that the causal variable level of education influences the poverty process in a specific way: It reduces the probability of entering poverty, but has no impact on the probability of leaving poverty. Since the poverty process is in a steady state, a logit and/or log-linear analysis of cross-sectional data only sheds light on the difference in these two directional effects on the poverty process (Coleman 1981). In other words, cross-sectional logit and/or log-linear models can only show the net effect of the causal variables on the steady state, and this can be misleading, as the following example demonstrates.

Let us assume that we examine a process with two states ("unemployed" and "employed") that is in equilibrium (i.e., the unemployment rate is trendless over time), and let us further assume that the covariate "educational attainment" determines the probability of movement from unemployment to employment (UE $\rightarrow$ E) and increases the probability of movement from employment to unemployment (E $\rightarrow$ UE) for each individual. In a cross-sectional logistic regression analysis using the probability of being employed as the dependent variable, the estimated coefficient for "educational attainment" only gives the net effect of both directional effects. Therefore, if the effect of educational attainment on UE $\rightarrow$ E offsets the effect on E $\rightarrow$ UE, the net effect of "educational attainment" in the logistic regression on the steady-state probability will be about zero and not significant. This means that a zero effect of a covariate in a logistic cross-sectional regression analysis could have two very different meanings: that there is no effect at all of "educational attainment" on UE $\rightarrow$ E and on E $\rightarrow$ UE, or that the directional effects on UE $\rightarrow$ E and on E $\rightarrow$ UE cancel each other out. Thus, a nonsignificant effect in the cross-sectional logit model should not be taken as evidence that a variable is irrelevant for the process, but only that it has no net effect on the equilibrium distribution (Tuma and Hannan 1984).

If the net effect of "educational attainment" in a cross-sectional logistic regression on the probability of becoming employed is positive, then, in principle, the following four different interpretations are possible: (1) that the positive effect on UE $\rightarrow$ E is greater than the positive effect on E $\rightarrow$ UE, (2) that the negative effect on UE $\rightarrow$ E is smaller than the negative effect on E $\rightarrow$ UE, (3) that there is only a positive effect on UE $\rightarrow$ E and no effect on E $\rightarrow$ UE, and (4) that there is no effect on UE $\rightarrow$ E and only a negative effect on E $\rightarrow$ UE. Conversely, in the case of negative effects in the cross-sectional logistic regression, the four interpretations must be reversed.

If there is no equilibrium in the process, however, cross-sectional coefficients may not only be ambiguous but also give a completely misleading picture. In a study on unemployment incidence, Rosenthal (1991), for example, demonstrated how confusing cross-sectional estimates can be when the proportion of unemployed in a particular region is increasing or decreasing and the process of change is, therefore, not in equilibrium.

In the social sciences, it can be assumed that stability is very rare. For example, life history studies (Blossfeld 1989, 1995; Mayer 1990; Blossfeld and Hakim 1997; Blossfeld and Drobnič 2001; Blossfeld and Müller 2002/2003; Blossfeld and Timm 2003; Bernardi et al. 2019; Settersten et al. 2021; Settersten et al. 2024) show that change across age, cohort, and historical period is an enduring and important feature in all domains of modern individuals' lives (Mayer and Tuma 1990); organizational studies demonstrate that most social organizations appear to follow a program of growth and not of stability (Carroll and Hannan 2000, 2018; Schecter et al. 2018; Surdu et al. 2018; Furr 2029; Momtaz 2021; Agarwal and Kapoor 2023); and most modern societies exhibit an accelerating rate of change in almost all of their subsystems (cf. the rapid changes in family systems, job structures, educational systems, etc.; see Mayer 1990; Heinz 1991a, 1991b, 1992; Huinink et al. 1995; Blossfeld and Hakim 1997; Leisering and Leibfried 1998; Blossfeld and Drobnič 2001; Blossfeld and Müller 2002/2003; Blossfeld and Timm 2003). But even in areas considered to be relatively stable, the crucial methodological question must be asked: To what extent is the process under study close to an equilibrium (Tuma and Hannan 1984)? This question can only be answered if longitudinal data are used, because these are the only type of data that indicate whether a steady state actually exists, or how long it will take until a system returns to a new equilibrium after some external upheaval.

Apart from the crucial assumption of process stability, cross-sectional data have several other inferential limitations with regard to causal modeling. We want to address at least some of the more important problems here.[19]

Direction of Causality. There are only a few situations in which the direction of causality can be established on the basis of cross-sectional data (Davies 1987). For example, consider the strong positive association between parental socioeconomic characteristics and educational attainment of sons and daughters, controlling for other important influences (Shavit and Blossfeld 1993; Erikson and Jonsson 1996; Blossfeld et al. 2014, 2016a, 2016b, 2017, 2019a, 2019b; Triventi et al. 2020; Blossfeld and Blossfeld 2021; Becker et al. 2024, Weinert et al. 2023, 2024). A convincing interpretation of this effect could be that being born into a middle-class family increases the probability of a university degree, because no other plausible explanation for the statistical correlation can be imagined. However, such recursive relationships in which all the causal linkages run "one way" and have no "feedback" effects, are rare in social science research. For example, in modern industrial societies there is very often a correlation between the age of the youngest child and the labor force participation of women (Blossfeld and Hakim 1997; Blossfeld, Drobnič, and Rohwer 1998; Blossfeld and Drobnič 2001, Blossfeld 2022). The common interpretation is that there is a one-way causality with young children tending to keep mothers at home. However, it is quite possible that the lack of jobs encourages women to enter into marriage and motherhood, suggesting a inverse relationship (Davies 1987).

The ambiguity of causation seems to be particularly important for modeling the *relationship between attitudes and behavior*. There are two interesting aspects of this relationship: There is a direct effect in which behavior affects attitudes, and there is a "feedback" process in which attitudes change behavior (Davies 1987).[20] The well-known disputes among sociologists as to whether value change causes change in social behavior or whether structural change in behavior leads to changing values of individuals often originate from the fact that cross-sectional surveys are used that can assess the net association of these two processes.

Various Strengths of Reciprocal Effects. The impossibility of establishing the direction of causality in cross-sectional surveys has the disadvantage that these data cannot be used to discover the different strengths of reciprocal effects. For example, many demographic studies have shown that first marriage and first motherhood are closely interrelated (Blossfeld and Huinink 1991; Blossfeld et al. 1999; Blossfeld and Mills 2001; Blossfeld 2022). In order to understand how family formation has evolved in modern societies, it may be of interest to know more than just the effects of marriage on fertility, but also the effect of pregnancy or first birth on marriage (Blossfeld and Huinink 1991; Blossfeld 1995; Blossfeld et al. 1999; Blossfeld and Mills 2001; Mills and Trovato 2001), and, perhaps, how these effects have changed over historical time (Manting 1994, 1996).

Observational Data. Most sociological research is based on nonexperimental observations of social processes, and these processes are highly selective. For example, in a study examining the influence of type of school (private vs. public) on student test performance, Lieberson (1985) distinguished at least three types of nonrandom processes: (1) *self-selectivity*, in which the units of analysis sort themselves out by choice (e.g., certain students choose specific types of schools); (2) *selective assignment by the independent variable* itself, which determines, for example, which members of a population are exposed to certain levels of the independent variable (e.g., schools select their students based on their past achievement); and (3) *selectivity due to forces exogenous to variables under consideration* (socioeconomic background, ethnicity, gender, previous school career, changes in intelligence over time, etc.). Many of these sources are not only *not observed* but also practically *unmeasurable*. Of course, no longitudinal study is able to overcome all the problems of identifying these various effects; however, cross-sectional data offer the worst of all data types to disentangle the effects of the causal factors of interest on the outcome from other forces operating at the same time, as these data are the least informative about the process of change. Cross-sectional analysis therefore requires a particularly careful justification, and the substantive interpretation of results must always be qualified appropriately (Davies 1987; Pickles and Davies 1989).

Previous History. One aspect of observational data deserves special attention in the social sciences. Life courses of individuals (and other units of analysis such as organizations, etc.) involve complex and cumulative time-related layers of selectivity (Mayer and Müller 1986; Mayer and Schöpflin 1989; Mayer and Tuma 1990; Mayer 1991; Huinink et al. 1995; Mayer and Baltes 1996; Mayer 2015; Blossfeld and Blossfeld 2021). Therefore, there is a high probability that certain individuals have entered a specific origin state (see, e.g., the discussion in Blossfeld et al. 1999 on consensual unions). In particular, life-course research has shown that the past is an indispensable factor for understanding the present (Buchmann 1989; Mayer 1990, 2015; Heinz 1991a, 1991b, 1992; Allmendinger 1994; Huinink et al. 1995; Weymann 1995; Weymann and Heinz 1996). Cross-sectional analysis can be conducted with some proxy-variables and with assumptions about both the causal sequence and the interdependencies between the various explanatory variables. However, it is often not possible to adequately trace back the temporal selection processes occurring in the previous history because these data are simply not available. Thus the normal control approaches of cross-sectional statistical techniques will rarely succeed in isolating the influence of some specific causal force (Lieberson 1985).

Age and Cohort Effects. Cross-sectional data cannot be used to distinguish between age and cohort effects (Tuma and Hannan 1984; Davies 1987). In many social science applications, however, it is of substantive importance to know whether the people's behavior (e.g., their propensity to vote for a particular party) differs because they belong to different age groups or because they are members of different birth cohorts (Blossfeld 1986, 1989; Mayer and Huinink 1990).

Historical Settings. Cross-sectional data is also unable to take into account the fact that processes emerge in particular historical settings. Thus, in addition to individual resources (age, education, labor force experience, etc.), there are at least two ways in which a changing labor market structure affects career opportunities. The first is that people start their careers in different structural contexts. It has often been assumed that these specific historical conditions at entry into the labor market have an impact on people's subsequent careers. This type of influence is generally called a cohort effect (Glenn 1977). The second way in which a changing labor market structure influences career opportunities is that it improves or worsens the career prospects of all people within the labor market at a given point in time (Blossfeld 1986; Blossfeld and Becker

2017; Bukodi et al. 2017; Becker et al. 2024). For example, in a favorable economic situation with low unemployment, there is a relatively wide range of opportunities. This type of influence is generally referred to as a period effect (Mason and Fienberg 1985). Using longitudinal data, Blossfeld (1986) has shown that life-course, cohort, and period effects can be identified on the basis of substantively developed measures of these concepts (see also Rodgers 1982 as well as Blossfeld and Becker 2017, Becker et al. 2024) and that these effects represent key mechanisms of career mobility that must be distinguished in any valid account.

Multiple Clocks, Historical Eras, and Point-in-Time Events. From a theoretical or conceptual point of view, multiple clocks, historical eras, and point-in-time events very often influence the substantive process under study (Mayer and Tuma 1990). For example, in demographic studies of divorce, it is important to consider types of clocks, such as age of respondent, length of cohabitation, duration of marriage, ages of children, different phases of the business cycle, or changes in national (divorce or marriage) laws (Blossfeld et al. 1995; Blossfeld and Müller 2002/2003; Blossfeld 2022). With respect to cross-sectional data, such relationships can hardly be studied without making strong untestable assumptions.

Contextual Processes at Multiple Levels. Social scientists are often interested in the influences of contextual processes at multiple aggregation levels (Huinink 1989; Bukodi et al. 2017). Contextual process effects refer to situations where changes in the group contexts themselves influence the dependent variable. For example, career mobility of an individual can be conceptualized as dependent on changes in resources at the *individual level* (e.g., social background, educational attainment, experience, etc.), the success of the company in which she or he is employed (e.g., expansion or contraction of the organization) at the *intermediate level*, and changes in the business cycle at the *macro level* (Blossfeld 1986; DiPrete 1993; Blossfeld and Becker 2017; Becker et al. 2024). Cross-sectional data do not provide an adequate opportunity to study such influences at different levels (Mayer and Tuma 1990).

Duration Dependence. Another problem with cross-sectional data is that they are *inherently ambiguous when it comes to their interpretation at the level of the unit of observation*. Suppose we know that 30.6 % of employed women in West Germany worked part-time in 1970 (Blossfeld and Rohwer 1997a). At the one extreme, this could be interpreted to mean that *every* employed woman had a 30.6 % chance of being employed part-time in that year. But at the other extreme, one could conclude that 30.6 % of the employed women were always employed part-time and 69.4 % were exclusively employed full-time. In other words, the cross-sectional data do not provide any information about the *time women spent in these different employment forms*. They are therefore open to very different substantive interpretations (Heckman and Willis 1977; Flinn and Heckman 1982; Blossfeld and Hakim 1997; Blossfeld and Drobnič 2001). In the first case, each woman is expected to move back and forth between part-time and full-time employment. In the second case, there is no mobility between the two types of work, and the estimated percentages describe the proportions of two completely different groups of employed women. From an analytical point of view, it is therefore important to have data on the length of time spent in a state. Even repeated cross-sectional analyses using comparable samples of the same population (e.g., a series of micro censuses or cross-sectional surveys) can only show the net change, but not the flow of individuals.

Variability of State Dependencies. In many situations, cross-sectional data are problematic because the rate of change is strongly state dependent and *entries into and exits from these states are highly variable over time* (e.g., over the life course and historical period or across cohorts). For example, it is known that the roles of wife and mother (the latter in particular) have been central in women's lives. Therefore the *family cycle concept* has frequently been used in sociology to describe significant changes in circumstances that affect women's availability for paid work outside the home. The basic idea is that there is a set of ordered stages defined primarily by variations in family composition and size, which can be described with cross-sectional data. However, this view often leads to the tendency to assume that what happens to different women at a particular time in the various phases of the family cycle is similar to the pattern that women experience when they make these transitions in different historical times (which has been called the *life course fallacy*). Moreover, there is the well-known problem that individuals and families often do not conform to the assumption of a uniform trajectory through a given number of stages in a predetermined order (see, e.g., Blossfeld and Hakim 1997; Blossfeld and Drobnič 2001). There may be at least three reasons for this (Murphy 1991): (1) The chronology of the timing of events may not correspond to the ideal model (e.g., childbearing may start before marriage), (2) many stages are not reached (e.g., by individuals who never marry), and (3) the full set of stages may be truncated by events such as death or marital breakdown. Such complex constellations between the family cycle and women's labor-force participation can hardly be meaningfully described or examined on the basis of cross-sectional data.

Changes in Outcomes. In cross-sectional models, there is very often a *tendency to overpredict change* and consistently *overestimate the importance of the explanatory variables* (Davies 1987). The reason for this phenomenon is that these data analyses cannot be based on how *changes* in explanatory variables cause *changes* in outcomes. They are only concerned with how *levels* of explanatory variables "explain" an *outcome* at a particular point in time. However, if an outcome at time t (e.g., choice of mode of transport to get to work in June) depends on a previous outcome (e.g., established choice of mode of travel to work), and if both outcomes are positively influenced by an explanatory variable in the same way (e.g., merits of public transport), then the effect of the explanatory variable will reflect both the true positive influence of the explanatory variable on the outcome at time t and a positive spurious element due to that variable acting as a proxy for the omitted earlier outcome (established mode of travel to work). Hence, a cross-sectional analysis of mode choice for travel to work (e.g., public vs. private transport) would tend to overpredict the effect of policy changes (e.g., fare decreases or faster buses) as there is a strong behavioral inertia (Davies 1987).

In summary, all these examples show that cross-sectional data have many severe inferential limitations for social scientists. Therefore it is not surprising that causal inferences based on cross-sectional data have often been altered radically after the processes have been studied with longitudinal data (Lieberson 1985).

Longitudinal studies are also much more informative than cross-sectional studies, both in estimating bias from missing data and in the means for correcting it. This is because longitudinal studies often have data from previous points in time, so that the characteristics of non-responders or lost units can be assessed with some precision. It is worth noting that almost all the substantive knowledge about the biases associated with missing data, which all studies must seek to minimize, comes from longitudinal studies (Medical Research Council 1992).

Although longitudinal data are not a panacea, they are obviously more effective in causal analysis and have fewer inferential limitations (Mayer 1990, 2015; Mayer and Tuma 1990; Magnusson, Bergmann, and Törestad 1991; Clogg and Arminger 1993; Arminger, Clogg, and Sobel 1995; Blossfeld 1995; Blossfeld and Hakim 1997; Carroll and Hannan 2000; Blossfeld and Drobnič 2001; Blossfeld and Blossfeld 2021).

They are indispensable for the study of processes over the life course (of all types of units) and how these relate to historical change. Therefore research designs aiming at a causal understanding of social processes must be based on longitudinal data at the level of the units of analysis.

1.1.2 Panel Data

The temporal data commonly available to sociologists are *panel data*. In panel studies, the same persons or units are re-interviewed or observed at a series of discrete points in time (Chamberlain 1984; Arminger and Müller 1990; Engel and Reinecke 1994; Hsiao 2003; Andreß et al. 2013; Baltagi 2015; Blossfeld et al. 2016c; Laird 2020; Blossfeld 2025). Figure 1.1.1 shows a four-wave panel in which the family career of the respondent was observed at four different points in time. This means that there is only information about state of the units at pre-determined survey points, but the course of events between the survey points remains unknown.

Panel data usually contain more information than cross-sectional data, but are associated with well-known biases that arise from the method itself (see, e.g., Hunt 1985; Magnusson and Bergmann 1990).

Panel Bias. Respondents often answer the same questions differently in the second and later waves than they did the first time because they are less inhibited, or they have mulled over or discussed the issues between questioning dates.

Modification of Processes. Panel studies tend to influence the very phenomena they seek to observe—this sometimes alters the natural course of the observed processes.

Attrition of the Sample. In panel studies the composition of the sample normally diminishes selectively over time. Usually, these processes are particularly strong during the first panel waves and then gradually diminish. Therefore, what researchers observe in a long-term panel may not provide a good picture of what has actually happened to the process under study.

Nonresponses and Missing Data. In a cross-sectional analysis, one can afford to remove a small number of cases with non-response and missing data, but in a long-term panel study, removing incomplete cases in each observation round can eventually lead to a heavily pruned sample with very different characteristics than the original one.

Fallacy of Cohort Centrism. Very often, panel studies focus on members of a specific cohort (cf., e.g., the British National Child Development Study). In other words, these panels examine respondents who were born, grew up, and lived in a particular historical period. This creates a danger that researchers might assume that what happens to a particular group of people over time reveals general principles of the life course (fallacy of cohort centrism). Many events may simply be specific to that generation.

Fallacy of Period Centrism. Many panel studies include just a few waves and cover only a short historical period. At the time of these observations, special conditions may have prevailed (e.g., high unemployment), which can lead to individuals reacting differently than they would under other historical conditions (fallacy of historical period).

Confounded Age, Period, and Cohort Effects. In every long-term panel study in sociology, three causal factors—the individual's age, the cohort, and the historical period effect—are confounded (cf. the Panel Study of Income Dynamics). Methodological techniques are needed to unconfound these three factors and reveal the role of each. As will be discussed in more detail later, there are some specific problems with panel data in detecting the three main factors. However, in order to gain scientific insights into the interplay of processes governing life courses from birth to death, they appear to be a better approach than applying cross-sections. But a mechanical and atheoretical cohort analysis is a useless exercise, and statistical innovations alone will not solve the age-period-cohort problem (Blossfeld 1986; Mayer and Huinink 1990; Blossfeld and Becker 2017; Becker et al. 2024).

Most of the previously mentioned difficulties in panel studies can be overcome by better data collection methods, more sophisticated statistical procedures, or more panel waves. However, panel data also lead to a number of shortcomings with respect to the estimation of transition rates (Tuma and Hannan 1984; Blossfeld 2025): First, there is the problem of "embeddability," meaning that there may be difficulties in embedding a matrix of observed transition probabilities in a continuous-time Markov process (Singer and Spilerman 1976a); second, there is the problem that there may not be a unique matrix of transition rates that describes the data (Singer and Spilerman 1976b); and third, there is the drawback that the matrix of transition probabilities may be very sensitive to sampling and measurement error (Tuma and Hannan 1984). Multiple waves with irregular spacing or shorter intervals between waves can reduce these problems. However, as Hannan and Tuma (1979) noted, this only means that the more similar the panel and event history data are, the less problematic the modeling becomes (see also Blossfeld 2025).

Lazarsfeld (1948, 1972) was one of the first sociologists to propose a panel analysis of discrete variables. In particular, he wanted to find a solution to the problem of ambiguity in causation. He suggested that if one wants to know whether a variable X induces change in another variable Y, or whether Y induces change in X, observations of X and Y at two points in time would be required. Lazarsfeld applied this method to dichotomous variables whose time-related structure he analyzed in a resulting sixteenfold table. Later, Goodman (1973) applied log-linear analysis to such tables. For many years, such a cross-lagged panel analysis for qualitative and quantitative variables (Campbell and Stanley 1963; Shingles 1976) was considered a powerful quasi-experimental method for drawing causal inferences. It has also been extended to multiwave-multivariable panels to study more complex path models with structural-equation modeling (Jöreskog and Sörbom 1993). However, it appears that the strength of the panel design for causal inference has been greatly exaggerated (Davis 1978). Causal inference in panel approaches is much more complicated than is generally realized. There are several reasons for this.

Time Until the Effect Starts to Occur. It is important to recognize that the role of time in causal explanations is not only to indicate a *temporal order* in which the effect follows the cause in time. It also implies that there is a *temporal interval* between the cause and its impact (Kelly and McGrath 1988). In other words, if the cause has to precede the effect in time, it takes a certain amount of time for the cause to produce the effect. The time interval can be very short or very long, but it can never be zero or infinity (Kelly and McGrath 1988). Some effects occur almost instantaneously. For example, if the effect occurs at microsecond intervals, then the process must be observed within these small time units to reveal causal relations. However, some effects may occur in a time interval too small to be measured by any given methods, so that cause and effect *seem to occur* at the same point in time. Apparent simultaneity is often the case in social science applications where basic observation intervals are relatively crude (e.g., days, months, or even years), such as, for example, yearly data about first marriage and first childbirth (Blossfeld, Manting, and Rohwer 1993). For these parallel processes, the events "first marriage" and "first childbirth" may be interdependent, but whether these two events are observed simultaneously or successively depends on the degree of temporal refinement of the scale used in making the observations (Blossfeld 2016; 2021). Other effects need a long time until they start to occur. Thus, there is a *delay* or *lag* between cause and effect that must be specified in an appropriate causal analysis. However, in most current sociological theories and interpretations of research findings, this interval is left unspecified. In almost all cases, researchers assume, at least implicitly, that the effect takes place almost immediately and is then constant (Figure 1.1.2b). If this is the case, then there seems to be no need for theoretical statements about the time course of the causal effect. A single measurement of the effect at a specific point in time after the occurrence of a cause could be sufficient to catch it (see Figure 1.1.2a). However, if there is a reason to assume that there is a lag between cause and effect, then a single measurement of the outcome is insufficient to describe the process (see Figure 1.1.2b), and the interpretation based on a single measurement of the outcome then depends on the time chosen to measure the effect (the substantive conclusion based on p_3 or p_4 in the figure would of course be different; Blossfeld 2025).

Therefore, a *limiting assumption of panel designs* is that either cause and effect occur almost simultaneously, or the interval between observations is of approximately the same length as the true causal lag. The greater the discrepancy, the greater the likelihood that the panel analysis will fail to discover the true causal process. Hence, as expressed by Davis (1978), if one does not know the causal lag precisely, panel analysis is of little use when determining the causal direction or temporal sequence of causal effects. Unfortunately, we rarely have enough theoretically grounded arguments about the structure of a social process to accurately determine lags.

Temporal Shapes of the Unfolding Effect. In addition to the issue of the length of the delay between the timing of the cause and the beginning of the unfolding of the effect, it is also possible that *the way the effect develops over time takes different shapes*. Although the problem of time lags is widely recognized in social science literature, considerations on the temporal shape of the effect are quite rare (Kelly and McGrath 1988). Indeed, social scientists seem to be largely unaware of the fact that causal effects can be highly time-dependent.

Figure 1.1.2 illustrates several possible shapes these effects can take over time. In Figure 1.1.2a, there is an almost simultaneous change in the effect which is then maintained. In Figure 1.1.2b, the effect occurs with a certain time delay and is then time-invariant. In Figure 1.1.2c, the effect starts almost immediately and then gradually increases. In Figure 1.1.2d, an almost simultaneous increase can be observed, which reaches a maximum after some time before decreasing again. Finally, in Figure 1.1.2e, a cyclical effect pattern over time is described.

If the effect increases or decreases monotonically or linearly, oscillates in cycles, or has some other complicated temporal pattern, then the strength of the observed effect in a panel study depends on the timing of the panel waves. A panel design can be particularly problematic when there are non-monotonic cycles of the effect because completely opposite conclusions can be drawn about the effects of the explanatory variable depending on whether the panel places measurement points at a peak or an ebb of the curve (see Figures 1.1.2d and 1.1.2e).

Reciprocal Effects with Different Time Paths. With reciprocal causality, additional problems arise in panel studies if the time structure of the effects of X_1 on X_2 and of X_2 on X_1 differ in terms of lags and shapes. In these situations, a panel design might turn out to be completely useless for those wishing to detect such time-related recursive relationships.

Observational Data and Timing of Measurement of Explanatory Variables. Most sociological research is based on observational data, which means that manipulating the timing of the independent variables is generally not possible. For example, if a researcher wants to study the effects of job mobility on marriage behavior, it is impossible to force respondents to change their jobs, say, at the time of the first panel wave. Thus, the longer the interval between panel waves, the greater the uncertainty regarding the exact time at which a person changed jobs, and thus the *time at which we assess the time path of the effect* (Coleman 1981; Blossfeld 2026). The situation may be even more problematic when changes in independent variables are repeatable and multiple changes are possible between two consecutive panel waves, as may be the case for job exits observed at annual intervals (cf. Sandefur and Tuma 1987; Blossfeld 2025). In such panel studies, even the causal sequence of explanatory and dependent events may become ambiguous.

Observational Data and the Timing of Control Variables. Observational panel studies take place in natural settings and therefore offer little control over changes in other important variables and their timing. If these influences change arbitrarily and have time-related effect patterns, then panel studies are useless in disentangling the effect of interest from time-dependent effects of other parallel exogenous processes (Blossfeld 2026).

Continuous Changes of Explanatory and Control Variables. In observational studies, explanatory and control variables can change not only stepwise from one state to another, but often *continuously* over time. For example, individuals age continuously, constantly acquire general labor force experience or job-specific human capital when they are employed (Blossfeld and Huinink 1991), are exposed to continuously changing historical conditions (Blossfeld 1986; Blossfeld and Becker 2017; Becker et al. 2024), are steadily changing their social relationships in marital or consensual unions (Blossfeld et al. 1995; Blossfeld et al. 1999; Blossfeld 2021), and so on. Even in cases when these continuous changes are not connected with lags or time-related effect patterns, panel data reveal deficiencies in their ability to detect time dependence in substantive processes. For this reason, panel analyses are often unable to appropriately identify age, period, and cohort effects (Blossfeld 1986).

Therefore the use of panel data causes an identification problem due to omitted factors whose effects are summarized in a disturbance term. These factors are not stable over time, which means that the distur-

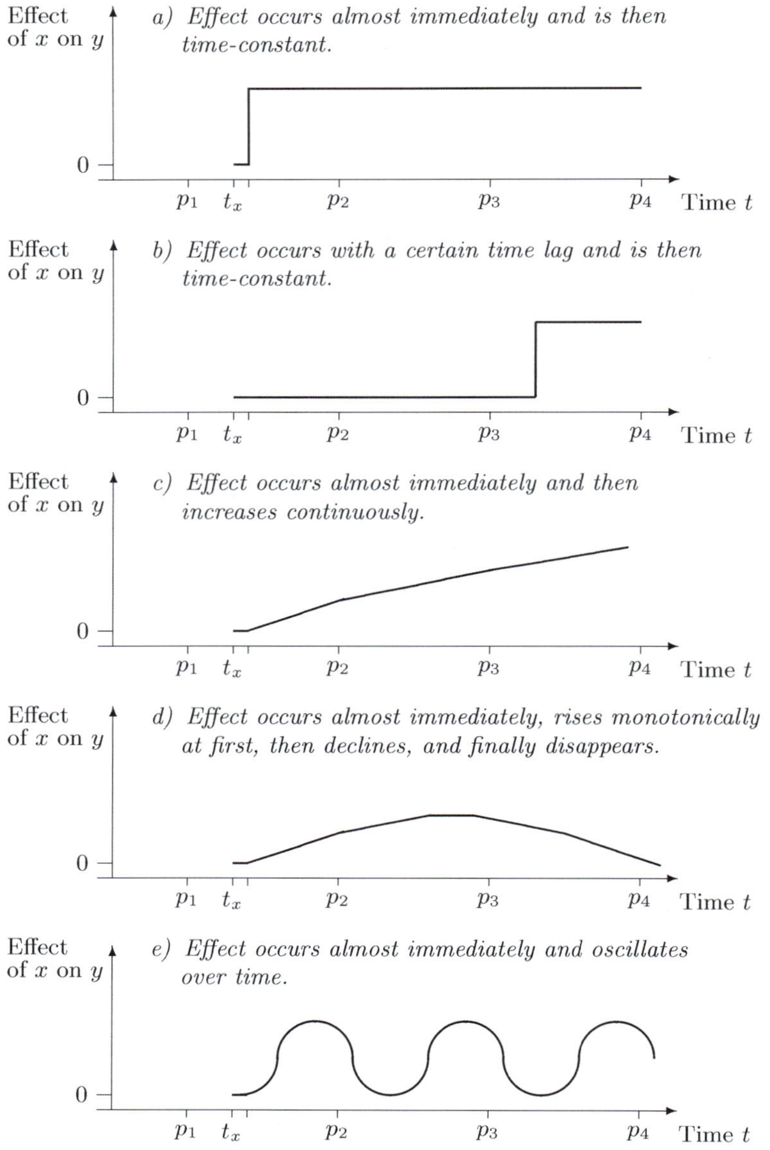

Figure 1.1.2 Different temporal lags and shapes of how a change in a variable x, occurring at point in time t_x, effects a change in a variable y.

bance term cannot be uncorrelated with the explanatory variables. Panel analysis thus depends critically on solutions to the problem of autocorrelation. This problem can be tackled reasonably well by increasing the number of panel waves and modifying their spacing. The panel analysis is particularly sensitive to the *length of the time intervals between waves relative to the speed of the process* (Coleman 1981). The intervals can be too short, so that too few events will be observed, or too long, so that it is difficult to establish a time order between events (Sandefur and Tuma 1987; Blossfeld 2026). A major advantage of the continuous-time observation design in event history analysis is that it makes the timing between waves irrelevant (Coleman 1968; Blossfeld 2026).

1.1.3 Event History Data

For many processes in the social sciences, a continuous measurement of qualitative variables seems to be the only adequate method of assessing empirical change. This is achieved by utilizing an *event oriented observation design* that records all the changes in qualitative variables and their timing. As shown in Figure 1.1.1, the major advantage of event history data is that they provide the *most complete data possible on changes in qualitative variables that can occur at any point in time*. Therefore, the observation of events is an attractive alternative to the observation of states for social scientists.

Event history data, which are usually *collected retrospectively* in *life history studies*, cover the entire life course of individuals. One example for such a study is the German Life History Study (GLHS; Mayer and Brückner 1998; Brückner and Mayer 1998; Mayer 2015). Retrospective studies have the advantage of being generally *less expensive* than collecting data with a long-term panel study. In addition, they are also systematically *coded to one framework of codes and meanings* (Dex 1991). However, retrospective (in contrast to prospective) studies suffer from several limitations that are increasingly recognized (see, e.g. Medical Research Council 1992).

Nonfactual Data. It is well known that retrospective questions on motivational, attitudinal, cognitive, or affective states are particularly problematic, because respondents can hardly recall the timing of changes in these states with any great accuracy (Hannan and Tuma 1979). This type of data is not verifiable even in principle, as these states exist only in the minds of the respondents and, if at all, are only directly accessible to the respondent concerned (Sudman and Bradburn 1986). For these non-factual data, panel studies have the advantage that they can repeatedly capture the current states of the same person over time. For studies that aim to model the relationship between attitudes and behavior over time, panel observations of attitudinal states, combined with retrospective information on behavioral events since the last sweep, appear to be an appropriate design.

Recall Problems with Regard to Behavior or Facts. Behavioral or factual questions ask the respondents about characteristics, things they have done, or things that have happened to them that are, in principle verifiable by an external observer. Most (cross-sectional, panel, or event oriented) surveys collect retrospective information on behavior or facts (e.g., by asking people about their education, social origin, etc.), so that the disadvantages of retrospection are only a matter of degree. However, event history studies are particularly ambitious (see Mayer and Brückner 1989; Mayer 2015). They attempt to collect continuous records of qualitative variables that have a high potential for bias because of their strong reliance on (autobiographical) memory. However, research on the accuracy of retrospective data shows that individuals' marital and fertility histories, family characteristics and education, health service usage, and employment history can be collected with reasonable accuracy. Dex (1991) has presented a very good overview of the types of data that can be collected retrospectively, the factors that influence recall accuracy, and the methods for improving recall.

Unknown Factors. Retrospective designs cannot be used to study factors involving variables that are not known to the respondent (e.g., emotional and behavioral problems when the respondent was a child). In such cases, panel studies are essential (Medical Research Council 1992).

Limited Capacity. There is a limit to respondents' tolerance for the amount of data that can be collected on one occasion (Medical Research Council 1992). A carefully corseted panel design can therefore provide a broader coverage of variables (if these are not unduly influenced by variations at the time of assessment).

Survivors Only. Due to their nature, retrospective studies must be based on survivors. Thus, people who have died or migrated out of the geographical area being studied will necessarily be omitted. If either is related to the process (which can often be the case), biases arise. This problem is particularly important for retrospective studies that include a broad range of birth cohorts, such as the German Life History Study (GLHS) or international migration studies (Blossfeld 1987b).

Misrepresentation of Specific Populations. Retrospective studies can also lead to a misrepresentation of certain population groups. For example, Duncan (1966) has shown that if men are asked about their fathers, men from earlier generations who had no sons, or whose sons died or emigrated will not be represented in a retrospective father-son mobility table. So, part of the population at risk may not be included in an analysis (see Blossfeld and Timm 2003, who discuss this problem with regard to studies of educational homogamy, in which singles are often excluded at the time of the interview).

To avoid these problems with retrospective event history data, a mixed design with a follow-up (or "catch-up") and a follow-back strategy appears to combine the strengths of traditional panel designs with the virtues of retrospective event history studies. In modern panel studies, event histories are therefore collected retrospectively for the period before the start of the panel and between successive panel waves. Sometimes complete administrative records also contain time-related information about events in the past. All of these methods (retrospective, combined follow-up and back-up, or available registers) offer a comparatively superior opportunity to model social processes, regardless of which method is selected.

One aim of our book is to show that event history models are a useful approach for uncovering causal relationships or mapping a system of causal relationships. As will become clear later in the book, event history models are linked very naturally with a causal understanding of social processes, as they relate changes in future outcomes to conditions in the past and attempt to predict future changes on the basis of past observations (Aalen 1987).

1.2 Event History Analysis and Causal Modeling

The investigation of causal relationships is an important but difficult scientific endeavor. As already shown, the possibilities for evaluating causal inferences depend heavily on the type of observation available to the social scientist. This is because they determine the extent to which the researcher is forced to make untested assumptions. Therefore, the overarching goal of research design should not only be to produce data, but to obtain the most appropriate data about the intended aspects of the social world.

1.2.1 Causal Explanations

Correlation and Causation

First of all, statements about causality should be distinguished from statements about association. In making correlational inferences, one can be satisfied with observing how the values of one variable are associated with the values of other variables in the population under study and perhaps also over time (Hernán 2018). In this context, time is important only insofar as it determines the population under analysis or specifies the operational meaning of a particular variable (Holland 1986). Although data are always viewed through the medium of an implicit or explicit theory (Fox 1992), statements about associations are quite different from causal statements because the latter are designed to say something about how events are caused or conditioned by other events. In causal analysis, we are often interested in causal mechanisms that we think lie behind the correlation in the data and specify how causal events affect effects (see Popper Shaffer 1992).

Sometimes social scientists argue that the units of sociological analysis are constantly learning and involve actors with goals and beliefs, so that sociology can at best provide systematic descriptions of phenomena at various points in history. This position is based on the view that causal statements about events are only possible if they are governed by "eternal," timeless laws (Kelly and McGrath 1988). Of course, the assumption that such laws can be established with regard to social processes can reasonably be disputed (see, e.g., Giere 1999). In particular, we are not forced to accept a simple contrast: either to describe contingent events or to assume "eternal" laws. Many social phenomena exhibit *systematic temporal variations* and *patterned regularities* under specific conditions which are themselves a legitimate focus of our efforts to understand social change (Kelly and McGrath 1988; Goldthorpe 2000). Thus, sociology can do more than simply describe the social world. This book emphasizes the usefulness of techniques of event history modeling as "new" approaches to the investigation of causal explanations (see also Blossfeld 2017a, 2017b).[21]

Role of Causal Models in Empirical Research

There is a long-standing philosophical debate on the question whether causality is a property of the real world, so that crucial variables can be identified in causal models and the functional form of their relationships can be discovered (this position might be called the realist perspective); or whether it is just a concept by humans, so that the variables and relationships are only constructions embedded in a theoretical framework (this position might be called the social constructivist perspective). This kind of debate was enriched by an intriguing proposal by Giere (1999), who suggested that one could think about the representation in scientific models in analogy to road maps. A map may, for example, be more or less accurate, more or less detailed, and of smaller or larger scale. Maps (like causal models) require a large background of human convention for their production and use (Giere 1999, p. 24). Much deliberate care is required to ensure that a map accurately represents the intended space. Mistakes are easy to make. It is also possible to deliberately construct incorrect maps, or even maps of completely fictional places. Map-making and map-using take advantage of similarities in spatial structures. But one cannot understand map-making solely in terms of abstract, geometrical relationships. Interpretative relationships are also required. One must be able to understand that a certain area on a map is supposed to represent something specific. These two features of representation using maps, similarity of structure and similarity of interpretation, can be applied to understanding the way social researchers use causal models to represent aspects of the social world. That is, thinking about causal models in terms of maps combines what is valuable in both constructivism and realism, but requires abandonment of the universal applicability of both views (Giere 1999). One can agree that scientific representations are socially constructed, but then one must also agree that some socially constructed representations may turn out to be a good picture of aspects of the world, while others are mere constructions with little genuine connection to the world (Giere 1999).

This epistemological perspective suggests that causal models provide only partial access to social reality (Giere 1999). Some ways of constructing models of the social world do provide resources for capturing some aspects of the social world more or less well. Other ways may provide resources for capturing other aspects more or less well. Both ways, however, can capture some aspects of social reality and are thus candidates for a realistic understanding of the social world. That is, there is no such thing as a perfect causal model, complete in all details. The fit of causal models is always partial and imperfect. However, this does not prevent causal models from providing us with deep und useful insights into the workings of the social world. In this view, the empirical question is not whether causal models about the social world, as ontologically well-defined entities, are empirically true or not, but how well causal models fit the intended aspects of the social world. In other words, there may be multiple valid theoretical models and statistical analyses, and a causal model may fit the social world more or less well in something like the way maps fit the world more or less well (Giere 1999; see also Pötter and Blossfeld 2001). In such a framework, it is enough that empirical evidence sometimes helps us to decide that one type of model fits better than another type in some important respect. Sometimes this means that a fortunate combination of data and research design makes us justifiably confident that a particular model has a good fit, i.e. that this model has a structure that is similar to the social world itself. However, often the situation in nonexperimental social science research is less ideal (Popper Shaffer 1992), and scientific consensus may then be based on shared values rather than empirical data. We will return to these ambivalences of empirical social research in chapter 10.

Causal Mechanisms and Substantive Theory

The identification of causal mechanisms has been one of the classic concerns in sociology (Weber 1972). Causal statements are made to explain the occurrence of events, to understand why particular events happen, and to make predictions when the situation changes (Marini and Singer 1988). Although sociologists sometimes seem reluctant to use the word *cause*, they are far less opposed to apply very similar words such as *force*, *agency*, or *control* when trying to understand social phenomena.

As discussed earlier, there seems to be a consensus that causal inferences cannot be made simply and directly from empirical data, regardless of whether they are collected through sophisticated research designs or summarized by particularly advanced statistical models. Therefore, we cannot prove causality with event history observation plans and event history models per se, as is the case with all other statistical methods. However, as already shown in section 1.1, event-oriented observation designs offer richer information and, as we attempt to demonstrate in this book, event history models provide more appropriate techniques for exploring causal relations.

If we view causality as a property of theoretical statements rather than the empirical world itself (Goldthorpe 1996, 2000), then causal statements are based primarily on substantive hypotheses that the researcher develops about the social world. In this sense, causal inference is theory-driven (Freedman 1991) and always reflects the changing state of sociological knowledge in a field.[22] Of course, descriptive statements also depend on theoretical views that guide the selection processes and provide the categories that underly any description. The crucial point about causal statements, however, is that they require a *theoretical argument* that specifies the *particular mechanism by which a cause produces an effect* or, more generally, *the way in which interdependent forces in a particular setting influence each other over time*.

The important task of event history modeling is therefore not to demonstrate causal processes directly, but to establish relevant empirical evidence that can serve as a link in a chain of reasoning about causal mechanisms (Goldthorpe 1996, 2000). In this respect, event history models can be particularly helpful tools, as they allow for a time-related empirical representation of the structure of causal arguments.

Three Models of Causal Inference

The goal of finding scientifically sound evidence for causal relationships leads to design questions, such as which inference model is appropriate to specify the relationship between cause and effect, and which statistical procedures can be used to determine the strength of this relationship (Schneider et al. 2007). Two distinct models of causal inference have dominated the work of social science practitioners in recent decades: (i) the model of *causality as robust dependence* and (ii) the model of *causality as consequential manipulation* (Blossfeld 2017a). A third conception of causal inference is the idea of *causality as a generative process*, which was proposed by David R. Cox (1990).

Causation as a robust dependence. The approach "causation as robust dependence"—which in multiple regression or path analysis is referred to as the control variable approach (Duncan 1966) and in the econometric analysis of time series data as "Granger causality" (Granger 1969)—assumes that correlation does not necessarily imply causation, but that causation must imply correlation in one way or another. From this perspective, the main problem of causal inference is to determine whether an observed correlation of the variable X with the variable Y, where X precedes Y in time, can be established as a "genuine causal relationship." X is said to be a true cause of Y if the dependence of Y on X cannot be removed by additional variables introduced into the statistical analysis. In this approach, causality is thus established essentially by removing spurious (or non-causal) influences. Although this approach has dominated the social sciences for several decades, many social scientists today consider it too limited. First, they believe that causal inference should not be restricted to a matter of statistical predictability alone, but should also include predictability in line with theory (Goldthorpe 2001). Second, it is not possible to be certain that all other important variables have actually been controlled for, since scientists rarely know all the causes of the observed effects or how they are related (Shadish et al. 2002). A variable X can therefore never be considered causally significant for Y, except in a preliminary sense.

Causation as Consequential Manipulation. The second approach, "causation as consequential manipulation", seems to have emerged as a reaction to the limitations of "causation as robust dependence". Instead of "establishing the causes of effects", Holland (1986) and Rubin (1980) deal with "establishing the effects of causes." They make clear that it is more to the point to simply accept causes as given and then focus on the question of how their effects can be reliably measured. According to this approach, causes can only be factors that could serve as treatments or interventions in well-designed experiments or quasi-experiments. Thus, when a causal factor X is manipulated, with appropriate experimental control, a systematic effect on the response variable Y is produced. The particular strength of this design is that "[...] while statements in the form 'X is a cause of Y' are likely to be proved wrong as knowledge advances, statements in the form 'Y is an effect of X', once they have been experimentally verified, do not subsequently become false: 'Old, replicable experiments never die, they just get reinterpreted'." (Goldthorpe 2000, p. 198).

Understood in this way, causation is always relative in the sense that the specific treatment of X_{tr} and its observed outcome Y_{tr} are compared to what would have happened to the same unit if it had not been exposed to that treatment (counterfactual presentation of causality). Since it is not possible in the same experiment for an entity to both be exposed to the treatment and not be exposed, the view of "causality as a consequence of manipulation" leads to what Holland (1986) called the fundamental problem of causal inference. For example, a student taking a math program cannot go back in time and take a different one so that the two test scores can be compared, since individuals learn from their past experiences. The question, then, is how researchers can ensure that they obtain convincing measures of the outcome of Y_{con} when the unit was not exposed to the treatment X_{con} in an experiment.

In natural sciences such as physics, it is often relatively easy to conduct tightly controlled laboratory experiments and to show, based on the properties of the objects studied, what would have happened to the same entity if it had not been subjected to the treatment. In other disciplines, such as biology, medicine, economics, sociology, or psychology, however, it is often not possible to assume this kind of temporal stability and causal transience at the level of each unit, and it is usually impossible to eliminate the influence of confounding factors at the level of the unit. For these sciences, Rubin (1980) and Holland (1986) proposed a statistical approach to the fundamental problem of causal inference: Rather than focusing on specific units of study, this approach estimates an average causal effect (ACE) or average treatment effect (ATE) for a population of units: $ACE = E(Y_{tr} \mid X_{tr}) - E(Y_{con} \mid X_{con})$, where $E(Y_{tr} \mid X_{tr})$ is the expected value for treatment group participants and $E(Y_{con} \mid X_{con})$ is the expected value for control group participants. Using randomised experiments, in which participants are randomly assigned to treatment and control conditions, one can assume that assignment to treatment is on average independent of all pre-treatment characteristics, measured and unmeasured, across repeated experiments (Cox 1958). In randomised experiments, treatment assignment and unit response are therefore statistically independent of each other, and any kind of selection or confounding bias is eliminated. Today, the randomized control trial (RCT) is considered the "gold standard" for causal inference (Holland 1988). Of course, there are many practical research questions that could undermine the credibility of a causal effect estimated with an RCT (Blossfeld 2017a).

In the context of this book, we will limit ourselves to three important restrictions of the "Effect of Causes Design" that are particularly relevant for answering questions in long-term life course research:

Contexts of RCTs. A well-designed RCT maximises internal validity, i.e. it provides an unbiased ATE and "gets causation right." However, one cannot simply assume that a causal effect established in this way is immutable in different contexts (external validity). Rather, our findings from observational studies over the life course suggest that causal effects in modern societies vary between different population groups and societal settings. These differences between contexts can only be implemented empirically in RCTs with great effort.

Changes of Contexts in RCTs. Furthermore, macroeconomic changes (e.g. unemployment rates, growth rates and changes in population structure) could be relevant for certain causal effects, but in practice are difficult to integrate into a (long-term) experimental design. However, it is necessary to control for changing contexts in long-term RCTs. In other words, causal effects established by experiments are (at least) implicitly based on highly specialised economic, cultural or historical social structures (Deaton and Cartwright 2016). It is therefore necessary to determine whether the causal effect of an experiment is only applicable in a specific historical context (Deaton and Cartwright 2016) or whether it is also credible outside the original experimental setting.

Long-Term Effect Shapes in RCTs. Another important point in RCTs is the question of whether the effect is actually time-constant or time-dependent. In many long-term studies in the social sciences, the effects are time-dependent. These temporal forms of the unfolding effect are difficult to study with RCTs.

Causation as a Generative Process. For Cox (1990, 1992), the idea of "causation as a generative process" is crucial for causal inference. From this perspective, the underlying causal process needs to be elaborated in time and space. In this book, we show how event history data can be used to examine the effect shape over time, to analyze the effects of parallel and interdependent processes, and to cope with changing contexts of causal processes. Event history models are very closely related to the understanding of "causation as a generative process", because—as we will see below—the transition rate provides a local description of the development of the causal process over time (Blossfeld 2017b). These models attempt to predict future changes in the transition rate of the dependent process at each point in time based on events of possibly interdependent processes in the past. In the following, we will elaborate on these advantages of event history data for causal analysis. Furthermore, we describe in chapter 10 how unobserved heterogeneity and selection processes can be handled in event history models.

Attributes, Causes, and Time-Constant Variables

Holland (1986) tried to establish some links between causal inference and statistical modeling. In particular, he emphasized that it is essential for a conception of causality that each unit of a population must be exposable to any of the various levels of a cause—at least hypothetically. He argued, for example, that the schooling a student receives may be a cause of that student's performance on a test, whereas the student's race or sex cannot. In the former case, it seems possible to consider measuring the causal effect, whereas in the latter cases, where we have the enduring attributes of a student, only association can be discussed (Yamaguchi 1991).

We agree with Holland that causal mechanisms imply a counterfactual reasoning: *If* the cause *had been* different, there *would have been* another outcome, at least with a certain probability. In this sense, counterfactual statements reflect imagined situations. However, it is not always clear which characteristics of a situation can sensibly be assumed to be variable (i.e., can be used in counterfactual reasoning) and which characteristics should be regarded as fixed. At least to some degree, this distinction depends on the field of investigation. For example, from a sociological point of view, what is important with regard to gender is not the biological attributes per se, but the social meaning attached to these attributes. The social meaning of gender can change regardless of whether their biological basis changes or not. For example, societal rules might change to create more equality between the races or sexes. We therefore think that, in sociological applications, counterfactuals can also be applied meaningfully to such attributes. They can be represented as *time-constant* "variables" in statistical models to examine their possible impact on some outcome to be explained. It is, however, important to be quite explicit about the sociological meaning of causal statements that involve references to biological or ethnic attributes. There is, for example, no eternal law connecting gender and/or race with wage differentials. But there are probably *social* mechanisms that link gender and ethnic differences to different opportunities in the labor market.

Causes and Time-Dependent Variables

A key point in causal statements is counterfactual reasoning, which means that causes are states that could be different from what they actually are. However, the consequences of states that could be different from their actual state are obviously not observable. To find an empirical approach to causal statements, the researcher must look at conditions that actually do change over time. These changes are events. In the context of variables, one could say that an event is a change of the value of a variable, and this change must happen at a specific point in time. This implies that the most obvious empirical representation of causes is in terms of time-dependent variables which can change their states over time. Formal notations will be introduced in section 1.2.2.

The role of a time-dependent covariate in event history models is to indicate that a (qualitative or metric) causal factor has changed its state at a specific time and that the unit under study is exposed to a different causal condition. For example, in the case of gender, the causal events could be the steps in the acquisition of gender roles over the life course or the exposure to gender-specific opportunities in the labor market at a particular historical point in time. Thus, a time-constant variable "gender," should ideally be replaced in an empirical analysis by time-varying events assumed to produce gender-specific differences in the life histories of men and women. Of course, that is not always possible in empirical research, so one often has to rely on time-constant "variables" as well. However, it is important to recognize that for these variables, the implied *causal relation* is not examined longitudinally. For example, if we observe an association among people with different levels of educational attainment and their job opportunities, then we usually draw the conclusion that changes in job opportunities are a result of changes in educational attainment level. The implied idea is the following: If we started having people with the lowest educational attainment level and followed them over the life course, they would presumably differ in their rates of attaining higher levels of educational attainment and this would lead to changes in job opportunities. Whether this would be the case for *each* individual is not very clear from a study based on people with *different* levels of education. In particular, one would expect that the causal relationship between education and job opportunities would be altered radically if all people attained a higher (or the highest) level of education.[23] The two statements—the first about associations across different members of a population and the second about life course dependencies for each individual member of the population—are thus quite different; one type of statement may be empirically true, while the other one may be empirically false. Therefore statements of the first type cannot be regarded as substitutes for statements of the second type. However, since all causal statements have consequences for longitudinal change (see Lieberson 1985), only time-changing variables provide the most convincing empirical evidence for causal relationships.[24]

Time Order and Causal Effects

We can summarize our understanding of causal statements as follows:

> A change in a variable X at time t contributes causally to a change in a variable Y at a later time t'.

Of course, this does not mean that a change in X is the only cause that might affect a later change of Y. We therefore also speak of *causal conditions* to stress that there can be, and usually are, a number of complex causes.[25]

So when causal statements are examined empirically, they must be closely related to time. There are three important aspects: Firstly, to speak of a change in variables necessarily implies *reference to a time axis*. We need at least two points in time to observe that a variable has changed its value. Of course, we can say, at least approximately, that a variable has changed its value *at a certain point in time*.[26] This will be used below when we formally define time-dependent variables.

Secondly, there is a *temporal order between causes and effects*. The cause must *precede* the effect in time: $t < t'$, in the formal representation given earlier. This seems to be generally accepted.[27] As an implication, there must be a *temporal interval* between the change in a variable representing a cause and a change in the variable representing a corresponding effect. This time interval may be *very short* or *very long*, but it can never be *zero* or *infinity* (Kelly and McGrath 1988). Thus, logically *the cause and its effect cannot occur at the same point in time*. Any appropriate empirical representation of causal effects in a statistical model must therefore take into account that there may be various *delays* or *lags* between the events assumed to be causes and the unfolding of their effects (see a and b in Figure 1.1.2).

This leads directly to a third point: There may be a variety of different *temporal shapes (functional forms) in which the occurrence of a change in Y_t unfolds over time*. Some of these possibilities have been illustrated in Figure 1.1.2. A proper understanding of causal relationships between variables should therefore take into account that the *causal relationship itself may change over time*. This seems particularly important in sociological applications of causal reasoning. In these applications, we generally cannot rely on the assumption of eternal, timeless laws, but have to recognize that causal mechanisms may change during the development of social processes.

It is sometimes argued that sociologists should abandon the causal analytical view in favor of a systems view, because the operation of causal forces is mutually interdependent, and variables change more or less simultaneously in many systems (Marini and Singer 1988). However, even in systems of interdependent processes, time does not run backwards, and the change in one of the interdependent variables takes (at least a little) time to produce a change in another one. In systems of variables, therefore, there cannot be any simultaneity of causes and their effects. This allows us to demonstrate in chapter 6 that a causal approach to interdependent systems is possible using the concept of transition rate. In other words, the systems view is not a substitute for a proper causal approach in our field (Kelly and McGrath 1988).

Actors and Probabilistic Causal Relations

It is generally agreed that social phenomena are always based directly or indirectly on the actions of individuals. This clearly separates the social from the natural sciences. Sociology is therefore not concerned with associations among variables per se, but with variables that are linked to each other through the actions of individuals. This has at least three consequences for causal relationships. First, from a methodological point of view, this means that if individuals relate causes and effects through their actions, then research on social processes should be based at best *on individual longitudinal data* (Coleman and Hao 1989; Coleman 1990; Blossfeld and Prein 1998; Goldthorpe 2000). This is why life history data on individuals, and not aggregated longitudinal data, provide the most appropriate information for analyses of social processes. Only with these data can one trace the courses of action at the level of each individual over time. Second, in theoretical terms, it means that the explaining or understanding of social processes requires a time-related specification of (1) the past and present conditions under which people act,[28] (2) the many and possibly conflicting goals that they pursue at the present time, (3) the beliefs and expectations guiding their behavior, and (4) the actions that probably will follow in the future.[29] Third, if it is people who act, then causal inference must also take the free will of individuals into account (Blossfeld and Prein 1998). This introduces an essential element of indeterminacy into causal inferences. It means that in sociology, we can only reasonably consider and model the *generality* but not the determinacy of behavior. The aim of substantive and statistical models must therefore be to capture common elements in the behavior of people or patterns of action that recur in many cases (Goldthorpe 1996, 2000). This means that in sociological applications randomness has to enter as a defining characteristic of causal models. We can only hope to make meaningful causal statements about how a given (or hypothesized) change in variable X_t in the past affects the *probability of a change* in variable $Y_{t'}$ in the future. Accordingly, the basic form of a causal statement is:

> A change in a variable X at time t contributes causally to a change in the probability of a change in a variable Y occurring at a later time

In sociology, this interpretation seems more appropriate than the traditional deterministic approach. The essential difference is not that our knowledge about causes is insufficient because it only permits probabilistic statements; but that *causal explanations concern probabilities*. In this context, probability is no longer just a technical term, but is regarded as a theoretical one: It is the propensity of social actors to produce and experience events.

Causal Statements and Limited Empirical Observations

A quite different type of randomness related to making inferences occurs when causal statements are applied to real-world situations in the social sciences. There are at least four additional reasons to expect further randomness in empirical studies. These are essentially the same ones as those that occur in deterministic approaches and are well known from traditional regression modeling (Lieberson 1991, Lee 2022). The first reason is *measurement error*, a serious problem in empirical social research, meaning that the observed data deviate somewhat from the predicted pattern without invalidating the causal proposition. The second reason is particularly important in the case of nonexperimental data. It is often the case that *complex multivariate causal relations* operate in the social world. For example, a certain result may occur due to the presence of more than one influencing factor. In addition, an outcome may not occur because the influence of an independent variable is outweighed by other influences acting in the opposite direction. In such situations, the observed influence of the cause is only approximate unless the other important factors can be controlled. The third motive is that sociologists often do *not know* or are *not able to measure* all the important factors. Hence, social scientists must abandon the idea of a complete measurement of causal effects, even if they would like to make a deterministic proposition. Finally, *chance* sometimes affects observed outcomes in the social world. It is not important here to decide whether chance per se exists or whether it is merely a surrogate for the poor state of our knowledge of additional influences and/or inadequate measurement.

In summary, these problems imply that social scientists can only hope to make empirical statements of a probabilistic nature. This situation can lead to problems, as will be discussed in chapter 10. Without strong assumptions about missing information and errors in the available data, it is generally not possible to make unambiguous statements about causal relationships (see, e.g., Arminger 1990).

A Simplistic Conception of Causal Relationships

At this point, it is important to stress that the concept of causal relation is a rather special abstraction implying a time-related structure that does not immediately follow from our sensory impressions. Consider the following example in Figure 1.2.1 in which we characterize the necessary time-related observations of a unit that is affected by a causal effect. This figure shows that an empirical representation of the most simple causal effect (i.e., [1] where the condition X_t changes—from one state $X_{t_1} = x_1$ to another one $X_{t_2} = x_2$—and [2] is then constant afterward, [3] the change in Y_t—from $Y_{t_2} = y_1$ to $Y_{t_3} = y_2$—takes place almost instantaneously and [4] is also time-constant thereafter) requires at least three points in time at which the researcher must note the states of the independent and dependent variables, respectively.[30] This is because, if we assume that a change in the independent variable X_t has taken place at t_2, then we need to know the

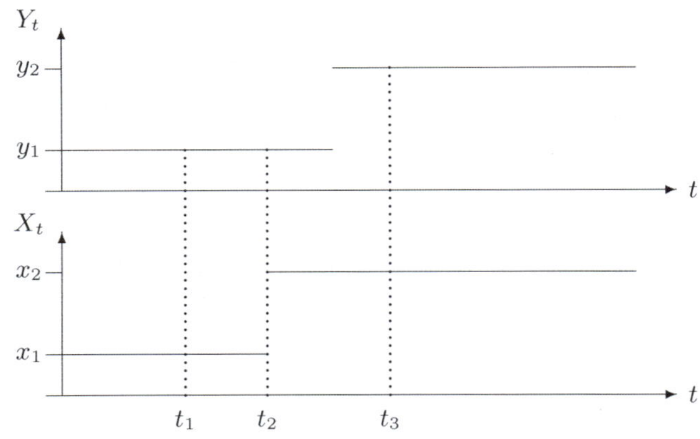

Figure 1.2.1 Observation of a simple causal effect.

state of the independent variable X_t at an earlier time, t_1, in order to be able to fix the particular change in the condition in the past, (see Figure 1.2.1). For the dependent variable Y_t we need an observation before the effect occurred. Assuming that everything else is constant, this observation can be made at the latest at point t_2, as the effect must *follow* the cause in time. To evaluate whether the hypothesized effect has indeed taken place at a later time, t_3, we must again record the state of the dependent variable Y_t. Thus, a *simple representation of a causal effect* exists if we compare the change in the observations for the independent variable in the past and the present with the change in the observations for the dependent variable in the present and in the future and link both changes with a substantive argument.[31]

However, as already demonstrated in section 1.1, this is only a simple and rather unrealistic example of a causal effect. In the case of observational data in the social sciences in which there are many (qualitative and metric) causal variables that can change their values at any point in time and whose causal effects can have different lags and various shapes in time (see Figure 1.1.2), the quantity of the observed causal effect as shown in Figure 1.2.1 depends strongly on when the measurements are taken at the three points in time.

Thus, what can we say about the causal effect(s) at any given point in time if the situation is more complex? A paradox arises: The concept of causal effect depends on a comparison of the changes in the independent and dependent variables at at least three points in time. Yet the concept of causal effect should itself reflect a state of a unit of observation at any point in time that is appropriate in real empirical situations. Thus, what is still needed in our discussion is a concept that represents the quantity of the causal effect at any point in time.

1.2.2 Transition Rate Models

The Concept of the Transition Rate

The formulation of causal statements proposed above refers to *a change in the probability of a change in a variable Y occurring at time t*. The concept of transition rate can help to make this formally more precise. This requires a notation for time-dependent variables. Here, we focus on variables representing the occurrence of a single event (often interpreted as a transition from a state j into a state k), and assume a continuous time axis starting at $t = 0$. To represent the occurrence of an event of a specified type (e.g., the transition from "being unmarried" to "being married"), it suffices to use a single variable, say T_y, whose value is the point in time when the event of type y occurs ($T_y = \infty$ if the event never occurs). With this notation, one can abbreviate *a change in a variable Y at time t " to "$T_y = t$*.

The term transition rate is based on T_y as a random variable. This allows us to consider the expression

$$\Pr(t \leq T_y < t' \mid T_y \geq t). \tag{1.1}$$

This is the probability that an event of type y occurs in the time interval from t to t', given that no event of type y has occurred before (i.e., in the interval from 0 to t). This probability is well-defined and is well-suited to describe the temporal evolution of the process. The definition refers to each point in time at which the process develops and can thus express the idea of change during its development. Moreover, the definition relies only on information about the past of the process: what has happened up to the present time (t). Therefore the concept can be used meaningfully applyed to describe the process before it is completed for all individuals in the population. Suppose we know the probabilities defined in (1.1) for all time points up to a certain point t^*. Then, we have a description of the process up to this point in time, and this description is possible without knowing how the process will develop in the future (i.e., for $t > t^*$).

The *ratio* of the transition probability to the length of the time interval represents the probability of future changes in the dependent variable per unit of time (Coleman 1968); that is, we consider

$$\Pr(t \leq T_y < t' \mid T_y \geq t) \,/\, (t' - t). \tag{1.2}$$

As t' approaches t, we define the central notion of a transition rate as

$$r(t) = \lim_{t' \to t} \frac{\Pr(t \leq T_y < t' \mid T_y \geq t)}{t' - t}. \tag{1.3}$$

Due to the different origins of the transition rate framework in different disciplines, the transition rate is also referred to as *hazard rate*, *intensity rate*, *failure rate*, *transition intensity*, *risk function*, or *mortality rate*.

The concept of transition rate obviously provides the possibility of giving a local, time-related description of how a process evolves over time. We can interpret $r(t)$ as the *propensity* to change states, from origin j to destination k, at t. But note that this propensity is defined with respect to a *risk set* at moment t (i.e., the set of individuals who may experience the event because they have not experienced it before t).

The transition rate is also a suitable tool for modeling the *time arrow* (Conveney and Highfield 1990) of social processes and for conceptually distinguishing at each point in time the *presentness* from a *closed past* and an *open future*. The transition rate enables us to link the events of the closed past with the intensity of *possible future changes* at any point in time. Thus, the transition rate is particularly well suited for causal analysis, because stochastic processes require an approximation of the *underlying distribution* from which future realizations of events are "drawn" (March, Sproull, and Tamuz 1991). The appropriate focus of causal analysis based on stochastic processes is the *distribution of possible future events* (or their potentiality), *not their specific "realizations."*

Additional Statistical Concepts

Because the transition rate is indeed an abstraction, we need to relate it to directly observable quantities; that is, to frequencies of state occupancies at particular points in time. To support such inferences, some additional statistical concepts are useful: As before, we refer to a variable T_y that captures the duration until the occurence of a change from state j into state k. We use the word "episode" to denote the process that begins at $t = 0$ in state j until it ends with a transition to the new state k.

We begin with the basic concept for describing the probability distribution of T_y; that is, the distribution function

$$F(t) = \Pr(T_y \leq t).$$

This is the probability that the episode's duration is less than or equal to t; or, in other words, the probability that an event happens in the time interval from 0 to t. Equivalently, we can describe the probability distribution of T_y by a *survivor function*, defined by

$$G(t) = 1 - F(t) = \Pr(T_y > t).$$

This is the probability that the episode's duration is at least t, and that the event that ends the current episode occurs later than t.

Both concepts, the distribution function and the survivor function, are mathematically equivalent. However, when describing event histories, we generally prefer the survivor function, as it provides a more intuitive description. We can imagine a population of individuals (or other units of analysis) that all begin a particular episode with origin state j at the same point in time $t = 0$. Then, as time goes on, events occur (i.e., individuals leave the given origin state). It is exactly this process that is described by the survivor function. If N is the size of the population at $t = 0$, then $N \cdot G(t)$ is the number of individuals who have not yet left the origin state up to t. Sometimes this is called the "risk set" (i.e., the set of individuals who remain at "risk" of experiencing the event that ends the current episode).

Since T_y is a continuous random variable, its distribution can also be described by a density function:

$$f(t) = \lim_{t' \to t} \frac{\Pr(t \leq T_y < t')}{t' - t}.$$

The definition is similar to the definition of the transition rate (1.3). On the right-hand side, before going to the limit, we have the probability that the event occurs in the time interval from t to t'. $f(t)$ is approximately proportional to this probability, when the time interval becomes very short. The density is related to the distribution function by

$$F(t) = \int_0^t f(\tau)\,\mathrm{d}\,\tau. \tag{1.4}$$

The distribution function, survivor function, and density function are quite common concepts to describe the probability distribution of a random variable. However, these functions do not make explicit that our random variable T_y has a quite specific meaning: the duration of an episode. Our mathematical concepts are intended to describe a *process evolving in time*. In defining such a process, we refer to a population of individuals (or other units of analysis) who are considered as "bearing" the process. These individuals evolve over time, and their behavior generates the process. In relation to these individuals, and as the process evolves, a distinction is always made between past, present and future. This is particularly important for a causal view of the process. The past conditions the present, and what happens in the present shapes the future. The question is how these temporal aspects of the process can be made explicit in the concepts we use to describe the process. As we have seen, the development of an episode can be represented by a random variable T_y, and statistics offers well-known concepts to describe the distribution of this variable. However, these concepts have hidden the temporal nature of the process. This becomes clear if we ask the question *when* does a description of the distribution of T_y become available. At the earliest, this is the case when the current episode has ended for all individuals in the population. Therefore, although a description of the distribution of T_y provides a description of the process *as it had evolved*, we need a quite different description in order to make a causal assessment of the evolution of the process. We need a concept that allows us to describe the development of the process at every point in time, while the process is running, and that relies only on information about the past development of that process for its definition.

Now we can re-examine the relationship with the transition rate. By definition, we have

$$\Pr(t \leq T_y < t' \mid T_y \geq t) = \frac{\Pr(t \leq T_y < t')}{\Pr(T_y \geq t)}.$$

Therefore, definition (1.3) can also be written as follows

$$r(t) = \lim_{t' \to t} \frac{\Pr(t \leq T_y < t')}{t' - t} \, \frac{1}{\Pr(T_y \geq t)} = \frac{f(t)}{G(t)}. \tag{1.5}$$

This shows that the transition rate is a conditional density function; that is, the density function $f(t)$ divided by the survivor function $G(t)$.

The transition rate enables a local description of the development of a process. To calculate $r(t)$, we need information about the local probability density for events at t, given by $f(t)$, and about the development of the process up to t, given by $G(t)$. Of course, if we know the transition rate for a time interval, say t to t', we have a description of how the process evolves during this time interval. And if we know the transition rate for all (possible) points in time, we eventually have a description of the entire process. This is mathematically equivalent to a complete description of the distribution of T_y.

There is a simple relationship between the transition rate and the survivor function. First, given the survivor function $G(t)$, we can easily derive the transition rate as (minus) its logarithmic derivative:[32]

$$\frac{\mathrm{d}\log(G(t))}{\mathrm{d}t} = \frac{1}{G(t)}\frac{\mathrm{d}G(t)}{\mathrm{d}t} = \frac{1}{G(t)}\frac{\mathrm{d}}{\mathrm{d}t}(1 - F(t)) = -\frac{f(t)}{G(t)} = -r(t).$$

With this relationship, the other direction is provided by integration. We have

$$-\int_0^t r(\tau)\,\mathrm{d}\,\tau = \log(G(t)) - \log(G(0)) = \log(G(t)),$$

because $G(0) = 1$. It follows that the basic relationship, often used in subsequent chapters, is as follows:

$$G(t) = \exp\left(-\int_0^t r(\tau)\,\mathrm{d}\,\tau\right). \tag{1.6}$$

The expression in brackets,

$$\int_0^t r(\tau)\,\mathrm{d}\,\tau,$$

is often referred to as the *cumulative hazard rate*.

Finally, note that $r(t)$ is a transition *rate* and not a transition probability (Blossfeld 2026). As shown in (1.5), $r(t)$ is similar to a density function. To derive correct probability statements, we need to integrate over a time interval, as follows:

$$\Pr(t \leq T_y < t' \mid T_y \geq t) = \frac{G(t) - G(t')}{G(t)} = 1 - \frac{G(t')}{G(t)}$$

$$= 1 - \exp\left(-\int_t^{t'} r(\tau)\,\mathrm{d}\,\tau\right). \tag{1.7}$$

It is easy to verify, however, that $1 - \exp(-x) \approx x$ for small values of x. Therefore, the probability that an event happens in a *small* time interval (t, t') is approximately equal to $r(t)$ multiplied by the length of that interval (see also Blossfeld 2026):

$$\Pr(t \leq T_y < t' \mid T_y \geq t) \approx (t' - t)\, r(t). \tag{1.8}$$

Transition Rate Models

We have tried to show that transition rates are well suited to formally represent an explanandum: the propensity of social actors to change their state. Transition rate models can be used to investigate how a transition rate $r(t)$ depends on values of covariates representing the situation at time t and may be relevant for the propensity to change states.

We begin with a time-constant covariate, X, whose values do not change after the beginning of the process at $t = 0$. In order to make the transition rate $r(t)$ dependent on values of X, we can simply use a conditional transition rate:

$$r(t \mid X = x) = \lim_{t' \to t} \frac{\Pr(t \leq T_y < t' \mid T_y \geq t, X = x)}{t' - t}. \tag{1.9}$$

In order to think of effects, we need to refer to different values of X. We use the notion of a contrast (x', x'') where x' and x'' are two possible values of X. Given a contrast, we can compare two transition rates, $r(t \mid X = x')$ and $r(t \mid X = x'')$. The term effect, as it is usually used in the field of statistical models, is intended to represent this comparison by a quantity. This can be done in different ways. A simple definition that will be used in the remainder of this introduction uses the difference

$$r(t \mid X = x'') - r(t \mid X = x'). \tag{1.10}$$

Another definition that will be used in chapter 4 employs the ratio

$$r(t \mid X = x'')/r(t \mid X = x'). \tag{1.11}$$

Of course, since X is a time-constant variable, a change from x' to x'' does not actually take place. However, as explained above, we nevertheless might be willing to engage in counterfactual considerations and then think about possible causal meanings (Dickerman and Hernán 2020).

Although X is a time-constant variable, the effect defined in (1.10) can change during the process. This will be the case if the transition rate also depends on the process time, t. In any case, the effect is a temporally local concept. It refers to a relationship between a transition rate and values of covariates in the situation at time t. The definition is therefore well suited to capture the fact that effects of covariates almost always depend on a context that can change over time (Blossfeld 2017a, 2017b).

In order to represent a changing context, we can use time-dependent covariates. As an example, we use a variable T_z that captures the time at which a person gets married ($T_z = \infty$ if a marriage never occurs). Since $T_z = t_z$ is to be used to formulate a condition, it is helpful to use variables Z_t that are defined for each point in time, t, as follows:

$$Z_t = \begin{cases} 0 & \text{if } t < t_z \\ 1 & \text{if } t \geq t_z. \end{cases}$$

Consequently, $Z_t = 0$ means that a person is not married at t, and $Z_t = 1$ means that she or he is married at t. These variables can be used formally in the same way as time-constant covariates.[33] In the notation (1.9), we can simply replace $X = x$ by $Z_t = z_t$ (or add $Z_t = z_t$ as a further condition). The reason for this is that effects are locally defined in time, which means that the transition rate at t depends only on covariate values at this point in time. It is of course possible to include further information. For example, it is quite possible that a transition rate does not only depend on whether a person is married at t, but also on the past duration of the marriage. We would then define a further time-dependent covariate, say D_t, which indicates the previous duration of the marriage at t. Including both variables simultaneously, the model becomes

$$r(t \mid Z_t = z_t, D_t = d_t).$$

This is still a temporally local approach that makes the transition rate at t dependent on covariate values at this point in time.

The fact that information from time-dependent covariates is used in a temporally local way is important for understanding effects. In the case of time-dependent covariates, the effects as defined in (1.10) or (1.11) are based on a comparison between different groups of persons, as is the case for time-constant covariates. For example, we can compare persons who are married at t with persons who are not married at t, or compare two groups of married persons with different marriage durations. Effects as defined in (1.10) or (1.11) are therefore called *comparative effects*.

Parametric Model Specifications

Given sufficient data, it would be possible to estimate effects for each contrast (x', x'') and temporal location t separately. However, due to data limitations, we mostly use parametric model specifications. A general notation is

$$r(t \mid X = x) = g(t, x; \theta). \tag{1.12}$$

The expression on the right-hand side represents a mathematical function with the arguments t and x (possible values of the covariate X), which depends on a parameter vector θ. Many different parametric models can be defined. This will be discussed later in this book. At this point, we would like to emphasize that these models are not just innocuous means for estimating numerical values of the conditional transition rates postulated on the left-hand side, but often entail strong assumptions arising from the chosen mathematical form. For example, if X is a single variable, we might use:

$$r(t \mid X = x) = \alpha + x\,\beta, \tag{1.13}$$

which entails the assumption that the transition rate does not depend on the process time. Because transition rates can assume only non-negative values, we might also use

$$r(t \mid X = x) = \exp(\alpha^* + x\,\beta^*), \tag{1.14}$$

where exp( ) denotes the exponential function. If X is a binary variable, both specifications provide identical estimates of an effect of X when defined according to (1.10).[34] If, on the other hand, X is a quantitative variable such as age, both models entail different assumptions about the dependency on age. The first formulation assumes a linear relationship; the second a nonlinear relationship. Consequently, estimates of effects will also be different.

We also learn that it is important to distinguish between model parameters and effects.[35] Model parameters are merely a means for calculating transition rates and effects of covariates. Consequently, there is no reason why we should be interested in the "interpretation" of parameters. Only in very simple linear models without interactions can parameters directly represent an effect if they are defined according to (1.10). When transition rate models are specified with an exponential link function, as in, for example, (1.14), the definition (1.11) might be preferred, because parameters directly provide information about effects (this will be illustrated in chapter 4). However, this assumes that there are no interactions between covariates. In any case, whenever there is no direct correspondence between parameters and effects, it is more informative to present estimates of effects.

It should be noted that a transition rate model has no dependent variable in the usual sense; and consequently, there is no "error term." It is possible, of course, to consider a regression model for T_y such as:

$$T_y = \alpha + x\,\beta + \epsilon,$$

where ϵ denotes a residual variable. However, this model serves a different purpose: to investigate how durations lasting until the occurrence of a state change depend on values of covariates. Consequently, effects of covariates also have a different meaning. They refer to differences between conditional expectations:

$$\mathrm{E}(T_y \mid X = x'') - \mathrm{E}(T_y \mid X = x').$$

In contrast, the effect defined in (1.10) concerns the propensity to change a state and is conceptually associated with a temporally local situation. In fact, since a regression model for T_y assumes completed durations (even if observations are censored), it is not compatible with the idea that a value of T_y is an outcome of a process that evolves through a sequence of temporally local situations. Consequently, it is not easy to include covariates that may change their value as the process that generates a value of T_y is evolving. Therefore, regression models for T_y are no longer considered in this book.

Context-Dependent Effects

Apart from artificially constructed examples, the effects of covariates almost always depend on a context; that is, on values and distributions of other covariates. Let us consider a transition rate model

$$r(t \mid X = x, Z = z),$$

where X and Z are two time-constant covariates. For a contrast (x', x''), the corresponding effect is

$$r(t \mid X = x'', Z = z) - r(t \mid X = x', Z = z). \tag{1.15}$$

In general, this effect will depend on the value of Z used in the evaluation. This means that the effect cannot be attributed to X alone.

The dependence of effects on a context should be taken seriously; and, if possible, it should be made explicit by, for example, by listing the effects for different contexts separately. The use of marginal effects can instead hide important information. In our example, the marginal effect is based on a reduced model that containing only X:

$$r(t \mid X = x) = \sum_z r(t \mid X = x, Z = z) \; \Pr(Z = z \mid T_y \geq t, X = x).$$

This shows that the marginal effect, $r(t \mid X = x'') - r(t \mid X = x')$, is not just a simple mean of the conditional effects given in (1.15), because the distribution of Z is different for $X = x'$ and $X = x''$.

The conclusion is not that effects derived from reduced models are in any way "wrong." But they arise both from a focal variable (X in this example) and from all other variables whose distributions contribute to marginalization (Z in the example). A marginal effect therefore hides information that could be revealed by a consideration of conditional effects. Of course, this is only possible insofar as variables are observed and included in a model. Therefore, conditional effects can also be considered as marginal effects with respect to omitted variables. In any case, effects are derived from a comparison of groups of units that have different distributions of other relevant variables (see also Kuha and Mills 2018).

Even if X and Z are independent, we have to deal with the fact that the distribution of Z also depends on $T_y \geq t$. This entails that the distribution of Z also changes due to changes in the composition of the risk set (an example will be discussed in section 10.1). However, as long as we can explicitly condition on Z, this only affects the relative number of cases in the two groups that differ by x' and x''. This is also an argument in favor of conditional effects.

Temporally Remote Effects

Finally, we consider effects of a time-dependent variable, say T_z, which indicates the time at which an event occurs. The corresponding state variables are Z_t: $Z_t = 1$ means that the event occurred at, or before, t; and $Z_t = 0$, otherwise. Based on a contrast $(0, 1)$, a temporally local effect can be defined as

$$r(t \mid Z_t = 1, X = x) - r(t \mid Z_t = 0, X = x) \tag{1.16}$$

if $X = x$ is used as a context. This is exactly analogous to the definition of effects of time-constant covariates.

However, referring to an event that occurs at $T_z = t_z$ allows us, at least formally, to represent temporally remote effects of the event for points in time $t \geq t_z$. For example, we can think of a transition rate for the occurrence of marriages in consensual unions, and the question concerns effects of a pregnancy $T_z = t_z$ during a time period following t_z (Blossfeld et al. 1999).[36] A definition requires the reference to two groups to be compared during $t \geq t_z$. A treatment group can simply be defined as consisting of all cases who experienced the event at t_z. To define a control group, we use the following principles (Rohwer 2016): (a) The control group for time t should not contain any cases who experienced the event before or at t. (b) There should be no conditioning on the future. This requires that the control group for time t should not exclude cases who experience the event later than t.

This leads to the following definition of the effect for a point in time $t \geq t_z$:

$$r(t \mid T_z = t_z, X = x) - r(t \mid T_z > t, X = x). \tag{1.17}$$

For each point in time t, this is a temporally local effect. Viewed as a function of $t \geq t_z$, we call it an *effect shape* that provides information on how the effect changes over time.

Like other comparative effects, temporally remote effects also depend on a context. In part, this can be made explicit by including time-constant covariates (such as X, in our example). In general, we must also think of time-dependent covariates that can change their value at times greater than t_z. Conditioning on such events is difficult, if not impossible, as it would entail changing the definition of the effect we want to evaluate.[37] As always when calculating comparative effects, we need to think of unobserved variables whose distributions may develop in different ways in the treatment and control groups. The calculation and interpretation of effect shapes is therefore difficult, and will not be discussed further in this introduction to event-history analysis.

Notes

[1]See, e.g., Heckman and Borjas 1980; Andreß 1989; Galler and Pötter 1990; Huinink et al. 1995; Bernardi et al. 2000; McGinnity 2004; Blossfeld et al. 2005; Golsch 2005; Blossfeld, Buchholz, and Hofäcker 2006; Blossfeld and Hofmeister 2006; Blossfeld, Mills, and Bernardi 2006; Covizzi 2008; Gesthuizen and Wolbers 2010; Evertsson, Grunow, and Aisenbrey 2016; Hägglund and Bächmann 2017; Lott 2020; Kay 2023.

[2]See, e.g., Bernasco 1994; Blossfeld and Hakim 1997; Blossfeld, Drobnič, and Rohwer 1998; Courgeau and Guérin-Pace 1998; Cramm, Blossfeld, and Drobnič 1998; Smeenk 1998; Bernardi 1999a, 1999b; Blossfeld and Drobnič 2001.

[3]See, e.g., Sørensen and Tuma 1981; Blossfeld 1986; Carroll and Mayer 1986; Carroll and Mosakowski 1987; Mayer and Carroll 1987; Blossfeld and Mayer 1988; DiPrete and Whitman 1988; Diekmann and Preisendörfer 1988; Hachen 1988; Andreß 1989; Becker 1993; Brüderl, Preisendörfer, and Ziegler 1993; DiPrete 1993; Allmendinger 1994; Blau 1994; Esping-Andersen, Leth-Sørensen, and Rohwer 1994, Mach, Mayer, and Pohoski 1994; Jacobs 1995; Huinink et al. 1995; Halpin and Chan 1998; Blau and Riphahn 1999; Drobnič and Blossfeld 2004 Burton and Beckman 2007; Windzio 2008; Buchmann, Kriesi, and Sacchi 2009; Ruiter and De Graaf 2009; Jayawarna et al. 2021.

[4]See, e.g., Mulder and Smits 1999; Kurz 2000; Kurz and Blossfeld 2004; Lersch and Luijkx 2015.

[5]See, e.g., Hoem 1983, 1986, 1991; Rindfuss and John 1983; Rindfuss and Hirschman 1984; Hoem and Rennermalm 1985; Michael and Tuma 1985; Sørensen and Sørensen 1985; Papastefanou 1987; Huinink 1987, 1993, 1995; Diekmann 1989; Leridon 1989; Mayer and Schwarz 1989; Hannan and Tuma 1990; Wu 1990; Liefbroer 1991; Mayer, Allmendinger, and Huinink 1991; Grundmann 1992; Klijzing 1992; Diekmann and Weick 1993; Lillard 1993; Lillard and Waite 1993; Wu and Martinson 1993; Teachman 1983; Lauterbach 1994; Manting 1994; Bernasco 1994; Blossfeld 1995; Blossfeld et al. 1995; Courgeau 1995; Huinink et al. 1995; Lillard, Brien, and Waite 1995; Yamaguchi and Ferguson 1995; Manting 1996; Blossfeld and Timm 1997; Li and Choe 1997; Defo 1998; Ostermeier and Blossfeld 1998; Timm, Blossfeld, and Müller 1998; Blossfeld et al. 1999; Brian, Lillard, and Waite 1999; Diekmann and Engelhardt 1999; Mills 2000b; Blossfeld and Müller 2002/2003; Blossfeld and Timm 2003; Nazio and Blossfeld 2003; Kravdal and Rindfuss 2008; Lyngstad and Prskawetz 2010; Bernardi and Martínez-Pastor 2011; Erola, Härkönen, and Dronkers 2012; Smith, Maas, and van Tubergen 2012; Carlson, VanOrman, and Pilkauskas 2013; Begall and Mills 2013; Tach and Edin 2013; Balbo and Barban 2014; Kneip, Bauer, and Reinhold 2014; van Damme and Kalmijn 2014; Wagner, Schmid, and Weiß 2015; Yu, Xie, and Xie 2015; Boertien and Mortelmans 2018; Cantalini 2017; Thorsen 2017; Blossfeld, 2022; Neels et al. 2024.

[6]See, e.g., Blossfeld 1986; Carroll and Mayer 1986; Carroll and Mosakowski 1987; Mayer and Carroll 1987; DiPrete and Whitman 1988; Blossfeld and Mayer 1988; Allmendinger 1989a, 1989b; Featherman, Selbee, and Mayer 1989; Mayer et al. 1989; Sørensen and Blossfeld 1989; Esping-Andersen 1990, 1993; Blossfeld and Hakim 1997; Blossfeld, Drobnič, and Rohwer 1998; Drobnič, Blossfeld, and Rohwer 1998; Cramm, Blossfeld, and Drobnič 1998; Blossfeld and Drobnič 2001 Wilson and Roscigno 2010; Wilson and Maume 2014; Sacchi, Kriesi, and Buchmann 2016; Bihagen et al. 2017.

[7]See, e.g., Carroll and Delacroix 1982; Freeman, Carroll, and Hannan 1983; Preisendörfer and Burgess 1988; Hannan and Freeman 1989; Brüderl, Diekmann, and Preisendörfer 1991; Brüderl 1991a; Hedström 1994; Brüderl and Diekmann 1995; Greve 1995; Lomi 1995; Popielarz and McPherson 1996; Hannan et al. 1998a, 1998b; Carroll and Hannan 2000; Wooley and MacGregor 2022.

[8]See, e.g., Strang 1991; Strang and Meyer 1993; Popielarz and McPherson 1996; Chaves 1996; Box-Steffensmeier and Bradford 1997, 2004; Ramirez, Soysal, and Shanahan 1997; Soule and Zylan 1997; Longhofer and Schofer 2010; Wimmer and Feinstein 2010; Walder and Lu 2017; Dudzevičiute et al. 2021; Boldrini and Grimaldi 2023.

[9]See, e.g., Bilsborrow and Akin 1982; Pickles and Davies 1986; Wagner 1989a, 1989b, 1990; Courgeau 1990; Baccaïni and Courgeau 1996; Courgeau, Leliévre and Wolber 1998; Kalter 2011; Liu 2013; Lu, Liang, and David 2013.

[10]See, e.g., Eder 1981; Krempel 1987; Snyder 1991; Peng et al. 2021; Butts et al. 2023.

[11]See, e.g., Blossfeld 1989, 1990; Meulemann 1990; Singer and Willett 1991; Willett and Singer 1991; Becker 1993; Schömann and Becker 1995; Plank, DeLuca, and Estacion 2008; Reisel and Brekke 2010; Roksa and Velez 2012; Müller and Schneider 2013; Blossfeld et al 2014, 2016a, 2016b, 2017; Nieuwenhuis, Hooimeijer, and Meeus 2015; Minello and Blossfeld 2017; Myers et al. 2020.

[12]See, e.g., Olzak and West 1991; Olzak 1992; Olzak, Shanahan, and West 1994; Olzak and Shanahan 1996, 1999; Olzak, Shanahan, and McEneaney 1996; Rasler 1996; Krain 1997; Minkoff 1997, 1999; Myers 1997; Olzak and Olivier 1998a, 1998b; Soule and Van Dyke 1999.

[13]See, e.g., Yamaguchi 1991; Hser et al. 1995; Diekmann et al. 1996.

[14]See, e.g., Yamaguchi and Jin 1999; Mills 2000a.

[15]See, e.g., Eerola 1994.

[16]See, e.g., Allmendinger 1994; Mayer and Baltes 1996; Leisering and Leibfried 1998; Leisering and Walker 1998; Zwick 1998; Hansen 2009; Mooi-Reci and Mills 2012; Mood 2013; Visser et al. 2016; Schels 2018.

[17]See, e.g., Andersen et al. 1993; Maggs et al. 2019; van den Hout et al. 2019.

[18]See, e.g., Cox and Oakes 1984; Tuma and Hannan 1984; Hutchison 1988a, 1988b; Kiefer 1988.

[19]These problems are, however, not mutually exclusive.

[20]The relationship between attitudes and behavior suggests that there is some kind of inertia (or positive feedback), meaning that the probability of a specific behavior increases as a monotonic function of attitudes, and attitudes depend on previous behavior (Davies and Crouchley 1985).

[21]We speak of a "new" approach to emphasize the difference to the traditional "causal analysis" based on structural equation models, which are basically timeless models. See the discussion in Campbell, Mutran, and Nash Parker (1987), Bollen (1989), or Faulbaum and Bentler (1994). Structural equation modeling typically fits a deterministic structure across the observed time points and does not distinguish between a completed past, the present, and a conditionally open future.

[22]Causal relationships are always identified against the background of a particular field, and what is to be considered as background and field is always relative to the conceptual framework under consideration (Marini and Singer 1988). Thus, the observed relationships between stochastic processes generally depend on the number of processes considered. If further processes are taken into account, the causal relationships between them may change. Since the theoretical background in the social sciences is rarely specific enough to determine exactly which processes need to be considered, several valid causal analyses may exist based on different sets of stochastic processes (see Pötter and Blossfeld 2001).

[23]However, a longitudinal approach would provide the opportunity to study this type of change in causal relationships over time.

[24]Another aspect is also important in this context (see Lieberson 1985): Causal relationships can be symmetrical or asymmetrical. In examining the causal influence of a change in a variable X on a change in a dependent variable Y, it must be considered whether shifts to a given value of X have the same consequences for Y from both directions. For example, researchers rarely consider whether an upward shift on the prestige scale, say from 20 to 40, leads to a different outcome of Y (say family decisions) than a downward shift of X from 60 to 40. In other words, most researchers assume symmetry. However, even if a change is reversible, the causal *process* may not be. The question is, if a change in a variable X causes a change in another one, Y, what happens to Y if X returns to its earlier level? "Assuming everything else is constant, a process is *reversible*, if the level of Y also returns to its initial condition; a process is irreversible if Y does not return to its earlier level. Observe that it is the *process*—not the *event*—that is being described as reversible or irreversible" (Lieberson 1985, p. 66).

[25]It is important to note here that the effect of a variable X is always measured relative to other causes. A conjunctive plurality of causes occurs if various factors must be jointly present to produce an effect. In contrast, a disjunctive plurality of causes exists, when the effect is caused by each of several factors alone, and the joint occurrence of two or more factors does not alter the effect (see the extensive discussion in Marini and Singer 1988; and the discussion in regard to stochastic processes in Pötter and Blossfeld 2001).

[26]A statement such as this refers implicitly to some specification of "point in time." The meaning usually depends on the type of event to be described, for instance, a marriage, the birth of a child, or becoming unemployed. In this book, we always assume a continuous time axis for purposes of mathematical modeling. This should, however, be understood as an idealized way of representing social time (Blossfeld 2026). Since we use mathematical concepts to speak about social reality, we disregard the dispute about whether time is "continuous" (in the mathematical sense of this word) or not.

[27]See, for instance, the discussion in Eells (1991, chapter 5).

[28]These conditions are, of course, strongly influenced by social structural regularities in the past and the present. Sociology must always be a historical discipline (Goldthorpe 1991, 2000).

[29]It is sometimes argued that the intentions or motives of actors to bring about an effect in the future cause the actor to behave in a certain way in the present, because human actors act intentionally and behavior is goaloriented (e.g., Marini and Singer 1988). However, this does not contradict a causal view. One simply has to distinguish intentions, motives, or plans as they occur in the present from both their impact on the behavior that follows their formation temporally, and the final result, as an outcome of the behavior. An expectation about a future state should be clearly distinguished from what eventually happens in the future. Therefore the fact that social actors can behave intentionally on the basis of expectations, does not reverse the time order underlying our causal statements.

[30]This example is instructive because Lazarsfeld (1948, 1972) and many others after him have argued that for causal inferences two observations of the units would be sufficient.

[31]Indeed, such a simplistic idea of the causal effect is the basis of all panel designs, as shown in section 1.1.

[32]In this book, we use log(.) to denote the natural logarithm.

[33]Note that Z_t is a single variable defined for the point in time t. As with time-constant variables, values of Z_t cannot change. When we call Z_t a time-dependent variable, we implicitly refer to a sequence of variables Z_t, defined for each $t \geq 0$.

[34]But not if you use the definition (1.11).

[35]If this distinction is not made, a lot of confusion arises. As an example, see Mood (2010). For a critique that emphasizes the distinction between parameters and effects, see Rohwer (2012).

[36]We discuss this example in chapter 6 (example 3).

[37]For a further discussion of these difficulties, see Rohwer (2016).

Chapter 2

Event History Data Structures

This chapter deals with event history data structures. We first introduce the basic terminology for event history data and then give an example of an event history data file. Finally, we show how to use the file in Stata.

2.1 Basic Terminology

Event history analysis examines *transitions* across a set of discrete states, including the length of *time intervals* between entry to and exit from specific states. The basic analytical framework consists of a state space and a time axis. The choice of the *time axis* or *clock* (e.g., age, experience or marriage duration) used in the analysis must be based on theoretical considerations, and will influence the statistical model. In this book, only methods and models using a *continuous* time axis are discussed (for a comparison between continuous-time and discrete-time event history models see Blossfeld 2026). An *episode*, a *spell*, a *waiting time*, or a *duration*—terms that are used interchangeably—is the time span a unit of analysis (e.g., an individual) spends in a particular state. The *states* are *discrete* and usually small in number. The definition of a set of possible states, the so-called *state space* $\mathcal{Y}$, also depends on substantive considerations. Thus a careful, theoretically driven choice of the time axis and the design of the state space is important, as they are often serious sources of misspecification. In particular, model misspecification can occur because some of the important states are not observed. For example, in a study analyzing the determinants of women's labor market participation in West Germany, Blossfeld and Rohwer (1997a) showed that much more appropriate substantive conclusions can be reached if one differentiates the state "employed" into "full-time work" and "part-time work." Here, too, it should be noted that a small change in the focus of the content, which leads to a new definition of the state space, often requires a fundamental reorganization of the event history data file.

The most restricted event history model is based on a process with only a *single episode* and *two states* (one *origin* and one *destination* state). An example would be the duration of the *first* marriage until its end, for whatever reason. In this case, each individual who entered the first marriage (origin state) has started an episode that can be ended by a transition to the destination state "not married anymore." In the case of a *single episode*, each person who entered the first marriage is represented by an episode. If more than one destination state exists, we refer to these models as *multistate models*. Models for the special case, with a single origin state but two or more destination states, are also referred to as *models with competing events* or *risks*. For example, a housewife could become "unemployed" (i.e. entering into the state "looking for work") or start being "full-time" or "part-time employed." If more than one event is possible (i.e., if there are repeated events or transitions over the observation period), we use the term *multiepisode models*. For example, an employment career usually consists of a series of job exits. Figure 1.1.1c (p. 3) describes a *multistate–multiepisode* process. The individual moves repeatedly between several different states. As shown in Blossfeld, Hamerle, and Mayer (1989), most of the basic concepts for the one-episode and one-event case can be easily extended and applied to more complex situations with *repeated episodes* and/or *competing events*. In this book, we stick mainly to the more complex notation for the multistate–multiepisode case. Thus given a sample of $i = 1, \ldots, N$ multistate–multiepisode data, a complete description of the data[1] is given by

$$(u_i, m_i, o_i, d_i, s_i, t_i, x_i) \qquad i = 1, \ldots, N.$$

where u_i is the identification number of the individual or any other unit of analysis to which the ith episode belongs; m_i is the serial number of the episode; o_i is the origin state, the state occupied during the episode until the ending time; d_i is the destination state defined as the state reached at the ending time of the episode; and s_i and t_i are the starting and ending times, respectively. In addition, a covariate vector x_i is associated with the episode. We always assume that the starting and ending times are coded such that the difference $t_i - s_i$ is the duration of the episode and is positive and greater than zero. There is also an ordering of the episodes for each individual given by the set of serial numbers of the episodes. Although these serial numbers need not be contiguous, the starting time of an episode should not come before the ending time of a previous episode.

The observations of event histories are very often *censored* (Turkson et al. 2021). Censoring occurs when the recorded information about the duration in the origin state is incomplete. Figure 2.1.1 shows examples

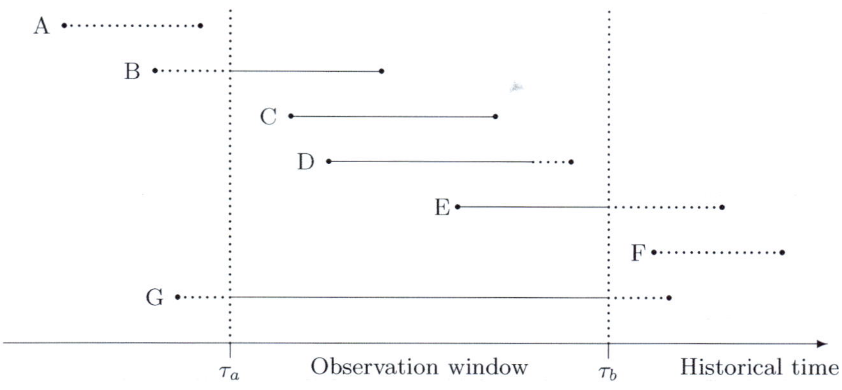

Figure 2.1.1 Types of censoring in an observation window.

of different types of censoring created by an observation window (see also Yamaguchi 1991; Guo 1993). The horizontal axis indicates historical time, and the observation period is usually of finite length with the beginning and end denoted by τ_a and τ_b, respectively.

1. *Episode A* is fully censored on the left, i.e. that the starting and ending times of this spell are located before the beginning of the observation window. Left censoring is usually a difficult problem, because it is not possible to take the effects of the unknown episodes in the past into account. It is only easy to cope with if the assumption of a Markov process is justified (i.e., if the transition rates do not depend on the duration in the origin state).

2. *Episode B* is partially censored on the left-hand side, so that it is not known how long a subject has already spent in the origin state. In this case, we have the same problems as for A-type episodes. However, sometimes (e.g., in a first panel wave), we have additional retrospective information about the time of entry into the origin state of Episode B. In this case, usually called a *left truncated* observation (see Guo 1993), we can reconstruct the entire duration of Episode B, but do not have information about A-type episodes. This leads to a sample selection bias for the period before the observation window. The earlier the starting time of the episode and the shorter the durations, the less likely it is that these episodes will appear in the observation window. One solution to this problem is to start analyzing the process at the beginning of the observation window and only evaluate the part of the duration that extends into the observation window beginning with time τ_a and ending with t_i. This means that the analysis is conditional on the fact that the individual has survived up to τ_a (see Guo 1993).

3. *Episode C* is complete. There is no censoring on the left or right.

4. *Episode D* is a special case. This episode is censored on the right-hand side within the observation window. If the censoring is a result of a random process, then the methods of event history analysis can easily take these episodes into account, as will be shown later (Kalbfleisch and Prentice 1980: Type II censoring). Technically speaking, it can be treated in the same way as Episode E. However, this type of censoring might occur because of attrition or missing data in a panel study. Such dropouts or missing data are usually not random, and the characteristics of the lost individuals are very often related to the process under study. Such a selection bias leads to problems and cannot be easily corrected in an event history analysis (see chapter 10).

5. *Episode E* is right-censored because the observation is terminated at the right-hand side of the observation window. This type of censoring typically occurs in life-course studies at the time of the retrospective interview or in panel studies at the time of the last panel wave. Since the end of the observation window, τ_b, is normally determined independently of the substantive process under study, this type of right censoring is unproblematic. It can be handled with event history methods (Kalbfleisch and Prentice 1980: Type I censoring).

6. *Episode F* is completely censored on the right. Entry into and exit from the duration occurs after the observation period. This type of censoring usually happens in retrospective life history studies in which individual members of different birth cohorts are observed over different life spans. To avoid sample selection bias, such models must take into account variables that control for selection—for example, by including birth cohort dummy variables and/or a time-dependent covariate age.

7. *Episode G* represents a duration that is left- and right-censored.[2] Such observations occur, for example, in panel studies recording job mobility. In such cases, one knows that a person is in a specific job at the time of the first sweep and holds the same job until the second sweep, but one has no information about the actual beginning and ending times.

In the examples in this book based on the German Life History Study (GLHS), we do not have left-censored data because all life histories of the 1929–31, 1939–41, and 1949–51 birth cohorts were collected retrospectively from the time of birth to the date of the interview (1981–1983, see Mayer and Brückner 1989; Mayer 1987, 1988, 1991, 2015). Thus, we have no data of Types A, B, and G. Type D censoring can only occur due to missing data, but not due to attrition, as the GLHS is a retrospective study. Because we examine different birth cohorts in our analyses (data of Types E and F), we must control for the fact that members of these birth cohorts could only be observed over different age spans (1929–31: up to the age of 50; 1939–41: up to the age of 40; and 1949–51: up to the age of 30).

This book offers numerous exercises, which you will find in the Appendix. They refer to a more recent data set from the German National Educational Panel Study (NEPS). NEPS is a research infrastructure project that provides data from several starting cohorts on educational processes and trajectories in Germany (Blossfeld and Roßbach 2019). One cohort focuses on the adult working population, and at the first interview, detailed information was collected retrospectively in the tradition of the German Life History Study (GLHS) (Allmendinger et al. 2019). In the following waves, information on labor market participation, partnership formation, partnership dissolution, and so forth were updated—as described at the end of section 1.1.3 in this book. The data version available for our book is a non-representative subsample restricted to West Germany, which was additionally anonymized by the Leibniz Institute for Educational Trajectories (LIfBi). The file includes data on 1,935 persons born between 1944 and 1986 with their last interview between 2009 and 2015. The data set is described in more detail in the Appendix.

2.2 Event History Data Organization

Event history data are more complex than cross-sectional data because, for each episode, information is given about an *origin state* and a *destination state*, as well as the *starting* and *ending times*. In most studies, there are also *repeated episodes* from *various parallel processes* (e.g., job, marital, or residential histories) at *different levels* (e.g., job history of an individual, histories of the firm where the individual worked at the meso level, and/or structural changes in the labor market at the macro level) for each unit of analysis. For this reason, large event history data sets have often been stored in data base systems. This book does not discuss the advantages and disadvantages of different data bank systems for event histories in terms of efficiency, convenience, data handling, and retrieval. However, we emphasize that event history data have to be organized as a *rectangular file* in order to analyze the data with standard programs such as SPSS, SAS, or GLIM (see e.g., Blossfeld, Hamerle, and Mayer 1989) or with the Stata program that is used throughout this book.

In an event-oriented data set, *each record of the file is associated with a duration in a state or episode* (see Carroll 1983). As shown previously, the type and number of states for each unit of analysis are dependent on the substantive question under consideration. Changes in the state space usually lead to a new definition of episodes and very often entail a fundamental reorganization of the data file. If only one episode is considered for each unit of analysis (e.g., entry into first marriage), the number of records in the file corresponds to the number of units. In an analysis that looks at repeated events (e.g., successive jobs in an individual's career), the number of which may vary between individuals, the sum of these person-specific episodes represents the number of records in the data set.

In the examples in this book, we use event history data from the GLHS. For our didactical task in this book, we use only an event history data file of 600 job episodes from 201 respondents (arbitrarily selected and anonymized). Each record in this file represents an employment episode, and the consecutive jobs in a respondent's career are stored successively in the file. For some individuals, there is only a single job episode, while for others, there is a sequence of two or more jobs.

The data file, `rrdat1.dta`, is a Stata system file containing 12 variables that are described briefly in Box 2.2.1.

`id` identifies the individuals in the data set. As the data file contains information about 201 individuals, there are 201 different ID numbers. The numbers are arbitrarily chosen and are not contiguous.

`noj` specifies the serial number of the job episode, which always begins with job number 1. For instance, if an individual in our data set had three jobs, the data file contains three records for this individual entitled job numbers 1, 2, and 3. Note that only job episodes are included in this file. If an individual has experienced an

Box 2.2.1 Variables in data file **rrdat1.dta**

```
Contains data from rrdat1.dta
  obs:            600
 vars:             12
 size:         11,000
-------------------------------------------------------------------
              storage  display
variable name   type   format     variable label
-------------------------------------------------------------------
id              int    %8.0g      ID of individual
noj             byte   %8.0g      Serial number of the job
tstart          int    %8.0g      Starting time of the job
tfin            int    %8.0g      Ending time of the job
sex             byte   %8.0g      Sex (1 men, 2 women)
ti              int    %8.0g      Date of interview
tb              int    %8.0g      Date of birth
te              int    %8.0g      Date of entry into the labor market
tmar            int    %8.0g      Date of marriage (0 if no marriage)
pres            byte   %8.0g      Prestige score of job i
presn           byte   %8.0g      Prestige score of job i+1
edu             byte   %8.0g      Highest educational attainment
-------------------------------------------------------------------
Sorted by:  id  noj
```

Box 2.2.2 First records of data file **rrdat1.dta**

```
    +------------------------------------------------------------------+
    | id  noj  tstart  tfin  sex   ti   tb   te  tmar  pres  presn  edu |
    |------------------------------------------------------------------|
 1. |  1    1     555   982    1  982  351  555   679    34     -1   17 |
    |------------------------------------------------------------------|
 2. |  2    1     593   638    2  982  357  593   762    22     46   10 |
 3. |  2    2     639   672    2  982  357  593   762    46     46   10 |
 4. |  2    3     673   892    2  982  357  593   762    46     -1   10 |
    |------------------------------------------------------------------|
 5. |  3    1     688   699    2  982  473  688   870    41     41   11 |
 6. |  3    2     700   729    2  982  473  688   870    41     44   11 |
 7. |  3    3     730   741    2  982  473  688   870    44     44   11 |
 8. |  3    4     742   816    2  982  473  688   870    44     44   11 |
 9. |  3    5     817   828    2  982  473  688   870    44     -1   11 |
```

interruption between two consecutive jobs, the difference between the ending time of one job and the starting time of the next job can be greater than 1 (see Figure 2.2.1b).

tstart is the starting time of the job episode in century months. (A century month is the number of months from the beginning of the century; 1 = January 1900.) The date given in this variable records the first month in a new job.

tfin is the ending time of the job episode in century months. The date given in this variable records the last month in the job.

sex records the gender of the individual coded 1 for men and 2 for women.

ti is the date of the interview in century months. This information can be used in the GLHS dataset to decide whether an episode is right-censored or not. If the ending time of an episode (**tfin**) is less than the interview date, the episode ended with an event (see Figure 2.2.1c); otherwise, the episode is right-censored (see Figures 2.2.1a and 2.2.1b).

tb records the birth date of the individual in century months. Therefore, **tstart** minus **tb** is the age in months at the beginning of a job episode.

te records the date of first entry into the labor market in century months.

tmar records whether/when an individual has married. If the value of this variable is positive, it indicates the date of marriage (in century months). For still unmarried individuals at the time of the interview, the value is 0.

pres records the prestige score of the current job episode.

presn records the prestige score of the consecutive job episode if there is a next job; otherwise, a missing value (-1) is coded.

edu records the highest educational attainment before entry into the labor market. In assigning school years to school degrees, the following values have been assumed (Blossfeld 1985, 1992): Lower secondary school qualification (*Hauptschulabschluss*) without vocational training is equivalent to 9 years; middle school qualification (*Mittlere*

Reife) is equivalent to 10 years; lower secondary school qualification with vocational training is equivalent to 11 years; and middle school qualification with vocational training is equivalent to 12 years. *Abitur* corresponds to 13 years, a professional college qualification (*Fachhochschulreife*) is equivalent to 17 years; and a university degree is equivalent to 19 years.

Box 2.2.2 shows the first nine records of data file `rrdat1.dta`. Note that all dates are coded in *century months*. Thus, 1 means January 1900; 2 means February 1900; 13 means January 1901; and so on. In general:

$$\texttt{YEAR} = (\texttt{DATE} - 1) \;/\; 12 + 1900$$
$$\texttt{MONTH} = (\texttt{DATE} - 1) \;\%\; 12 + 1$$

where `DATE` is given in century months, and `MONTH` and `YEAR` refer to calendar time. "/" means *integer division* and "%" is the *modulus operator*.[3] For instance, the first individual (Id = 1) has a single job episode. Its starting time is given in terms of century month 555, which corresponds to March 1946, and the ending time is 982 ≡ October 1981. Since this corresponds to the interview month, the episode is right-censored.

The panels in Figure 2.2.1 demonstrate the three basic types of job careers included in the example file:

a) For some respondents, the file `rrdat1.dta` contains an uninterrupted job history from the time of entry into the labor market to the time of the retrospective interview. If there is more than one job for the respondent, the ending time of job n (`tfin`) and the starting time of job $n + 1$ (`tstart`) are contiguous. For example, the first individual in the data set (`id = 1`, see Box 2.2.2), who is a man (`sex = 1`), had only one job from first entry into the labor market (`tstart = 555`) up to the time of the interview (`tfin = 982` equals `ti = 982`). This episode is therefore right-censored.

b) Some respondents' job histories were interrupted by an unemployment episode or because the respondents were out of the labor market for some period. In these cases, the example data set `rrdat1.dta` contains only the job episodes, and there may be gaps between the ending time of job n and the starting time of job $n + 1$ (see Figure 2.2.1b).

c) Finally, for some respondents, the observed job history may have ended before the time of the interview because the employment career stopped or was interrupted (due to unemployment or being out of the labor force), and re-entry did not take place before the time of the interview. For example, the second individual

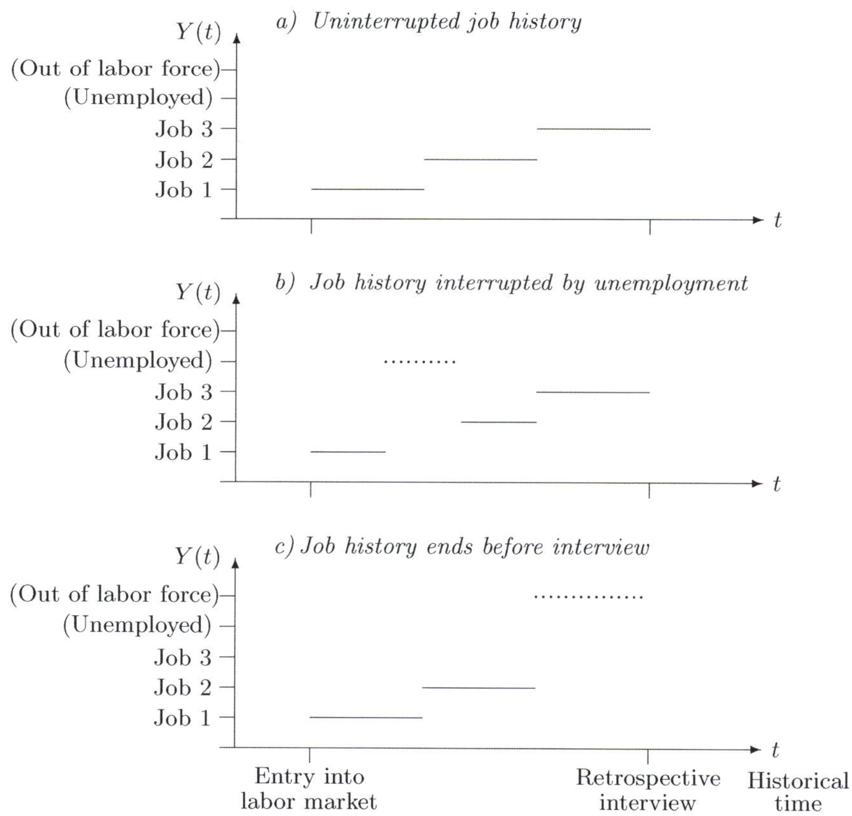

Figure 2.2.1 Examples of job histories included in the GLHS data set.

(`id = 2`), who is a woman (`sex = 2`), had an uninterrupted career of three jobs, then an interruption (`tfin = 892`), and did not re-enter the labor market before the time of the interview (`ti = 982`; see Figure 2.2.1c).

Using Event History Data Files with Stata

Here we use Stata to read the data file `rrdat1.dta`. For this purpose, we use a short do-file (`ehb1.do`) shown in Box 2.2.3.[4] As explained in the various examples in this book, a do-file is a short text file containing commands to be executed by Stata. A do-file can be created with the Stata Do-file Editor or any other editor such as NotePad or TextPad.

Our first example, shown in Box 2.2.3, contains eight commands. In Stata, there are several ways to terminate a command. By default, commands end with a line break. You may have to continue longer commands over several lines to make your do-file easier to read. There are three ways to allow for commands that run through several lines of a do-file. First, you can use the `#delimit` command, which defines the character that indicates the end of a command (carriage return or semicolon). If you change the delimiter to a semicolon, each section of text up to a terminating semicolon will be interpreted by Stata as a single command. Second, three slashes at the end of a line (///) can be used to continue commands across lines. Alternatively, you can also use /* at the end of a line and restart the following line with */. To include comments in a line, you either put an asterisk (*) at the beginning of a line or two slashes (//) somewhere in the line so Stata will ignore the text written in the remaining line.

Each do-file in this textbook begins with a series of commands. Stata is constantly being improved, and hence do-files written for earlier versions of Stata might produce an error message. To ensure that future versions of Stata will continue to execute your commands, specify the version of Stata for which the do-file was written at the top of your do-file. Once you submit a command, the results are displayed in the Stata Results Window. Stata pauses when the Results Window is full. With `set more off`, you ask Stata to run a do-file without interruption. Note that these results are only stored temporarily. You should therefore ensure in every Stata session that the commands and results of your do-file are saved in a so-called log file. In Stata, the command `log using` allows you to make a complete record of your Stata session. The command `log close` closes log files. If no log file is open, this command will produce an error message, and Stata will stop running the do-file. We therefore place `capture` before this command and instruct Stata to ignore any error message from this command.

Once you have set your working directory to the correct directory, you can load the file `rrdat1.dta` by simply typing `use rrdat1`.[5] You can open a Stata-format data set with the command `use`; for other data formats, you have to use other commands. Once you load the data file, you can view its contents by typing `describe`. Using the command `list`, you can look at all the contents and every single observation of the data file. The `in` qualifier in do-file `ehb1.do` tells Stata to list only the first nine observations; with `sepby(id)`, you draw a separator line whenever the `id` changes to make the output easier to read (see Box 2.2.2).

Box 2.2.3 Do-file `ehb1.do`

```
version 19
capture log close
set more off
log using ehb1.log, replace

use rrdat1, clear

describe

list id noj tstart tfin sex ti tb te tmar pres presn edu in 1/9, sepby(id)

sort id noj
by   id: gen noj2 = _n
tab  noj noj2, missing

by   id: egen te2 = min(tstart)
list id noj tstart te te2 in 1/9, sepby(id) noobs
cor te te2  /* cor(relelation) */

sort id noj
by   id: gen presn2 = pres[_n+1]
tab presn presn2              , miss
tab presn presn2 if presn <= 22, miss

log close
```

Box 2.2.4 Results of using do-file `ehb1.do` (Box 2.2.3)

```
. sort id noj
. by   id: gen noj2 = _n
. tab  noj noj2, missing
Serial     |
number of  |                          noj2
the job    |    1     2     3     4     5     6     7     8     9 | Total
-----------+------------------------------------------------------+------
1          |  201     0     0     0     0     0     0     0     0 |   201
2          |    0   162     0     0     0     0     0     0     0 |   162
3          |    0     0   107     0     0     0     0     0     0 |   107
4          |    0     0     0    62     0     0     0     0     0 |    62
5          |    0     0     0     0    32     0     0     0     0 |    32
6          |    0     0     0     0     0    20     0     0     0 |    20
7          |    0     0     0     0     0     0    11     0     0 |    11
8          |    0     0     0     0     0     0     0     4     0 |     4
9          |    0     0     0     0     0     0     0     0     1 |     1
-----------+------------------------------------------------------+------
Total      |  201   162   107    62    32    20    11     4     1 |   600

. by   id: egen te2 = min(tstart)
. list id noj tstart te te2 in 1/6, sepby(id) noobs
+-------------------------------+
| id    noj    tstart    te    te2 |
|-------------------------------|
|  1      1       555   555   555 |
|-------------------------------|
|  2      1       593   593   593 |
|  2      2       639   593   593 |
|  2      3       673   593   593 |
|-------------------------------|
|  3      1       688   688   688 |
|  3      2       700   688   688 |
+-------------------------------+

. sort id noj
. by   id: gen presn2= pres[_n+1]
(201 missing values generated)

. tab presn presn2                    , miss /* results not shown */
. tab presn presn2 if presn <= 22, miss
Prestige |
score of |                       presn2
job i+1  |       18        20        21        22         . |     Total
---------+--------------------------------------------------+----------
-1       |        0         0         0         0       201 |       201
18       |        6         0         0         0         0 |         6
20       |        0        14         0         0         0 |        14
21       |        0         0         4         0         0 |         4
22       |        0         0         0        16         0 |        16
---------+--------------------------------------------------+----------
Total    |        6        14         4        16       201 |       241
```

In Stata, it is possible to use abbreviations for both commands and variable names as long as the abbreviations are not ambiguous. For example, to create or change the contents of a variable you can use the command **generate**. A simple **g** suffice. Note, however, that many abbreviations in a do-file can also lead to ambiguity. In this book, we shall use abbreviations for several commands in do-files, but the full names of all commands are given in the text. One should also note that Stata is case-sensitive and that we use lowercase for variable names. In Stata, you can request for onscreen help. To display the online help system, type **help** at the command line. If you want to learn more about a specific command, enter **help command**, for example, **help generate** or **help list**, in the command line and press the return key.

Executing Stata with a Do-File

After you have prepared a do-file, you can make Stata execute the commands. There are several ways to do this. First, to run the do-file, enter

```
do filename
```

in the command line where `filename` is the name of a do-file and press the return key. If you have saved your file in your current working directory, you can omit any path specification. Another command that causes Stata to execute the commands of your do-file is `run`. In contrast to the command `do`, `run` will not display the commands and results in the Stata Results Window. Alternatively, you can use the pull-down menu, Files > Do. If you are using the Stata Do-File Editor, you can select Tools > Do and Tools > Run from the top menu bar or hit one of the two icons on the toolbar (by default the last two icons). Stata tries to find the do-file, reads the commands given in the file, and tries to execute the commands, possibly producing error messages. Any output is written in the Results Window, although these results are only stored temporarily. To create a full record of all commands and results generated in your Stata session you need to open a Stata log file. When executing do-file `ehb1.do`, Stata opens the data set `rrdat1.dta`, which consists of 600 cases and 12 variables, describe the data, and list values of all variables for the first nine records of the data set (see Boxes 2.2.1 and 2.2.2). If you want any further actions, you must specify these with additional commands. For instance, you can request a table of summary statistics of the variables by using the command `tabulate`. As an example, add the command

```
tab noj
```

to the do-file to obtain a frequency distribution of the variable `noj` (i.e., the distribution of the number of job episodes in the data file). Note that `tab` ignores system missing values if not specified in the command.

To generate the first variable, you need to sort your data by the individual identifier and the starting time of the job. Then you can use the prefix `by` followed by the variable `id`, a colon, and a command (see Box 2.2.3). The prefix `by` tells Stata that the subsequent command will be executed separately for all cases with the same value in the variable `id`. Thus `by id: gen noj2 = _n` leads to a variable that counts up the jobs for each person and starts every time with 1 in the first row of a person that equals the first job. `_n` is a system variable. A simple cross-tabulation shows that the variables `noj` and `noj2` contain exactly the same values (see Box 2.2.4). Typing `gen sn = _n` without any prefix would simply put the number of the row into the new variable `sn`. Another system information is `_N`. Which refers to the total number of lines or cases in the data set. If you use it in combination with the prefix `by` and the variable `id`, the new variable contains the maximum number of jobs a person has held. Such a variable would be constant within one person. As there are diverse internal variables in Stata that begin with an underscore, you should not start any variable name with an underscore.

Putting the date of first labor market entry into one variable, which has a constant value within one person but separate dates between persons, also works with the prefix `by` followed by the variable `id` and a colon. Now you need the command `egen`, which is a very practical and useful extension of the ordinary generate command (for details, type `help egen` in the command line). To check whether the new variable `te2` contains the same information as the original variable `te`, you can list some cases (see Box 2.2.4); and, in addition, you can ask Stata to estimate a correlation matrix that will lead to a perfect correlation between both variables and report the full number of observations, namely 600.

To determine the prestige score of the following job, you have to sort the data by the individual identifier and the variable on starting time of a job (or the serial number which would produce the same result). Then you can use the prefix `by` as already described. What is new is that you tell Stata to use the value from the following line by adding the suffix `[_n+1]` to the variable (see Box 2.2.3). If you use the suffix `[_n]`, the value of a variable is simply duplicated. Using `[_n-1]` or `[_n-2]` means copying the information from one or two rows above into the new variable. Because we are using the prefix `by` and `[_n+1]` in Box 2.2.3, there is no valid prestige score for the last job of a person in the next line. Consequently, Stata inserts a system missing into the variable. The default value for a system missing is a point ".". The results window or the log-file reports `201 missing values generated`. Since the data set contains 201 individuals, the same number of persons has one last observed job. The last job cannot be followed by a prestige score for the following job. Commands such as `tab(ulate)` or `table` drop all cases with system missing unless you specify the option `missing`. Before specifying such an option, a comma is needed. The cross tabulation of `presn` and `presn2` shows that if there is a valid case, both variables have the same value, whereas missing values have an a `1` in the original variable and a "." in the new variable. To avoid a huge cross tabulation, we restrict this to cases with values lower or equal to 22 in the variable `persn`. This can be done quite simply with the `if`-option.

If you intend to use the data as a set of event history data, you have to inform Stata explicitly about how to interpret the data matrix as a set of episodes. In fact, there are two different ways to interpret the data matrix as a set of episodes: single episode data and multiepisode data.

Single Episode Data

When working with single episode data, it is implicitly assumed that all individual episodes are statistically independent.[6] Before defining episode data, two basic dimensions must be specified: the state space and the

Box 2.2.5 Do-file `ehb2.do`

```
version 19
capture log close
set more off
log using ehb2.log, replace

use rrdat1, clear

gen org = 0                 /* origin state */
gen des = tfin ~= ti        /* destination state */
gen tf  = tfin - tstart + 1 /* ending time */

stset tf, f(des)    /* define single episode data */

describe _*

stdes  /* describe survival-time data */
stsum  /* summarize survival-time data */

log close
```

time axis. The state space is the set of possible origin and destination states. The values can be assigned arbitrarily. For instance, if you are only interested in job duration, the origin state of each job episode can be coded as 0, and the destination state can be coded as 1. This means that you do not distinguish between different ways to leave a job. There is only one transition: from 0 (being in a job) to 1 (having left that job). Of course, some episodes can be right-censored. In these cases, individuals are still in the origin state at the observed ending time of the episode. Note, however, that Stata does not explicitly use the concept of a state space. All calculations are based on just one origin and one destination state. Censored and uncensored episodes are distinguished by a censoring indicator.

In addition to a state space, you also need a time axis. For instance, in our example data file, jobs are coded in historical time defined in century months. The implied origin of this time axis is the beginning of January 1900. The easiest way to define the process time axis, for example, is to define the time of entry into the episode as zero and the ending time as the episode duration.[7] The do-file `ehb2.do`, shown in Box 2.2.5, illustrates the definition of single episode data with our example data set. The state space is $\mathcal{Y} = \{0, 1\}$: $0 \equiv$ being in a job episode, $1 \equiv$ having left the job. This is achieved by defining a new variable, `des`, which serves as a censoring indicator: `des` is assigned zero or one, depending on whether or not the episode is right-censored. An episode is right-censored in the GLHS data set if its ending time (`tfin`) is equal to the interview date (`ti`). Therefore, if `tfin` is equal to `ti`, the variable `des` becomes zero, otherwise it becomes one. As we are only interested in job duration in this example, starting and ending times are specified on a process time axis; that is, the starting time is always zero and the ending time corresponds to the (observed) duration of the episode. This is achieved in Stata by defining a variable `tf` for the ending time. The variable `tf` is calculated as the difference between the historical ending time (`tfin`) and the historical starting time (`tstart`). To avoid zero durations, we have added one month to the job duration. For example, it might be the case that an individual enters a job on the first day of a month and leaves it during that same month. As starting and ending times refer to the month in which an individual occupies the current job, starting and ending times would be equal, and the duration would be zero. Thus, we assume in our examples that any employment duration in the last month can be treated as a full month. The Stata command to create a new variable is `generate`. In Stata, you can use the following operators:

<table>
<tr><th colspan="2">Arithmetic</th><th colspan="2">Logical</th><th colspan="2">Relational</th></tr>
<tr><td>$+$</td><td>addition</td><td>$\sim$</td><td>not</td><td>$>$</td><td>greater than</td></tr>
<tr><td>$-$</td><td>subtraction</td><td>!</td><td>not</td><td>$<$</td><td>less than</td></tr>
<tr><td>$\star$</td><td>multiplication</td><td>$|$</td><td>or</td><td>$>=$</td><td>greater or equal</td></tr>
<tr><td>$/$</td><td>division</td><td>$\&$</td><td>and</td><td>$<=$</td><td>less or equal</td></tr>
<tr><td>$\wedge$</td><td>power</td><td></td><td></td><td>$==$</td><td>equal</td></tr>
<tr><td></td><td></td><td></td><td></td><td>$\sim=$</td><td>not equal</td></tr>
<tr><td></td><td></td><td></td><td></td><td>$!=$</td><td>not equal</td></tr>
</table>

For example, to create the variable `tf`, type: `gen tf = tfin - tstart + 1` (see Box 2.2.5). Note that a double equal sign (==) is used for equality testing. Next, use the `stset` command to declare the data set `rrdat1.dta` to be single-record event-history data: `stset timevar, failure(failvar)`. The variable `timevar` measures the duration before either a particular event ("failure") occurs or the observation is

Box 2.2.6 Result of using do-file **ehb2.do** (Box 2.2.5)

```
. stset tf, f(des = 1)    /* define single episode data */

failure event:  des == 1
obs. time interval:  (0, tf]
exit on or before:  failure
------------------------------------------------------------------------------
600  total observations
0  exclusions
------------------------------------------------------------------------------
600  observations remaining, representing
458  failures in single-record/single-failure data
40,782  total analysis time at risk and under observation
at risk from t =          0
earliest observed entry t =          0
last observed exit t =       428

. describe _*
storage   display    value
variable name    type    format     label      variable label
-------------------------------------------------------------------------------------
_st             byte    %8.0g                 1 if record is to be used; 0 otherwise
_d              byte    %8.0g                 1 if failure; 0 if censored
_t              int     %10.0g                analysis time when record ends
_t0             byte    %10.0g                analysis time when record begins


. stdes   /*describe survival-time data*/

failure _d:  des == 1
analysis time _t:  tf

|------------- per subject -------------|
Category                   total        mean          min      median          max
------------------------------------------------------------------------------
no. of subjects              600
no. of records               600           1            1           1            1

(first) entry time                         0            0           0            0
(final) exit time                      67.97            2          37          428

subjects with gap              0
time on gap if gap             0
time at risk               40782       67.97            2          37          428

failures                     458    .7633333            0           1            1
------------------------------------------------------------------------------

. stsum   /* summarize survival-time data */

failure _d:  des == 1
analysis time _t:  tf

|               incidence       no. of     |------ Survival time -----|
| time at risk      rate        subjects          25%        50%        75%
---------+--------------------------------------------------------------------
total |      40782    .0112304           600          20          44          121
```

terminated due to right-censoring because of the end of the observation window. The variable **failvar** records whether an episode ends with an event or censoring. By specifying **failure(des)**, all episodes for which the variable **des** is equal to 0 or missing will be considered as censored. You can also specify a list of codes that are similiar to events: **failure(des==1)**. All other values of the variable **des** will be considered as indicators for censoring. For our example, type: **stset tf, f(des)**. There are several options for using the Stata command **stset**, although not all of them are relevant for this simple example. We assume that all individual episodes are statistically independent, so it is not necessary to note individual case identification numbers. We shall discuss repeated events in the next section on multiple record data. The option **origin** is used to identify when a subject becomes at risk. By default, Stata assumes that our variable **ts**, the onset of risk, is 0, and this is exactly what our data set looks like. When you declare your data as event history data using the **stset** command, Stata checks the data for consistency.[8] At the same time, the command **stset** automatically creates the following new variables:

- **_t0** analysis time when record begins
- **_t** analysis time when record ends
- **_d** 1 if failure, 0 if censored
- **_st** 1 if the record is to be used, 0 if ignored

The first two variables provide information about the process time. In our example, _t0 is equal to zero for all job episodes; _t is equal to our variable `tf`. The dummy _d contains the same type of information as our variable `des`. The variable _st is not relevant here, but we will come back to this issue in the next section on multiple-record data. It is important to note that most of Stata's event history analysis commands require the data to have been previously declared as single episode or multiepisode data by specifying the `stset` command. This command only needs to be executed once. If you want to save the data to preserve this information, type `save [filename], replace`. If there is already a file with the same name in your working directory, you will have to overwrite the existing data set. However, we strongly recommend that you never overwrite the original file rrdat1.dta. To see the setting for your current data set, simply type `st`. Once you have declared your data set to be single episode data, you can produce a summary of this data set using the Stata commands `stdes` and `stsum`. Executing the do-file `ehb2.do` with Stata can be done in the following way: `do ehb2`. The output of the Results Window is shown in Box 2.2.6. First, Stata shows how the data set was declared. You can see that 458 of all 600 job episodes end with an event. The total analysis time at risk is 40,782 months, and the last exit is observed after a duration of 428 months. The second part of the output in the Results Window yields a brief description of how our survival-time data are structured, though some of this information has already been generated by the `stset` command. In this simple example, we have only one record per subject, so the number of subjects is equal to the number of records and the minimum and maximum of records per subject is 1. The starting time (first entry time) is zero for all job episodes with the first exit occurring after two and the last after 428 months (final exit time). By specifying `stsum`, Stata presents summary statistics (third part of the output in Box 2.2.6). The hazard rate, here called incidence rate, is $458/40782 = 0.011$. The 25th, 50th, and the 75th percentile of survival times are derived from a Kaplan-Meier survivor function (see chapter 3). This function estimates the probability of ending a job episode within 20 months at around 25% and the probability of ending it within 44 months at 50%.

Multiepisode Data

For multiepisode data, you need two additional pieces of information. First, you have to know which episodes belong to which individual; second, you need to know the serial (sequence) number of the episodes.

To define a time axis, you have at least two options: You can reset the clock at the beginning of each new episode (e.g., use job-specific labor force experience as process time), or you can use a common process time axis in which the first episode for each individual begins at time zero (e.g., use general labor force experience over the life time as time axis). In the latter case, consecutive episodes are evaluated according to their starting and ending time on this common process time axis in, for example, time-dependent parametric models. We use this latter option to illustrate the setup of multiepisode data with our example data set. The example do-file `ehb3.do` is shown in Box 2.2.7.

To define the common process time axis "general labor force experience," we use the variable `te`, time of entry into the labor market. This variable has the same values as `tstart` in the first job. It is subtracted from the original starting and ending times to obtain their values on the process time axis. The new variables are `ts2` for the start and `tf2` for the end of a job episode with reference to labor market entry. Again, we have added one month to the ending times to get positive durations. For the individuals one, two, three, and six in our example data set, Box 2.2.8 shows how the common process time axis is created from the calendar time axis. Whereas the second and third person immediately have one job after another until the last one, followed by a complete labor market exit—at least during the observation period. The individual with id number 6 has several gaps in his/her employment history.

In the single-record-per-subject data set, each observation represents a different subject. For multiepisode data in which the same individual contributes more than one record to the data, you have to modify your `stset` command and include the option `id(idvar)`, where `idvar` records the subject-id variable. Yet, typing `stset tfp, f(des) id(id)` will not yield the expected result. Our example data `rrdat1` contains multiple events (see Box 2.2.8). If you do not take this into account, Stata will remove the subject from the risk pool after the first event has occurred. We therefore specify `exit(time .)` to keep all records of a subject until its final exit. In our example, the variable `tf2` indicates the elapsed time before failure or censoring; `des` indicates whether an event (1) or censoring (0) occurred at this time. The variable `id` identifies the individuals in the data set. In addition, we have to use the `time0()` option and put the variable `ts2` in the brackets. Otherwise, Stata assumes that there are no gaps between episodes and the ending time of one episode is the starting time of the next episode.[9] To set the data for event history analysis, type:

```
stset tf2, f(des) id(id) time0(ts2) exit(time .).
```

Stata's output from the execution of do-file `ehb3.do` is shown in Box 2.2.8. The tables are similar to the tables shown in Box 2.2.6.

Box 2.2.7 Do-file `ehb3.do` for multiepisode data

```
version 19
capture log close
set more off
log using ehb3.log, replace

use rrdat1, clear

gen des = tfin ~= ti        /* destination state */
gen ts2 = tstart - te       /* starting time - for multi-episode concept */
gen tf2 = tfin - te + 1     /* ending time - for multi-episode concept */


list id noj tstart tfin te ti ts2 tf2 if id < 4 | id == 6, sepby(id) noobs

stset tf2, f(des) id(id) time0(ts2) exit(time .)
stdes
stsum

list id noj tstart tfin te ts2 tf2 _st _t0 _t   ///
     if (id == id[_n-1] & ts2 < tf2[_n-1])      ///
      | (id == id[_n+1] & tf2 > ts2[_n+1]), sepby(id) noobs

table noj if _st == 1, c(n _st min _t0 median _t max _t mean _d) f(%5.2g)

capture log close
```

The output presented in Box 2.2.8 contains important information. First, there are five cases where the end date of a job is after the start date of the next. Consequently, 5 job episodes are not considered in the event history analysis. See also the number of remaining episodes for the analysis, which is now 595. Producing a short extract from the data set with the `list` command reveals that there are overlaps from the fifth to the sixth job for the person with `id == 17`; from the first to the second job for the person with `id == 113`; and from the third to the fourth job for `id == 140` (see Box 2.2.8). For the persons with `id == 100` and `150`, the end of a job and the start of the next one take place in the same month. As we added one month to the job duration to ensure, that jobs last at least one month, Stata also identifies these two cases as overlapping. The Stata-generated variable `_st` indicates that in the case of an overlap, the earlier job episode remains in the sample, while the later one is discarded. In a research project, one would have to decide how to deal with these overlaps. In our example, we use Stata's default option. There are not only overlaps but also gaps in our data. The ouput in Box 2.2.8 also shows that for 91 of the 201 persons, there is at least one gap between employment spells and the total time of the gaps is 4,717 months.

Finally, we provide some further information on the multi-episode data using the ordinary command `table`. We restrict the output to the 595 records remaining in the pool for survival time data analysis (`if _st == 1`). For each consecutive job number, the table reports the number of episodes in the data set, the earliest time a job starts, as well as the median and the maximum time of leaving a job. Since the time information refers to labor market entry, from the second job onward, the minimum time is always greater than zero; and in most cases, the median time increases by job order. In addition, the last column of the table shows the proportion of episodes ending with an event. The format-option at the end of the command `table` determines the number of digits after the decimal point.

To ensure that you have understood the basic concepts presented in chapter 2 we recommend to work through the exercises in the Appendix. There, you can familiarize yourself with the second data set (rrdat2.dta) used in this book from the German National Educational Panel Study (NEPS).

Box 2.2.8 Example of multiepisode data for four individuals and further results of using do-file `ehb3.do` (Box 2.2.7)

```
. list id noj tstart tfin te ti ts2 tf2 if id < 4 | id == 6, sepby(id) noobs

  +--------------------------------------------------+
  | id   noj   tstart   tfin    te    ti   ts2   tf2 |
  |--------------------------------------------------|
  |  1     1      555    982   555   982     0   428 |
  |--------------------------------------------------|
  |  2     1      593    638   593   982     0    46 |
  |  2     2      639    672   593   982    46    80 |
  |  2     3      673    892   593   982    80   300 |
  |--------------------------------------------------|
  |  3     1      688    699   688   982     0    12 |
  |  3     2      700    729   688   982    12    42 |
  |  3     3      730    741   688   982    42    54 |
  |  3     4      742    816   688   982    54   129 |
  |  3     5      817    828   688   982   129   141 |
  |--------------------------------------------------|
  |  6     1      691    716   691   982     0    26 |
  |  6     2      728    753   691   982    37    63 |
  |  6     3      771    846   691   982    80   156 |
  |  6     4      859    982   691   982   168   292 |
  +--------------------------------------------------+

. stset tf2, f(des) id(id) time0(ts2) exit(time .)
                id:  id
     failure event:  des != 0 & des < .
obs. time interval:  (ts2, tf2]
 exit on or before:  time .

------------------------------------------------------------------------------
      600  total observations
        5  overlapping records (tf2[_n-1]>ts2)                  PROBABLE ERROR
------------------------------------------------------------------------------
      595  observations remaining, representing
      201  subjects
      455  failures in multiple-failure-per-subject data
   40,578  total analysis time at risk and under observation
                                                at risk from t =         0
                                     earliest observed entry t =         0
                                          last observed exit t =       465
. stdes
         failure _d:  des
   analysis time _t:  tf2
  exit on or before:  time .
                 id:  id
                                   |-------------- per subject --------------|
Category                   total        mean         min     median        max
------------------------------------------------------------------------------
no. of subjects              201
no. of records               595    2.960199           1          3          8

(first) entry time                         0           0          0          0
(final) exit time                   225.3483           7        188        465

subjects with gap             91
time on gap if gap          4717    38.04032           1       18.5        299
time at risk               40578    201.8806           7        163        465

failures                     455    2.263682           0          2          7
------------------------------------------------------------------------------
```

Box 2.2.8 (cont.)

```
. stsum

failure _d:  des
analysis time _t:  tf2
exit on or before:  time .
id:  id

        |               incidence     no. of    |------ Survival time -----|
        | time at risk     rate      subjects       25%        50%       75%
--------+---------------------------------------------------------------------
  total |        40578   .011213          201        17         34        72

. list id noj tstart tfin te ts2 tf2 _st _t0 _t   ///
>      if (id == id[_n-1] & ts2 < tf2[_n-1])      ///
>      | (id == id[_n+1] & tf2 > ts2[_n+1]), sepby(id) noobs

  +-----------------------------------------------------------------+
  | id   noj   tstart   tfin    te   ts2   tf2   _st   _t0      _t |
  |-----------------------------------------------------------------|
  | 17     5      831    900   595   236   306     1   236     306 |
  | 17     6      891    918   595   296   324     0     .       . |
  |-----------------------------------------------------------------|
  | 110    3      672    673   664     8    10     1     8      10 |
  | 110    4      673    681   664     9    18     0     .       . |
  |-----------------------------------------------------------------|
  | 113    1      787    882   787     0    96     1     0      96 |
  | 113    2      877    982   787    90   196     0     .       . |
  |-----------------------------------------------------------------|
  | 140    3      887    940   832    55   109     1    55     109 |
  | 140    4      939    952   832   107   121     0     .       . |
  |-----------------------------------------------------------------|
  | 150    3      879    936   809    70   128     1    70     128 |
  | 150    4      936    982   809   127   174     0     .       . |
  +-----------------------------------------------------------------+

. table noj if _st == 1, c(n _st min _t0 median _t max _t mean _d) f(%5.2g)

----------------------------------------------------------------------
Serial     |
number of  |
the job    |     N(_st)    min(_t0)      med(_t)      max(_t)    mean(_d)
-----------+----------------------------------------------------------
         1 |        201           0           33          428         .92
         2 |        161           3          106          450         .78
         3 |        107           8          162          463         .64
         4 |         59          30          175          465         .61
         5 |         32          39          216          441         .72
         6 |         19          46          210          379         .58
         7 |         11         112          265          427         .36
         8 |          4         129          337          440         .25
         9 |          1         285          319          319           0
----------------------------------------------------------------------
```

Notes

[1]A complete history of state occupancies and times of changes is often called a "sample path" (see Tuma and Hannan 1984).

[2]A special case of such a data type is "current-status" data. This data includes information on whether or not an event has occured at the time of a survey, as well as information on age at the time of the survey. If the event has occurred, one has incomplete information on when it occurred. Conversely, we do not know when it will occur (if at all) for those respondents who have not yet experienced the event at the time of the survey (see Diamond and McDonald 1992).

[3]In Stata, you specify `mod(x,y)` to ask for the modulus of x with respect to y. For two integer numbers, x and y, `mod(x,y)` is the remainder after dividing x by y. For instance: `display mod(13,12) = 1`.

[4]The naming convention to identify do-files is: `ehxn.do`, where the single letter `x` is one of the series `b,c,d,...` referring to the successive chapters `2,3,4,...`, and `n` numbers the do-files in each chapter.

[5]We omit path information and assume that you have copied our example data set and all do-files to your working directory. To change the working directory to the correct drive and directory, use the command `cd drive:directory_name`. The path of your current working directory is displayed in the bottom of your Stata window and is also be shown in the Results Window if you type `cd`.

[6]In the case of multiepisode data, this may not be true because some individuals may contribute more than a single episode and these dependencies are not sufficiently controlled for by covariates (see sections 4.3, 9.4, and 10).

[7]It should be noted that the time of entry into the origin state is sometimes hard to determine. For example, consider the case when people start a consensual union, begin a friendship, or are looking for a marriage partner.

[8]Once you have declared your data to be event history data, Stata may warn you about inconsistent records. For example, Stata will recognize when the event or entry time is missing: Stata will mark records if the observed entry is on or after the observed exit, and it will warn you about any overlapping records. As a result,observations may be excluded by Stata (the automatically created variable `_st` will be equal to zero for these observations).

[9]The `stset` command offers many helpful options that can be somewhat confusing at first glance. Cleves et al. (2016) provide an excellent introduction to these options. The option `time0()` takes the beginning of an episode into account.

Chapter 3

Nonparametric Descriptive Methods

This chapter discusses nonparametric estimation methods that can be used to describe the characteristics of the process under study. Since these methods make no assumptions about the distribution of the process, they are particularly suited for first exploratory data analyses. Stata includes procedures to calculate life tables and Kaplan-Meier (or product-limit) estimates. Both of these methods are useful because they provide graphical presentations of both the survivor function (and its transformations) and the transition rate. The life table method is the more traditional procedure and has been used for large data sets as it requires less computing time and space. However, compared to the Kaplan-Meier estimator, the life table method has the disadvantage that the researcher has to define discrete time intervals—as will be shown later. Given the capabilities of modern computers, there no longer seems to be any reason to prefer the life table method for reasons of computer time or memory. We therefore only give a few examples of the life table method, while we discuss the Kaplan-Meier estimator in more detail. For multiple destination states, we also demonstrate the limitations of the standard Kaplan-Meier estimator. A suitable way to deal with such data is to use cumulative incidence functions (CIF). Therefore, we also provide information on how the results of CIF can be calculated and presented.

3.1 Life Table Method

The life table method enables the calculation of nonparametric estimates of the survivor function, the density function, and the transition rate for durations given in a set of episodes.[1] This method has two disadvantages: First, it is necessary to group the durations into fixed intervals. The results therefore depend on these arbitrarily defined time intervals. Second, this method is only useful if there is a relatively large number of episodes, so that estimates for each interval are reliable. However, if this second requirement is fulfilled, the method provides good approximations that are easy to calculate.

Time intervals are defined by split points on the time axis

$$0 \leq \tau_1 < \tau_2 < \tau_3 < \ldots < \tau_L$$

with the convention that $\tau_{L+1} = \infty$ and that there are L time intervals, each including the left limit, but not the right one:

$$I_l = \{t \mid \tau_l \leq t < \tau_{l+1}\} \qquad l = 1, \ldots, L.$$

Given these time intervals, Stata always calculates life tables by using episode durations. In the following description, we therefore assume that all episodes have starting time zero. In addition, we assume that the time intervals start at zero (i.e., $\tau_1 = 0$).

The calculation depends somewhat on the type of input data. Stata recognizes the following possibilities:[2] (1) If the input data are split into groups, a separate life table is calculated for each group. (2) If there is more than one origin state, the life table calculation can be done separately for each subset of episodes having the same origin state. (3) If there is only a single destination state for a given origin state, an ordinary life table is calculated. However, if there are two or more destination states, a so-called multiple-decrement life table is produced.

To explain the formulas used to calculate the life table, let us first consider the case of a single transition (i.e., only a single origin and a single destination).

Single Transitions

All formulas used in the calculation of single transition life tables are based on the following quantities, defined for each interval I_l, $l = 1, \ldots, L$.

$$\begin{aligned} E_l &= \text{the number of episodes with events in } I_l, \\ Z_l &= \text{the number of censored episodes ending in } I_l. \end{aligned}$$

The next important point is the definition of a *risk set*, $\mathcal{R}_l$, for each of the time intervals; that is, the set of units (episodes) that are at risk of an event occurring during the lth interval.[3] Two steps are required to take into account episodes that are censored during the interval: First the number of episodes, N_l that enter the lth interval is defined recursively by

$$N_1 = N, \quad N_l = N_{l-1} - E_{l-1} - Z_{l-1}.$$

Second, one has to decide how many of the episodes that are censored during an interval should be contained in the risk set for that interval. A standard assumption is that half of the episodes should be included, but this is a somewhat arbitrary assumption.[4] To provide the possibility to change this assumption, we assume a constant ω ($0 \leq \omega \leq 1$) for the definition of the proportion of censored episodes that should be included in the risk set.[5] The number of elements in the risk set is then defined by

$$R_l = N_l - \omega\, Z_l.$$

Using these basic quantities, it is easy to define all other concepts used in the life table setup. First of all, the conditional probabilities for the occurrence of an event in the lth interval, q_l, and for the survival of this interval, p_l, are as follows

$$q_l = \frac{E_l}{R_l} \quad \text{and} \quad p_l = 1 - q_l.$$

This results in the following estimator for the survivor function:

$$G_1 = 1, \quad G_l = p_{l-1} G_{l-1}.$$

Note, however, that the output of the Stata life table provides values of the survivor function for end points of time intervals.

After estimating the survivor function, the density function is evaluated approximately at the midpoints of the intervals as the first derivative

$$f_l = \frac{G_l - G_{l+1}}{\tau_{l+1} - \tau_l} \qquad l = 1, \ldots, q-1.$$

If the last interval on the right-hand side is open, it is of course not possible to calculate the survivor function for this interval. In addition, estimates of the transition rate, r_l, are calculated at the midpoints of the intervals. They are defined by

$$r_l = \frac{f_l}{\bar{G}_l} \qquad \text{where} \qquad \bar{G}_l = \frac{G_l + G_{l+1}}{2},$$

and this can also be written as

$$r_l = \frac{1}{\tau_{l+1} - \tau_l} \frac{q_l}{1 - q_l/2} = \frac{1}{\tau_{l+1} - \tau_l} \frac{E_l}{R_l - E_l/2}.$$

Finally, it is possible to calculate approximate standard errors for not only the estimates of the survivor and density functions, but also for the transition rates using the following formulas

$$\begin{aligned}
\mathrm{SE}(G_l) &= G_l \left[\sum_{i=1}^{l-1} \frac{q_i}{p_i\, R_i} \right]^{1/2}, \\
\mathrm{SE}(f_l) &= \frac{q_l\, G_l}{\tau_{l+1} - \tau_l} \left[\frac{p_l}{q_l R_l} + \sum_{i=1}^{l-1} \frac{q_i}{p_i R_i} \right]^{1/2}, \\
\mathrm{SE}(r_l) &= \frac{r_l}{\sqrt{q_l R_l}} \left[1 - \left[\frac{r_l\,(\tau_{l+1} - \tau_l)}{2} \right]^2 \right]^{1/2}.
\end{aligned}$$

For large samples, it can be assumed that the values of the survivor, density, and rate functions, divided by their standard errors, have an approximately standard normal distribution. In these cases, it is then possible to calculate confidence intervals.

As an example of life table estimation with Stata, we examine the length of durations until job exit. This means that there is only one type of event: "job exit" from the origin state "being in a job" to the destination state "having left the job." For didactical purposes, we start from a simple example and assume in this application that all job episodes in the data file `rrdat1` (see Boxes 2.2.1 and 2.2.2) can be considered

Box 3.1.1 Do-file `ehc1.do` (life table estimation)

```
version 19
capture log close
set more off
log using ehc1.log, replace

use rrdat1, clear

gen des = tfin ~= ti          /* destination state */
gen tf  = tfin - tstart + 1   /* ending time */

ltable tf des, intervals(30) su h f   /* command for life table estimation */

log close
```

Box 3.1.2 Result of using do-file `ehc1.do` (Box 3.1.1)

Interval		Beg. Total	Deaths	Lost	Survival	Std. Error	[95% Conf.	Int.]
0	30	600	223	28	0.6195	0.0201	0.5788	0.6574
30	60	349	113	23	0.4121	0.0208	0.3712	0.4524
60	90	213	51	15	0.3098	0.0199	0.2711	0.3492
90	120	147	25	16	0.2541	0.0192	0.2172	0.2924
120	150	106	24	15	0.1922	0.0182	0.1578	0.2291
150	180	67	9	5	0.1654	0.0177	0.1323	0.2016
180	210	53	4	9	0.1517	0.0175	0.1193	0.1878
210	240	40	3	5	0.1396	0.0175	0.1075	0.1758
240	270	32	0	5	0.1396	0.0175	0.1075	0.1758
270	300	27	2	7	0.1277	0.0179	0.0952	0.1651
300	330	18	2	5	0.1112	0.0190	0.0775	0.1517
330	360	11	2	1	0.0900	0.0205	0.0552	0.1352
360	390	8	0	3	0.0900	0.0205	0.0552	0.1352
390	420	5	0	4	0.0900	0.0205	0.0552	0.1352
420	450	1	0	1	0.0900	0.0205	0.0552	0.1352

Interval		Beg. Total	Deaths	Lost	Cum. Failure	Std. Error	[95% Conf.	Int.]
0	30	600	223	28	0.3805	0.0201	0.3426	0.4212
30	60	349	113	23	0.5879	0.0208	0.5476	0.6288
60	90	213	51	15	0.6902	0.0199	0.6508	0.7289
90	120	147	25	16	0.7459	0.0192	0.7076	0.7828
120	150	106	24	15	0.8078	0.0182	0.7709	0.8422
150	180	67	9	5	0.8346	0.0177	0.7984	0.8677
180	210	53	4	9	0.8483	0.0175	0.8122	0.8807
210	240	40	3	5	0.8604	0.0175	0.8242	0.8925
240	270	32	0	5	0.8604	0.0175	0.8242	0.8925
270	300	27	2	7	0.8723	0.0179	0.8349	0.9048
300	330	18	2	5	0.8888	0.0190	0.8483	0.9225
330	360	11	2	1	0.9100	0.0205	0.8648	0.9448
360	390	8	0	3	0.9100	0.0205	0.8648	0.9448
390	420	5	0	4	0.9100	0.0205	0.8648	0.9448
420	450	1	0	1	0.9100	0.0205	0.8648	0.9448

Interval		Beg. Total	Cum. Failure	Std. Error	Hazard	Std. Error	[95% Conf.	Int.]
0	30	600	0.3805	0.0201	0.0157	0.0010	0.0137	0.0177
30	60	349	0.5879	0.0208	0.0134	0.0012	0.0110	0.0158
60	90	213	0.6902	0.0199	0.0094	0.0013	0.0069	0.0120
90	120	147	0.7459	0.0192	0.0066	0.0013	0.0040	0.0092
120	150	106	0.8078	0.0182	0.0092	0.0019	0.0056	0.0129
150	180	67	0.8346	0.0177	0.0050	0.0017	0.0017	0.0083
180	210	53	0.8483	0.0175	0.0029	0.0014	0.0001	0.0057
210	240	40	0.8604	0.0175	0.0028	0.0016	0.0000	0.0059
240	270	32	0.8604	0.0175	0.0000	.	.	.
270	300	27	0.8723	0.0179	0.0030	0.0021	0.0000	0.0071
300	330	18	0.8888	0.0190	0.0046	0.0032	0.0000	0.0110
330	360	11	0.9100	0.0205	0.0070	0.0049	0.0000	0.0167
360	390	8	0.9100	0.0205	0.0000	.	.	.
390	420	5	0.9100	0.0205	0.0000	.	.	.
420	450	1	0.9100	0.0205	0.0000	.	.	.

independent of each other (single episode case) and that there is no substantial heterogeneity among the individuals.[6] Thus, we will estimate "average" survivor and transition rate functions across all the job spells and individuals.

Box 3.1.1 shows the do-file (`ehc1.do`) for the life table estimation with Stata. The upper part of this file is identical to do-file `ehb2.do`, which is shown in Box 2.2.5, that was used to define single episode data. To request life table estimations, we have added the `ltable` command. We also need a specification of time intervals. This is done with the option `intervals`:

```
ltable tf des,intervals(30).
```

This defines time intervals, each having a duration of 30 months and beginning at time zero: $0 \le t < 30$, $30 \le t < 60$, and so on. You can also use the `intervals` option to define cutoff points by specifying more than one number as the argument, such as `intervals(0 30 60 120 240 360)`.[7]

By default, `ltable` displays the estimates of the survivor function. Specifying the option `failure` results in the cumulative failure table (1 - survival). To obtain estimates for the transition rate, we type `hazard`.

If we run the do-file `ehc1.do` with Stata, we get the survival and failure table and the estimates of the transition rate aggregated in 30-month intervals. The results of do-file `ehc1.do` are shown in Box 3.1.2.

We give a short example of how the numbers in the life table of our example are related to the formulas developed previously. The third column of the life table in Box 3.1.2 gives the number of episodes entering into the successive intervals. In the first interval, all 600 episodes are recorded: $N_1 = N = 600$. The numbers of the following intervals $l = 2, 3, \ldots$ are calculated as

$$N_l = N_{l-1} - E_{l-1} - Z_{l-1}$$

where E_l, the number of events in the lth interval, is printed in Column 4, and Z_l, the number of censored episodes in the lth interval, is printed in Column 5. In our example:

$$\begin{aligned} N_1 &= 600, \\ N_2 &= 600 - 223 - 28 = 349, \\ N_3 &= 349 - 113 - 23 = 213. \end{aligned}$$

Under the assumption that censored episodes are equally distributed within each interval ($w = 0.5$), we can estimate the number of episodes at risk in each interval. For the lth interval:

$$\hat{R}_l = N_l - 0.5\, Z_l.$$

In our example:

$$\begin{aligned} \hat{R}_1 &= 600 - 0.5 \cdot 28 = 586.0, \\ \hat{R}_2 &= 349 - 0.5 \cdot 23 = 337.5. \end{aligned}$$

The conditional probability that an event occurs in the lth interval is given by

$$\hat{q}_l = \frac{E_l}{\hat{R}_l}.$$

In our example:

$$\begin{aligned} \hat{q}_1 &= \frac{223}{586.0} = 0.38055, \\ \hat{q}_2 &= \frac{113}{337.5} = 0.33481. \end{aligned}$$

The conditional probabilitythat no event occurs in the lth interval is then

$$\hat{p}_l = 1 - \hat{q}_l.$$

In our example:

$$\begin{aligned} \hat{p}_1 &= 1 - 0.38055 = 0.61945, \\ \hat{p}_2 &= 1 - 0.33481 = 0.66519. \end{aligned}$$

Based on these estimates, we can calculate estimates of the survivor function (Column 6 of the upper panel of the life table):

$$\begin{aligned} \hat{G}_1 &= 1, \\ \hat{G}_l &= \hat{p}_{l-1} \cdot \hat{p}_{l-2} \cdots \hat{p}_1. \end{aligned}$$

Box 3.1.3 Do-file `ehc2.do` to plot a survivor function

```
version 19
set scheme sj
capture log close
set more off
log using ehc2.log, replace

use rrdat1, clear

gen des = tfin ~= ti          /* destination state */
gen tf  = tfin - tstart + 1   /* ending time */

ltable tf des, intervals(30) graph ci title("Life Table Survivor Function")   ///
               ysc(r(1)) ylabel(0(0.2)1) xtitle("analysis time")               ///
               saving("Figure 3_1_1",replace)

log close
```

In our example:

$$\begin{aligned}\hat{G}_1 &= 1,\\ \hat{G}_2 &= 0.61945 \cdot 1 \;=\; 0.61945,\\ \hat{G}_3 &= 0.66519 \cdot 0.61945 \cdot 1 \;=\; 0.41205.\end{aligned}$$

Finally, the length of the intervals must also be taken into account, $\tau_{l+1} - \tau_l$ (for the lth interval). The duration density function is given as

$$\hat{f}_l = \frac{\hat{G}_l - \hat{G}_{l-1}}{\tau_{l+1} - \tau_l}.$$

In our example:

$$\begin{aligned}\hat{f}_1 &= \frac{1.00000 - 0.61945}{30 - 0} = 0.01268,\\ \hat{f}_2 &= \frac{0.61945 - 0.41205}{60 - 30} = 0.00691.\end{aligned}$$

The "average" transition rate, which is evaluated at the midpoint of each interval, is printed in Column 6 of the bottom panel in the life table:

$$\hat{r}_l = \frac{1}{\tau_{l+1} - \tau_l}\,\frac{E_l}{\hat{R}_l - 0.5\,E_l}.$$

In our example:

$$\begin{aligned}\hat{r}_1 &= \frac{1}{30-0}\,\frac{223}{586.0 - 0.5 \cdot 223} = 0.01567,\\ \hat{r}_2 &= \frac{1}{60-30}\,\frac{113}{337.5 - 0.5 \cdot 113} = 0.01340.\end{aligned}$$

The standard errors for the survivor function, cumulative failure table, and the rate function are printed in Column 7 of each table, and the 95% confidence interval is printed in Columns 8 and 9.

Life tables, as shown in Box 3.1.2, are very complex and not easy to interpret. It is therefore better to plot the interval-related information of the survival or failure function. Box 3.1.3 shows an example of a Stata do-file, `ehc2.do`, that can be used to generate a plot of the survivor function.[8] To draw a graph of the survivor function, you simply add the `graph` option to the `ltable` command.[9] The resulting plot is shown in Figure 3.1.1. A similar do-file can be used to generate plots for the failure function.

The plot of the survivor function shows estimates of the proportions of respondents who have not yet changed their jobs up to a specific duration. For example, after 10 years (or 120 months), about 25 % of respondents are still in their jobs, while about 75 % have already left.

Life Table Estimation for Different Groups

Life tables are particularly useful for comparing the behavior of subgroups. We therefore extend the example in Box 3.1.1 and demonstrate how separate life tables for men and women can be estimated with Stata. This can be achieved by adding the `by(varname)` option. Box 3.1.4 shows the do-file `ehc3.do`, which is simply a modification of the do-file `ehc2.do`. The variable `sex` is the indicator variable for men and women.[10] The

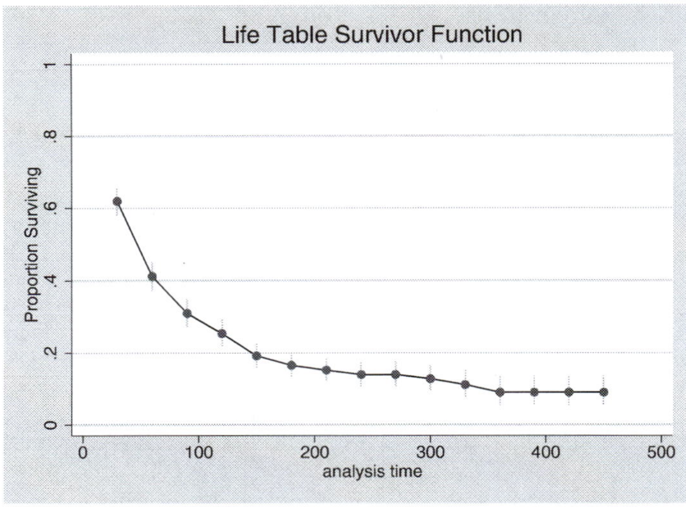

Figure 3.1.1 Plot of the survivor function generated with do-file `ehc2.do`.

command to request life table estimation is basically the same. We only added the option `by(sex)` in order to separate life tables for both groups.[11]

By using do-file `ehc3.do`, we obtain separate tables for men and women. To save space, these tables are not shown here. Instead, Figure 3.1.2 shows a plot of the survivor functions. It is easy to see that the process is quite similar for both men an women at the beginning. But after a duration of about 3 years, the survivor function decreases more steeply for women than that for men. Thus, women tend to leave their jobs sooner than men. After 20 years, about 20 % of men are still in their jobs, but only about 5 % of women. The median job durations are about 57 months for men and about 40 months for women.

Examples of the Application of Survivor Functions in Social Research

In longitudinal social research, survivor functions have been used increasingly to study social change. They are particularly suited to the analysis of how changing historical conditions affect life-course transitions of successive birth cohorts.[12] They enable a description of the age-graded character of roles and behaviors and, by documenting exits in the timing when individuals enter or leave specific social institutions or positions, the changes of life phases (Blossfeld and Nuthmann 1990; Becker and Blossfeld 1991).

Example 1: "Vulnerable" Phases in the Educational Careers of German Students

An example of the application of survivor functions in educational research can be found in Blossfeld (1990, 1992). He studied the process of entry into vocational training in Germany and showed that the particular organizational structure of the German education system creates what he calls a "vulnerable" life phase for

Box 3.1.4 Do-file `ehc3.do` (life table estimation)

```
version 19
set scheme sj
capture log close
set more off
log using ehc3.log, replace

use rrdat1, clear

gen des = tfin ~= ti          /* destination state */
gen tf  = tfin - tstart + 1   /* ending time */

label define sex 1 "Men" 2 "Women"
label value sex sex

ltable tf des, intervals(30) su h f by(sex)  /* command for life table estimation */

ltable tf des, intervals(30) by(sex) gr overlay              ///
               ysc(r(1)) ylabel(0(0.2)1) xlabel(0(100)500)   ///
               xtitle("analysis time") saving("Figure 3_1_2",replace)

log close
```

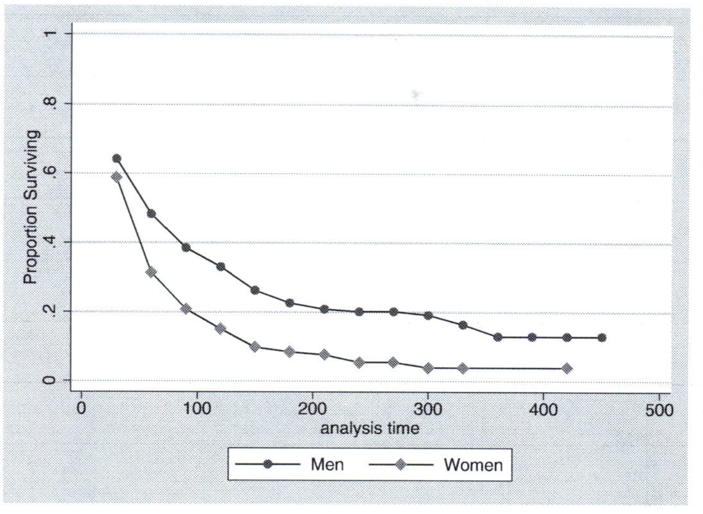

Figure 3.1.2 Plot of survivor functions for men and women, generated with do-file `ehc3.do`.

students. *Vulnerability* means that the time span between leaving the general school system and entering vocational training is limited to a short period of about 2 or 3 years, during which prevailing historical events, economic conditions, and demographic constellations strongly determine the opportunities of each generation to acquire vocational training.

The survivor functions in Figure 3.1.3 demonstrate this "vulnerable" phase for three different birth cohorts and for men and women. For every point in time after leaving the general education system, they show the proportions of school leavers who did not yet enter the German vocational training system. The curves are very different for the three birth cohorts and for men and women. The economic and political breakdown in the immediate postwar period (1945–50) had a particularly negative impact on the 1929–31 birth cohort's entry into vocational training. Confronted with the existing historical conditions, school leavers of this birth cohort did not rank entering vocational training highly as they had more pressing problems to deal with (e.g., earning a living), and it would have been very difficult to find any trainee positions at all (Mayer 1987, 1988, 1991). Compared to this immediate postwar period, the later social and economic development until the mid-1970s led to a steady rise in living standards, a decrease in unemployment, and a substantial growth in the number of trainee positions offered in the private sector.

The upper part of Figure 3.1.3 shows that about 50 % of men in the 1929–31 birth cohort started vocational training immediately after leaving the general education system. A further 27 % of these men began vocational training within three years of leaving the general education system. However, about 23 % never entered vocational training. In comparison, about 71 % and 79 % of the men in the 1939–41 and 1949–51 birth cohorts, respectively, began vocational training immediately, and an additional 14 % of the men in both cohorts started within 3 years. Thus, only about 15 % of the men in the 1939–41 cohort and about 7 % of the men in the 1949–51 cohort never entered vocational training.

These differences were even more pronounced for women (lower part of Figure 3.1.3). Within 3 years of leaving the general education system, only 40 % of the women in the 1929–31 cohort, but 60 % of the women in the 1939–41 cohort and 80 % of the women in the 1949–51 cohort, undertook vocational training. In other words, with regard to educational opportunities, women in particular carried the burden of the immediate postwar social and economic crises, but they also profited more than men from the rapid economic recovery in the 1960s and early 1970s.

In summary, this description based on survivor functions shows that in the German vocational training system, entries are hard to postpone beyond a "vulnerable" life phase of about 2 or 3 years after leaving the general education system. Individuals' careers are directed toward vocational training relatively early on and are difficult to change later on. The tremendous increase in the proportion of trainee positions until the early 1970s therefore had almost no effect on the process of entering vocational training for the 1929–31 cohort (Figure 3.1.3). It was not possible for members of this cohort to "stop" their life course and "resume" their educational careers when trainee positions finally became available. A temporary lack of trainee positions in the German education system, is therefore not just a short-term problem, but also leads to long-term life-course effects for many people. There is a long-term disadvantage, because it is difficult to acquire vocational degrees in later life stages when one is further away from the institutions of vocational training and has additional social commitments (such as maintaining one's own home, marriage, and children), and this makes entrance into the institutions of vocational training more and more unlikely. As a result, there are generations in Germany that are disadvantaged in terms of educational opportunities, such as those born

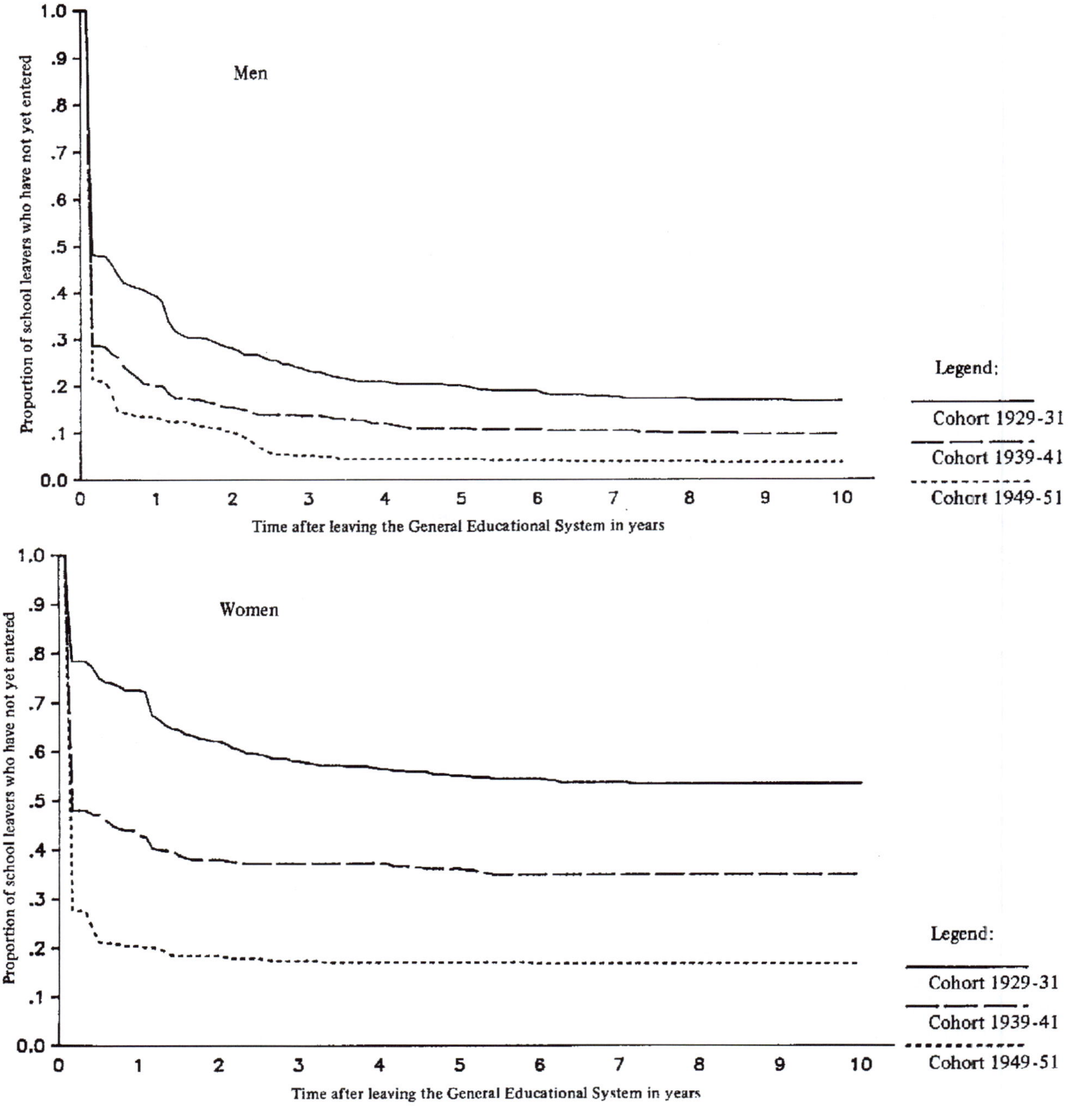

Figure 3.1.3 Process of entering vocational training (survivor functions).

around 1930, who completed their education in the immediate post-war period, or the baby boomers who entered vocational training at the beginning of the 1980s.

Example 2: Changes in Women's Ages at Family Formation

Blossfeld and Jaenichen (1992) provide a further illustration of the use of survival functions to describe social change. They discussed the changes in the process of women's entry into marriage and motherhood across successive birth cohorts in Germany. Tables 3.1.1 and 3.1.2 show the percentages of women who have not yet had their first marriage or first birth in each birth cohort at specific ages. These percentages are based on life table estimates of survivor functions for each cohort for the events of entry into marriage and first birth.

As shown in Table 3.1.1, the age at first marriage fell sharply from the 1919–23 cohort to the 1944–48 cohort, and has since been rising again until the youngest birth cohort. The greatest movements occurred among women aged 20–24, where the proportion of unmarried women fell from 46 % to 15 % and then rose again to 40 %. As a result, the youngest cohorts, 1964-68 and 1959-63, show more or less the same age pattern

Table 3.1.1 Changes in the timing of entry into marriage,as measured by proportions unmarried at specific ages (percentages).

Birth cohort	Proportion of unmarried women at age							
	20	24	28	32	36	40	44	48
1964–68	89	–	–	–	–	–	–	–
1959–63	78	40	–	–	–	–	–	–
1954–58	73	32	19	–	–	–	–	–
1949–53	65	24	11	7	–	–	–	–
1944–48	65	15	7	4	3	–	–	–
1939–43	80	21	8	5	3	3	–	–
1934–38	76	23	9	6	5	5	4	–
1929–33	86	32	13	7	6	5	5	4
1924–28	90	40	16	11	8	6	5	5
1919–23	90	46	20	13	10	9	7	7

Table 3.1.2 Changes in the timing of entry into motherhood, as measured by proportions childless at specific ages (percentages).

Birth cohort	Proportion of childless women at age							
	20	24	28	32	36	40	44	48
1964–68	92	–	–	–	–	–	–	–
1959–63	90	57	–	–	–	–	–	–
1954–58	84	55	30	–	–	–	–	–
1949–53	78	47	21	15	–	–	–	–
1944–48	77	32	16	11	9	–	–	–
1939–43	87	41	18	13	10	10	–	–
1934–38	83	45	18	13	11	11	11	–
1929–33	92	54	28	19	16	16	16	16
1924–28	87	56	26	19	15	14	14	14
1919–23	91	57	30	19	15	15	15	15

at entry into marriage as the oldest cohorts, 1924-28 and 1919-23.

Looking at ages at first birth in Table 3.1.2, a similar trend can be observed. Again, it is the 1944–48 cohort that entered motherhood at the youngest ages. In this cohort, not only marriages but also entries into motherhood were highly concentrated. And again, we find more or less the same temporal pattern of entry into motherhood for the youngest cohorts, 1964–68 and 1959–63, and the oldest cohorts, 1924–28 and 1919–23—at least as far as the youngest cohorts can be tracked.

Both tables show that the delay in entering marriage and motherhood seems to be less dramatic in Germany than in other countries such as those in Scandinavia and especially Sweden (see Hoem and Rennermalm 1985; Hoem 1986, 1991; Blossfeld 1995). In Germany, the pattern of the age of entry into marriage and motherhood is more or less the same as it was 50 years ago. However, it is also clear that the earlier trend toward younger and universal marriage and motherhood in Germany had come to a halt at the end of the 1960s and the beginning of the 1970s. But this reversal of the timing of marriage and motherhood is not consistent with the monotonic trend in women's educational attainment across cohorts (Blossfeld and Shavit 1993). It is therefore questionable whether changes in marriage and motherhood can be attributed mainly to women's growing economic independence (see Blossfeld 1995), as argued by, for example, Becker (1981).

3.2 Product-Limit Estimation

Another method for the nonparametric estimation of the survivor function and its derivatives is the product-limit, also known as the Kaplan-Meier (1958) method. One of the advantages of this approach over the life table method is that it is not necessary to group the episode durations according to arbitrarily defined time intervals. Instead, the product-limit method is based on calculating a risk set at each point in time at which least one event has occurred. In this way, the information contained in a set of episodes is optimally utilized. The only drawback of this method is that all episodes must be sorted by their ending (and starting) times; but with efficient sorting algorithms, the method can be used for quite large sets of episodes.

This section describes the product-limit estimation method and its implementation in Stata.[13] The options are essentially the same as with the life table method, depending on the type of input data: (1) If the input data are split into groups, separate product-limit estimates can be calculated for each group. (2) If there is more than a single origin state, one or more product-limit estimates can be calculated for each subset of episodes having the same origin state. (3) If there is more than a single destination state, separate product-limit estimates can be calculated for each transition found in the input data.

The following description proceeds in two steps: First, we consider the case of a single transition, then the case of two or more destination states.

Single Transitions

We assume a sample of N episodes that all having the same origin state and either have the same destination state or are right-censored. If groups are defined, it is assumed that all episodes belong to the same group. For the moment, we also assume that all episodes have the starting time zero.

The first step is to consider the points in time when at least one of the episodes ends with an event. There are, let's say, q such points in time.

$$\tau_1 < \tau_2 < \tau_3 < \ldots < \tau_q.$$

The second step is to define three basic quantities, all defined for $l = 1, \ldots, q$, with the convention that $\tau_0 = 0$.

$$\begin{aligned} E_l &= \text{the number of episodes with events at } \tau_l, \\ Z_l &= \text{the number of censored episodes ending in } [\tau_{l-1}, \tau_l), \\ R_l &= \text{the number of episodes in the risk set at } \tau_l, \text{ denoted } \mathcal{R}_l; \\ &\quad \text{that is, the number of episodes with starting time less than} \\ &\quad \tau_l \text{ and ending time} \geq \tau_l. \end{aligned}$$

Note that the implied definition of the risk set allows the handling of episodes with starting times greater than zero. Also note that the risk set at τ_l includes episodes that are censored at this point in time. It is assumed that a censored episode contains the information that there was no event up to *and including* the observed ending time of the episode. Therefore, censoring takes place at an infinitesimal amount to the right of the observed ending time.

Given these quantities, the product-limit estimator of the survivor function is defined as

$$\hat{G}(t) = \prod_{l:\tau_l < t} \left(1 - \frac{E_l}{R_l}\right).$$

This is a step function with steps at the points in time, τ_l. The usual formula to calculate estimates of standard errors for the survivor function is

$$\text{SE}(\hat{G}(t)) = \hat{G}(t) \left[\sum_{l:\tau_l < t} \frac{E_l}{R_l\,(R_l - E_l)} \right]^{1/2}.$$

In addition to survivor function estimates, the product-limit method gives a simple estimate of the cumulated transition rate:[14]

$$\hat{H}(t) = -\log\left(\hat{G}(t)\right).$$

This is again a step function. It is especially useful for simple graphical checks of distributional assumptions about the underlying durations (see chapter 8).

Unfortunately, unlike the life table estimation, the product-limit method does not provide direct estimates of transition rates. Of course, it is possible to obtain estimates by numerical differentiation of $\hat{H}(t)$, but this would require that you first apply a smoothing procedure to the cumulative rate.[15]

To illustrate the application of the product-limit estimator with Stata in Box 3.2.1, we again apply the job-exit example, where we assumed that there are only single episodes and two states ("being in a job" and "having left the job"). To estimate the Kaplan-Meier survivor function, type `sts list`. By default, the function is listed for all unique time values in the data. To reduce the output, you can also choose the points in time at which the estimated survivor function should be listed. This is achieved by using the option `at()`. In our example, the following command will produce the output shown in Box 3.2.2: `sts list, at(0 2/10 42/45 275 293 312 326 332 350 428)`.[16]

Box 3.2.1 Do-file `ehc4.do` (Kaplan-Meier estimation)

```
version 19
set scheme sj
capture log close
set more off
log using ehc4.log, replace

use rrdat1, clear

gen des = tfin ~= ti          /* destination state */
gen tf  = tfin - tstart + 1   /* ending time */

stset tf, f(des)   /* define single episode data */

sts list  /* list survivor and cumulative hazard functions */
sts list, at(0 2/10 42/45 275 293 312 326 332 350 428)

sts graph, title("Product-Limit Survivor Function") saving("Figure 3_2_1", replace)

log close
```

Box 3.2.2 Result of do-file `ehc4.do` (Box 3.2.1)

```
      failure _d:  des
 analysis time _t:  tf
          Beg.          Net     Survivor      Std.
Time     Total   Fail   Lost    Function     Error   [95% Conf. Int.]
-------------------------------------------------------------------
   2       600      2      1      0.9967    0.0024  0.9867    0.9992
   3       597      5      2      0.9883    0.0044  0.9757    0.9944
   4       590      9      0      0.9732    0.0066  0.9567    0.9835
   5       581      3      1      0.9682    0.0072  0.9506    0.9796
   6       577     10      0      0.9514    0.0088  0.9309    0.9660
   7       567      9      1      0.9363    0.0100  0.9136    0.9533
   8       557      6      3      0.9262    0.0107  0.9022    0.9446
   9       548      7      1      0.9144    0.0115  0.8889    0.9343
  10       540      8      4      0.9009    0.0123  0.8739    0.9223
                            output omitted
  42       273      1      0      0.5040    0.0209  0.4623    0.5442
  43       272      2      1      0.5003    0.0209  0.4586    0.5405
  44       269      5      1      0.4910    0.0210  0.4493    0.5313
  45       263      1      0      0.4891    0.0210  0.4474    0.5295
                            output omitted
 275        26      1      0      0.1345    0.0175  0.1025    0.1709
                            output omitted
 293        20      1      0      0.1278    0.0179  0.0953    0.1652
                            output omitted
 312        16      1      1      0.1198    0.0185  0.0866    0.1588
 326        14      1      0      0.1112    0.0190  0.0775    0.1518
                            output omitted
 332        11      1      0      0.1011    0.0198  0.0666    0.1440
                            output omitted
 350         9      1      0      0.0899    0.0205  0.0550    0.1353
                            output omitted
 428         1      0      1      0.0899    0.0205  0.0550    0.1353
```

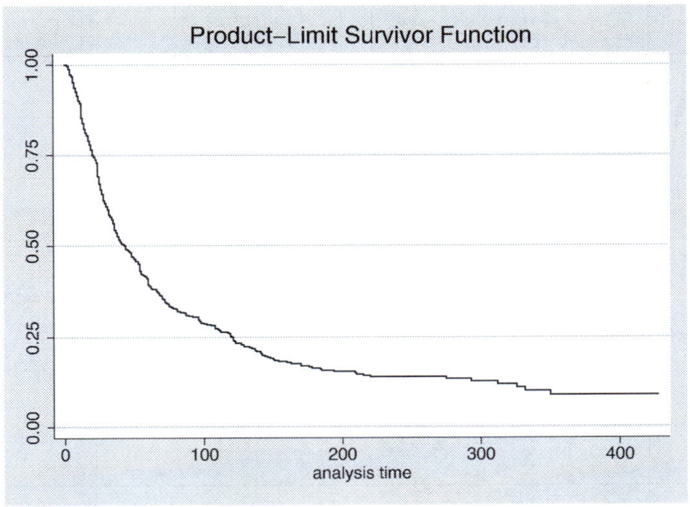

Figure 3.2.1 Plot of survivor function (product-limit estimation) generated with do-file `ehc4.do`.

The column labeled `time` shows the points in time at which at least one event takes occurs. The risk set is given in the second column. For example, at a job duration of 2 months, 600 episodes are still at risk. The number of events is given in Column 3. For example, at a job duration of 6 months, we observe 10 job moves, and at a job duration of 42 months we observe 1 event. The last four columns show estimates of the survivor function, its standard errors, and the 95 % confidence intervals. For example, after about 4 years (or 45 months), a proportion of 0.4891 of workers are still in the same job. The survivor function of the product-limit estimator is only defined up to the highest event time. The highest event time in Box 3.2.2 is 350 months. Eight additional censored cases with longer job durations follow. Under these circumstances, the estimated survivor function can no longer approach zero and can be interpreted only up until 350 months.

Again, the survivor function in Box 3.2.2 is not very comprehensible. The shape of the function is easier to assess if it is plotted against job duration. An example is provided at the end of `ehc4.do`, see Box 3.2.1. If you type `sts graph`, or simply `sts`, a graph of the Kaplan-Meier survival estimate is displayed. The resulting plot is given in Figure 3.2.1. A comparison with the life table estimate of the survivor function, shown in Figure 3.1.1, shows fairly identical results. This is consistent with our experience from practical research applications, which show that the difference between life table and product-limit estimates is usually very small.

3.3 Comparing Survivor Functions

When analyzing episode data, it is often necessary to compare survivor functions and test whether there are significant differences.There are basically two different methods available. The first relies on the calculation of confidence intervals for each of the survivor functions and then checking whether or not they overlap. This is possible with both the life table and the product-limit methods. Both methods provide estimates of standard errors for the survivor function. Another possibility is to calculate specific test statistics to compare two or more survivor functions. Both options are described in this section.

Defining Groups of Episodes

In order to make comparisons, you must have two or more groups of episodes. This is easy to achieve by using indicator variables that define membership in a group. In Stata, the syntax is

```
sts list, by(varlist).
```

In our example, we specify a single `by()` variable, `by(sex)`, but it is also possible to select several variables. The set of episodes specified in the current data matrix is then split into groups, and separate calculations are performed for each group characterized by equal values of the variables in `varlist`.

The example in Box 3.2.1 is extended to illustrate grouping in the case of estimating product boundaries. The new do-file, `ehc5.do` is a small modification of the do-file `ehc4.do` already shown in Box 3.2.1. The additional commands are shown in Box 3.3.1.

Box 3.3.1 Do-file `ehc5.do` (comparing survivor functions)

```
version 19
set scheme sj
capture log close
set more off
log using ehc5.log, replace

use rrdat1, clear

gen des = tfin ~= ti          /* destination state */
gen tf  = tfin - tstart + 1   /* ending time */

label define sex 1 "Men" 2 "Women"
label value sex sex

stset tf, f(des)   /* define single episode data */

sts list, by(sex)                    /* List survivor functions */
sts list, by(sex) at(12 24 48 120)   /* List survivor functions at 1,2,4,10 years */

sts graph, by(sex) gwood saving("Figure 3_3_1", replace)

sts test sex              /* Log-rank test for equality of survivor functions */
sts test sex, wilcoxon    /* Wilcoxon (Breslow) */
sts test sex, tware       /* Wilcoxon (Taron-Ware) */
sts test sex, peto        /* Wilcoxon (Prentice) */

log close
```

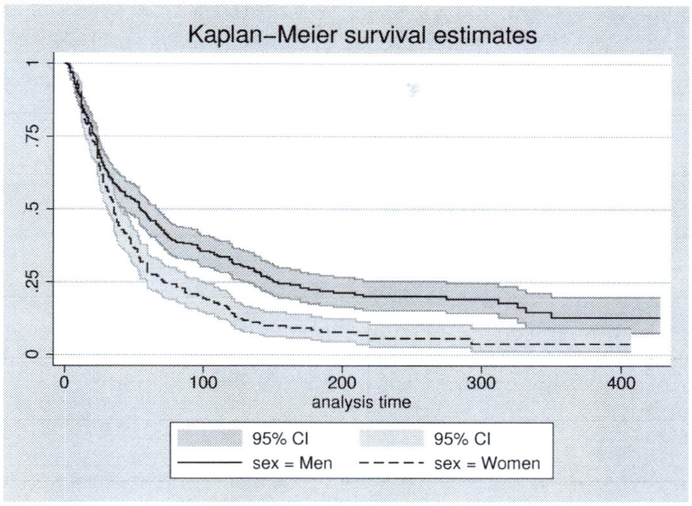

Figure 3.3.1 Plots of survivor functions (product-limit estimation) for men and women with 95 % confidence intervals (gray-scaled). The plot has been generated with do-file `ehc5.do`.

This modified file is used to estimate the product-limit estimation for men and women separately. The results window contains two tables: one with the estimates for men and the other for women. Figure 3.3.1 shows a plot of these two survivor functions. In Stata, you can obtain this plot with the following command: `sts graph, by(sex)`. The option `gwood` is used to show the pointwise Greenwood confidence bands around the survivor function. After about 3 years, the confidence bands of the survivor functions of men and women no longer intersect. There are therefore statistically significant differences in the job-exit behavior of men and women for longer durations.

Construction of Test Statistics

Many different test statistics have been proposed to compare two or more survivor functions. We describe four that can be calculated with Stata. All are based on product-limit estimates of survivor functions.

It is assumed that m groups have been defined that do not intersect. The total sample is implicitly defined as the set of all episodes contained in one of these groups. Then all calculations for each transition in the total sample are performed in exactly the same way as explained in the context of the product limit method. Therefore, we only consider a sample of episodes that have the same origin state and are censored or have the same destination state.

In general, a sample defined in this way consists of m groups, and the following table can be calculated:

$$
\begin{array}{cccccccc}
\tau_1 & R_{11} & E_{11} & R_{12} & E_{12} & \dots & R_{1m} & E_{1m} \\
\tau_2 & R_{21} & E_{21} & R_{22} & E_{22} & \dots & R_{2m} & E_{2m} \\
 & & & & & \vdots & & \\
\tau_q & R_{n1} & E_{q1} & R_{q2} & E_{q2} & \dots & R_{qm} & E_{qm}
\end{array}
$$

These are the basic quantities for the product-limit estimation for the entire sample and for each group separately. $\tau_1 < \tau_2 < \ldots < \tau_q$ are the points in time at which at least one episode contained in the sample has an event. E_{lg} is the number of episodes contained in group g with an event at τ_l; R_{lg} is defined as the number of elements in the risk set at τ_l for the episodes contained in group g (i.e., all episodes belonging to group g whose starting times are less than τ_l and whose ending times are equal to or greater than τ_l). Overall, these quantities are sufficient for a product-limit estimation in each of the m groups.

Given this, the four test statistics can be defined and are denoted as S_ν ($\nu = 1, \ldots, 4$). Since the calculations only differ due to different weights, we first give their definitions. The weights are denoted by

$W_l^{(\nu)}$; and they are defined for $l = 1, \ldots, q$ by

$$\begin{aligned} W_l^{(1)} &= 1, \\ W_l^{(2)} &= R_l, \\ W_l^{(3)} &= \sqrt{R_l}, \\ W_l^{(4)} &= \prod_{i=1}^{l} \frac{R_i - E_i + 1}{R_i + 1}. \end{aligned} \tag{3.1}$$

The next step is to construct one (m)-vector $U^{(\nu)}$ and one (m,m)-matrix $V^{(\nu)}$ for each of the four test statistics. The definitions are:[17]

$$\begin{aligned} U_g^{(\nu)} &= \sum_{l=1}^{q} W_l^{(\nu)} \, (E_{lg} - R_{lg} \frac{E_{l0}}{R_{l0}}), \\ V_{g_1 g_2}^{(\nu)} &= \sum_{i=1}^{n} W_l^{(\nu)2} \, \frac{E_{l0}\,(R_{l0} - E_{l0})}{R_{l0} - 1} \, \frac{R_{lg_1}}{R_{l0}} \left(\delta_{g_1 g_2} - \frac{R_{lg_2}}{R_{l0}} \right). \end{aligned}$$

Finally, the test statistics are defined by

$$S_\nu = U^{(\nu)'} \, V^{(\nu)^{-1}} \, U^{(\nu)}. \tag{3.2}$$

All of them follow a χ^2 distribution with $m - 1$ degrees of freedom given the null hypothesis that there are no significant differences. Note that, accordingly, the rank of $V^{(\nu)}$ is only $m - 1$. Therefore in the calculation of (3.2), you can use a generalized inverse or omit the last dimension without loss of generality. Stata chooses the latter of the two options.

Unfortunately, there is no standardized convention for naming the different test statistics, so we give the names used by Stata and make some comments on other naming conventions. In the order given by (3.1), we have:

1. *Log-rank.* Other names are *Generalized Savage Test* (Andreß 1985, p. 158, 1992). The same name is used by BMDP, with *Mantel-Cox* added. SAS calculates this test statistic under the name *Logrank.*
2. *Wilcoxon-Breslow-Gehan.* BMDP gives the name *Generalized Wilcoxon (Breslow).* SAS uses only the label *Wilcoxon.*
3. *Tarone-Ware.* This test statistic was proposed by Tarone and Ware (1977) and is named accordingly. It is also calculated by BMDP, using the label *Tarone-Ware.*
4. *Peto-Peto-Prentice.* Finally, there is a test statistic explained by Lawless (1982, p. 423) with reference to R. L. Prentice.

The Stata command for calculating test statistics to compare survivor functions is `sts test` as shown in Box 3.3.1. The `sts test` command performs the log-rank test by default. To compute the other test statistics, you have to select one of the following options: `wilcoxon` to specify the Wilxocon-Breslow-Gehan test, `tware` to conduct the Taron-Ware test, or `peto` to run the Peto-Peto-Prentice test.[18] The results of the do-file `ehc5.do` are shown in Box 3.3.2. All test statistics are based on the null hypothesis that the survivor functions of men and women do not differ. They are χ^2-distributed with $m - 1$ degrees of freedom (in the example we have two groups, men and women: $m = 2$). All four test statistics are significant. In other words, the null hypothesis that survivor functions of men and women do not differ must be rejected. However, it is easy to see that there is a great difference between the log-rank (or Savage) test statistic and the other test statistics. The reason for this is that the Wilcoxon test stresses differences in the survivor functions at the beginning of the duration, while the log-rank (or Savage) test statistic emphasizes the increasing differences at the end of the process time (see Figure 3.3.2).

Multiple Destination States: Pseudosurvivor Functions

We now turn to the case of multiple transitions. Here we are dealing with a situation of competing risks that depend on a given origin state. There are various concepts that can be used to describe such a situation. The simplest generalization of the single transition case leads to product-limit estimates for pseudosurvivor functions.[19]

Box 3.3.2 Results of do-file `ehc5.do` (Box 3.3.1)

```
. sts test sex
         failure _d:  des
   analysis time _t:  tf

  Log-rank test for equality of survivor functions
        |  Events          Events
  sex   |  observed        expected
  ------+-------------------------
  Men   |      245          290.89
  Women |      213          167.11
  ------+-------------------------
  Total |      458          458.00
                                        chi2(1) =       20.60
                                        Pr>chi2 =      0.0000

. sts test sex, wilcoxon
         failure _d:  des
   analysis time _t:  tf

  Wilcoxon (Breslow) test for equality of survivor functions
        |  Events          Events        Sum of
  sex   |  observed        expected       ranks
  ------+--------------------------------------
  Men   |      245          290.89       -11883
  Women |      213          167.11        11883
  ------+--------------------------------------
  Total |      458          458.00            0
                                                   chi2(1) =        9.36
                                                   Pr>chi2 =      0.0022

. sts test sex, tware
         failure _d:  des
   analysis time _t:  tf

  Tarone-Ware test for equality of survivor functions
        |  Events          Events        Sum of
  sex   |  observed        expected       ranks
  ------+--------------------------------------
  Men   |      245          290.89    -717.3311
  Women |      213          167.11     717.3311
  ------+--------------------------------------
  Total |      458          458.00            0
                                                   chi2(1) =       14.33
                                                   Pr>chi2 =      0.0002

. sts test sex, peto
         failure _d:  des
   analysis time _t:  tf

  Peto-Peto test for equality of survivor functions
        |  Events          Events        Sum of
  sex   |  observed        expected       ranks
  ------+--------------------------------------
  Men   |      245          290.89   -21.622585
  Women |      213          167.11    21.622585
  ------+--------------------------------------
  Total |      458          458.00            0
                                                   chi2(1) =       10.70
                                                   Pr>chi2 =      0.0011
```

The method is analogous to the single transition case. You start with N episodes that have the same origin state. Then, for each possible destination state k, you look at the points in time, $\tau_{k,l}$, when at least one transition to destination state k takes place. There are, say, $l = 1, \ldots, q_k$ such points in time.

Let $E_{k,l}$ denote the number of events at $\tau_{k,l}$, and let $\mathcal{R}_l$ denote the risk set at the same point in time.

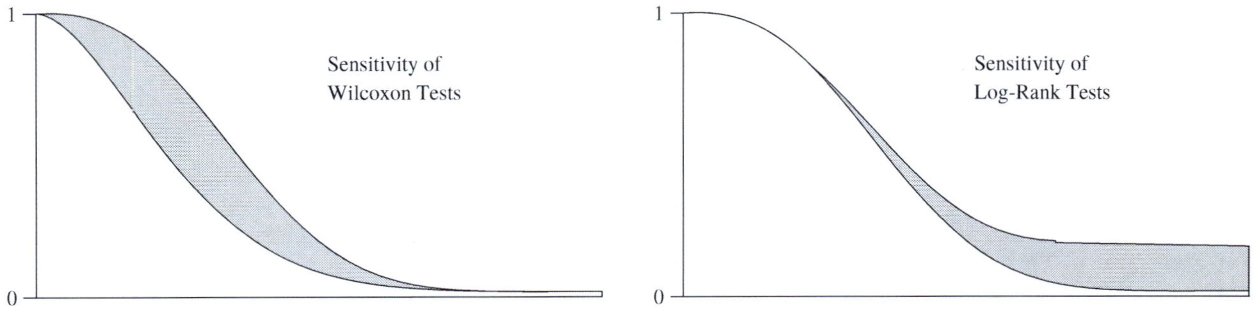

Figure 3.3.2 Regions of sensitivity for Wilcoxon and log-rank tests.

Box 3.3.3 Do-file `ehc6.do`

```
version 19
set scheme sj
capture log close
set more off
log using ehc6.log, replace

use rrdat1, clear

gen     des = 2                         /*destination state*/
replace des = 1 if (presn/pres -1)>0.2
replace des = 3 if (presn/pres -1)<0.0
replace des = 0 if presn==-1

gen tf = tfin - tstart + 1              /*ending time*/

stset tf, f(des=1) /*upward moves*/
sts list if pres <= 65, at(0 2 4 6/8 121 160 326 428)
sts gen surv1 = s

stset tf, f(des=2) /*lateral moves*/
sts list, at(0 3 4/7 121 184 209 350 428)
sts gen surv2 = s

stset tf, f(des=3) /*downward moves*/
sts list, at(0 2/6 121 275 312 332 428)
sts gen surv3 = s

for any surv1 surv2 surv3 \ any "upward" "lateral" "downward": ///
label var X "Y"

graph twoway line surv1 surv2 surv3 _t, sort ///
ysc(r(0 1)) ylabel(0(0.2)1) xtitle("analysis time")legend(row(1))

log close
```

Box 3.3.4a Part 1 (upward moves)

```
         failure _d:  des == 1
   analysis time _t:  tf
                Beg.                  Survivor      Std.
   Time        Total      Fail        Function     Error     [95% Conf. Int.]
-------------------------------------------------------------------------------
      0            0         0          1.0000         .          .          .
      2          591         1          0.9983    0.0017     0.9880     0.9998
      4          582         2          0.9949    0.0030     0.9842     0.9983
      6          569         3          0.9896    0.0042     0.9771     0.9953
      7          559         1          0.9879    0.0046     0.9747     0.9942
      8          549         1          0.9861    0.0049     0.9723     0.9930
    121          103        70          0.7720    0.0254     0.7175     0.8173
    160           61        74          0.7301    0.0316     0.6625     0.7864
    326           14         1          0.6663    0.0583     0.5382     0.7664
    428            1         0          0.6663    0.0583     0.5382     0.7664
-------------------------------------------------------------------------------
Note:  Survivor function is calculated over full data and evaluated at
indicated times; it is not calculated from aggregates shown at left.
```

Note that the risk set does not depend on the destination state, but is defined as in the single transition case as the set of all episodes with a starting time less than $\tau_{k,l}$ and with an ending time equal to or greater than $\tau_{k,l}$. The product-limit estimate of the pseudosurvivor functions can then be formally defined by

$$\tilde{G}_k(t) = \prod_{l:\tau_{k,l}<t} \left(1 - \frac{E_{k,l}}{R_l}\right).$$

Of course, the same algorithm can be used to calculate this estimate as for the case of a single transition. In the calculation for a specific destination state, you only have to treat *all* episodes that do not end in this destination as if they were censored.

To illustrate this, we use (in do-file `ehc6.do` in Box 3.3.3) our example data set, `rrdat1`, and construct a new variable, `des`, to distinguish three different destination states.[20] The Variable `des` takes the values 1, 2, and 3 for upward, lateral, and downward moves, respectively, or the value 0 for right-censored episodes.

Moreover, we change the content of this new variable `des` to zero for subjects who have left their last job prior the interview date, as these subjects are no longer at risk. In our example, we simply add the line `replace des = 0 if presn==-1`, since the prestige score of the consecutive job episode, `presn`, is -1 for both censored observations and all records with ending times before the date of the interview. Executing the do-file `ehc9.do`, shown in Box 3.3.3, with Stata generates three tables of product-limit estimates for the three destination states at specified times. We can also plot the results. First, we need to generate a new

Box 3.3.4b Part 2 (lateral moves)

```
         failure _d:  des == 2
   analysis time _t:  tf

              Beg.                      Survivor      Std.
  Time       Total       Fail           Function     Error     [95% Conf. Int.]
-----------------------------------------------------------------------------
     0           0          0             1.0000         .          .         .
     3         597          2             0.9966    0.0024     0.9867    0.9992
     4         590          4             0.9899    0.0041     0.9776    0.9954
     5         581          2             0.9865    0.0047     0.9732    0.9932
     6         577          4             0.9796    0.0058     0.9644    0.9884
     7         567          3             0.9745    0.0065     0.9580    0.9845
   121         105        187             0.5063    0.0276     0.4509    0.5590
   184          53        201             0.4181    0.0316     0.3557    0.4791
   209          41          1             0.3997    0.0328     0.3353    0.4632
   350           9          1             0.3553    0.0510     0.2571    0.4546
   428           1          0             0.3553    0.0510     0.2571    0.4546
-----------------------------------------------------------------------------
Note:  Survivor function is calculated over full data and evaluated at
indicated times; it is not calculated from aggregates shown at left.
```

Box 3.3.4c Part 3 (downward moves)

```
         failure _d:  des == 3
   analysis time _t:  tf

              Beg.                      Survivor      Std.
  Time       Total       Fail           Function     Error     [95% Conf. Int.]
-----------------------------------------------------------------------------
     0           0          0             1.0000         .          .         .
     2         600          1             0.9983    0.0017     0.9882    0.9998
     3         597          3             0.9933    0.0033     0.9823    0.9975
     4         590          3             0.9883    0.0044     0.9755    0.9944
     5         581          0             0.9883    0.0044     0.9755    0.9944
     6         577          3             0.9831    0.0053     0.9689    0.9909
   121         105         76             0.7541    0.0260     0.6988    0.8008
   275          26         84             0.6468    0.0445     0.5522    0.7263
   312          16          1             0.5760    0.0619     0.4458    0.6862
   332          11          0             0.5760    0.0619     0.4458    0.6862
   428           1          0             0.5760    0.0619     0.4458    0.6862
-----------------------------------------------------------------------------
Note:  Survivor function is calculated over full data and evaluated at
indicated times; it is not calculated from aggregates shown at left.
```

variable containing the estimated survivor function. The syntax is: `sts generate newvar = s`. Next, we use the graph command as shown in Box 3.3.3.

Figure 3.3.3 shows the survivor functions for these three directional moves. We can observe that workers move down faster than upwards. The estimates show that after a duration of 121 months, 22.8% of jobs end with an upward move, 49.4% with a lateral move, and 24.6% with a downward move (see Box 3.3.4a to c and calculate 1-G(t)).[21] This would mean that almost 100% of jobs end within the first 10 years. Since the

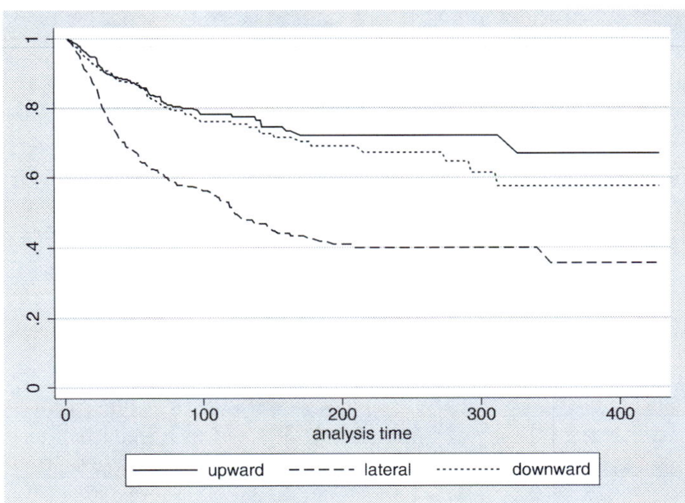

Figure 3.3.3 Plot of pseudosurvivor functions (product-limit estimation) for upward, lateral, and downward moves. The plot was generated with do-file `ehc6.do`.

pseudosurvivor curves are still falling after this (see also Figure 3.3.3), more than 100% of jobs "end," which is implausible. Hence, you should be very cautious using pseudosurvivor functions.

Multiple Destination States: Cumulative Incidence Function

An alternative is the cumulative incidence function (CIF), which can handle competing risks in an appropriate manner. In the previous sections, the focus was on the survivor function: the proportion of episodes still being at risk after some time. In contrast, the CIF has the perspective of cumulative probability distributions. How large is the proportion of already terminated episodes in a specific destination state at some point in time since the start of the process? The increase, the "jump," in the proportion ending in destination state k at $\tau_{k,l}$ is calculated by the ratio of the destination-specific events $E_{k,l}$ to all episodes still at risk R_l multiplied by the proportion of all survivors (see equation 3.3). The last part is the survivor function $\hat{G}(t-1)$, which is estimated by the Kaplan-Meier method. The sum of all destination-state-specific cumulative incidence functions is $\sum_j \widehat{CIF(t)} = 1 - \hat{G}(t)$ (Gooley et al. 1999; Rohwer 2006).

$$\widehat{CIF}_k(t) = \sum_{l:\tau_l \leq t} \frac{E_{k,l}}{R_l} \hat{G}(t-1). \tag{3.3}$$

Stata does not directly offer a command to estimate non-parametric CIFs. However, there is a program `stcompet` written by Coviello and Boggess that can be used to estimate these functions very easily in Stata.[22] Unlike in the file ehc6.do, in Box 3.3.3, we now treat all episodes containing an event as failures. Only episodes that are "truly" right-censored are treated as such. This means that we define a fourth destination state for those who exited their last job (and maybe the labor force) and have not started a new job by the time of the interview (see variable `lfexit` in Box 3.3.5). If you are only interested in upward mobility, you can place all other destination states in one single category. In a next step, we declare the data as survival time data and specify one destination state in the failure option, here `f(des=1)`. To estimate the CIF, we type `stcompet` followed by a new variable `(cif)` which should contain the cumulative indicator `(=ci)`. As options, we specify the risks that compete with the default already defined in the `stset` command. In our case, we define three additional destination states `compet1()` to `compet3()`, and place the values of the variable `des` in brackets indicating competing risks.

The new variable `cif` includes the results on all four destination states. We store the information in four separate variables containing the estimated value if an episode ends in the specified destination state and otherwise a system missing. Then we plot the incidence function for upward, lateral, and downward moves, the exit from the labor market (see Figure 3.3.4). After about 10 years (121 months), 14.1% of the jobs are over and followed by a more prestigious job, 36.7% by a similarly prestigious job, and 15.6% by a less prestigious job.

Incidence functions are sometimes presented in stacked forms. This requires some data preparation. First we have to sort the data by episode duration as well as by the variables containing the destination-state-specific CIF. Then we fill in missing values with the most recent valid information stored in some previous row. A look at Figure 3.3.4 should show what is happening here: The horizontal lines are periods with missing values in the CIF variables. Filling missing values with the last valid observation is identical to carrying forward the last valid observation until a new one is available. Before we do this, we have to replace system missings in the first row with 0 to ensure that all curves start from the beginning. Then we sum up the values for upward and lateral moves in one new variable and those for upward, lateral, and downward moves in a further new variable. We display the variable on upward moves and the two variables containing the sums (see Figure 3.3.5). We suppress the ordinary legend and place the labels in the appropriate space. In contrast to the previous figure, you can immediately see how large the proportion of jobs is that have been left and followed (sooner or later) by another one. The proportion of lateral moves is the difference between the middle and the lowest curve; the proportion of downward moves, is the difference between the between the highest and the middle curve.

Multiepisode Data (or Repeating Events)

We have limited the presentation of the life table method and the product-limit estimation in this book to the special case of single episodes (with competing risks). For didactic reasons, we assumed in our practical examples that the episodes in the data set `rrdat1` are statistically independent of each other and that the population is homogeneous. This is of course not the case, because the GLHS records complete job histories of men and women up to the time of the interview. Hence, individuals with a higher number of jobs are represented in the file `rrdat1` more often than individuals with a lower number. One solution to this problem is to only look at episodes with a specific job number. For example, one could study job behavior in only the

Box 3.3.5 Do-File ehc7.do

```
version 19
set scheme sj
capture log close
set more off
log using ehc7.log, replace

use rrdat1, clear

ssc install stcompet, replace  /* ado-file written by E. Coviello &  M. Boggess */

gen     des = 2                              /* destination state */
replace des = 1 if (presn/pres -1) > 0.2
replace des = 3 if (presn/pres -1) < 0.0
replace des = 4 if presn == -1 & tfin  < ti   /* labor force exit */
replace des = 0 if presn == -1 & tfin == ti   /* right censored */

gen tf = tfin - tstart + 1   /* ending time */

stset tf, f(des = 1)
stcompet cif = ci, compet1(2) compet2(3) compet3(4)

gen upward   = cif if des == 1
gen lateral  = cif if des == 2
gen downward = cif if des == 3
gen lfexit   = cif if des == 4

twoway line upward lateral downward _t, sort connect(step step step)       ///
  ylabel(0(0.2)1) ytitle("Cumulative Incidence") xtitle("analysis time")   ///
  saving("Figure 3_3_4", replace)

sort _t upward lateral downward
for X in var upward lateral downward: replace X = 0  if X == . & _n == 1   ///
  \ replace X = X[_n-1] if X == . & X[_n-1] <.
gen upw_lat      = upward + lateral
gen upw_lat_down = upward + lateral + downward

twoway line upward upw_lat upw_lat_down _t, sort connect(step step step)   ///
  ylabel(0(0.2)1) ytitle("Cumulative Incidence") xtitle("analysis time")   ///
  text(0.1 200 "upward moves") text(0.3 200 "lateral moves")               ///
  text(0.65 200 "downward moves") legend(off) saving("Figure 3_3_5", replace)

log close
```

first job, the second job, and so on. Another solution would be to compare survivor functions for several—say the first four—jobs (using the variable "serial number of job" as a group variable) and test whether the survivor functions are equal. If they are not significantly different, you can pool them and continue as in the single episode case.

In general, great differences in the number of episodes between observation units indicate that the sample is quite heterogeneous. However, homogeneous populations are also assumed in the case of individual episodes. Neglected (or unobserved) heterogeneity between observation units can lead to apparent time dependence (see chapter 10) and wrong substantive conclusions. You should therefore be cautious and only

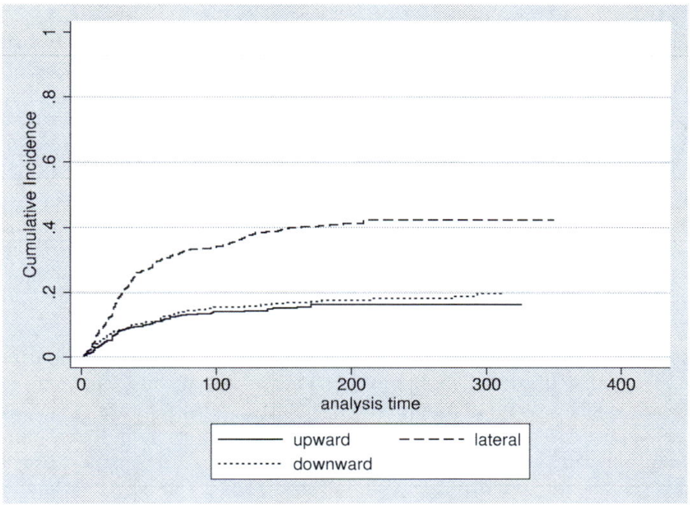

Figure 3.3.4 Plot of cumulative incidence function (CIF) for upward, lateral, and downward moves. The plot was generated with do-file ehc7.do

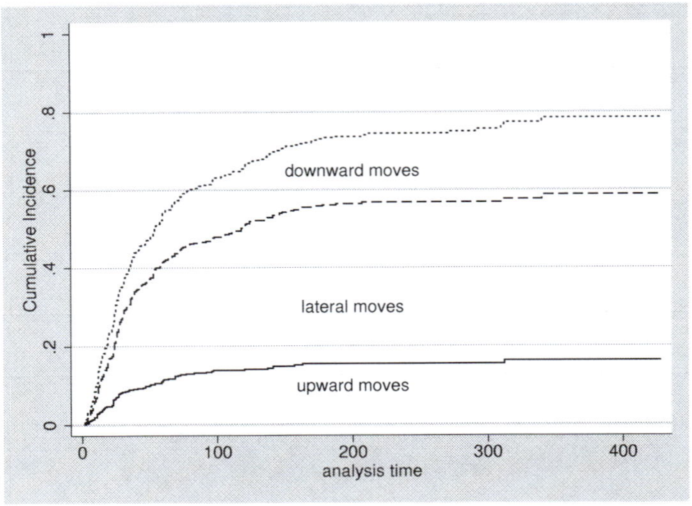

Figure 3.3.5 Plot of stacked CIF for upward, lateral, and downward moves. The plot was generated with do-file `ehc7.do`

estimate survivor functions for truly homogeneous groups by disaggregating the sample according to theoretically important variables. Unfortunately, this approach is usually of limited use in practical applications, as a large sample size would be required to study a great number of different subpopulations separately. Hence, comparisons of survivor functions between subgroups and the possibility to detect time dependence based on transformations of survivor functions (see chapter 8) usually have only a heuristic character. In many cases, it is easier to include population heterogeneity and the differences in the event histories of individuals into transition rate models—as will be shown in the following chapters.

Notes

[1] An detailed discussion of the life table method can be found in Namboodiri and Suchindran (1987).

[2]There are four types of weights in Stata: frequency weights (`fweights`), analytic weights (`aweights`), sampling weights (`pweights`), and importance weights (`iweights`). To learn more about these weighting types, enter the command `help weights` in the command line. The Stata command for life tables allows for `fweights`. For a discussion of the use of weights in longitudinal data analyses, see Hoem (1985, 1989).

[3]We generally denote the risk set by the symbol $\mathcal{R}$ and the number of units contained in the risk set by the symbol R.

[4]See the discussion in Namboodiri and Suchindran (1987: 58pp).

[5]Stata's default is $\omega = 0.5$. If the `noadjust` option is specified, $\omega = 0$ is assumed. Stata does not offer any further options for the weighting factor ω in the `ltable` command.

[6]This is of course an unrealistic assumption, as the individuals in the data file (`rrdat1`) are represented with varying numbers of job spells. Hence, there are dependencies between the episodes of each individual.

[7]If you do not specify an argument in the `intervals()`-option or if you do not write `intervals()` in the command line, Stata's `ltable` assumes Argument 1 and spans intervals with a length of 1 time unit. The output then displays only those intervals in which at least one episode ends, regardless of whether there is failure or censoring.

[8]Since there can be a considerable number of intervals in which no events occur and therefore the hazard rate is zero, a graph of the rate function is best created by using a kernel smooth. To plot the results of the hazard table, use the command `sts graph` (for more information see footnote 15 in this chapter).

[9]The appearance of a Stata graph in the Stata Graph Window is specified by a graph scheme. To ensure that you obtain the same results as in this textbook, enter `set scheme sj` to change to the Stata Journal scheme. We use different options to specify the title, scale, and labels of the graph. To learn more about the options available with `ltable`, type `help ltable` and `help twoway options`.

[10]In Stata, linking value labels and variables is a two-step process. First, you have to define a value label by using the command `label define`; then specify the `label value` to attach its contents to a variable.

[11]By default, Stata draws a separate graph within the same image for each group defined with the `by()` option. Sometimes it is useful to overlay these plots on the same graph. You can do this by specifying the option `overlay`.

[12]Elder 1975, 1978, 1987; Hogan 1978, 1981; Marini 1978, 1984, 1985; Mayer and Schwarz 1989.

[13]You must declare your data as event history data before you can use the Stata command for product-limit estimation. If you work with weighted data, `fweights`, `iweights`, and `pweights` may be specified at this point.

[14]For the relationship between the survivor function and the cumulated transition rate, see also equation 1.6 in chapter 1.2.2.

[15]Actually Stata is a very convenient program that performs some differentiation and smoothing procedure. To get a result, you simply have to add `hazard` after `sts graph`. However, note that the hazard rate is not derived from the survivor function estimated with the Kaplan-Meier method, but from the cumulative hazard rate $\hat{H}(t)$ estimated by a method proposed by Nelson (1972) and Aalen (1978), which is defined as $\hat{H}(t) = \sum_{l:\tau_l \leq t} \frac{E_l}{R_l}$. As already shown in chapter 1.2.2 and repeated in this section, there is a simple relationship between the survivor function and the cumulative hazard rate: $\hat{H}(t) = -log(\hat{G}(t))$ resp. $\hat{G}(t) = exp(-\hat{H}(t))$. Cleves et al. (2016, p. 110) point out that "in very large samples, it does not matter how you estimate the survivor function, whether by Kaplan-Meier or by transforming the Nelson-Aalen." Since the cumulative hazard rate is

a step function, some smoothing procedures based on a kernel function are used to produce a hazard rate instead of simple differentiation. For more details on, limitations of, and options for smoothed hazard rates, see Cleves et al. (2016, pp. 113–117).

[16]If you are not yet familiar with number lists in Stata, type `help numlist`.

[17]δ_{ij} is the Kronecker symbol, which is one if $i = j$ and zero otherwise.

[18]You may also obtain the test statistics for the Cox test and the Fleming-Harrington test, which are not covered here. Further details can be found in the Stata manual.

[19]This generalization is often used with the product-limit method. See, for instance, the discussions in Lawless (1982, p. 486f) and Tuma and Hannan (1984, p. 69f).

[20]We define upward shifts as job mobility that results in an increase in the prestige score of 20 % or more, downward shifts as job mobility that is associated with a decrease in the prestige score, and lateral shifts that have no impact or result in an increase in the prestige score of up to 20 %. It is important to note here that the prestige score in our example data varies between 18 and 78. Those already in good positions with a prestige score greater than 65 are no longer at risk of experiencing an upward career move. These records are therefore excluded from the analysis.

[21]These results differ from the results in Blossfeld and Rohwer (2002) because here we have taken into account the ceiling effect for upward moves (only people who can experience a 20 % or more increase in the prestige score from job n to job $n + 1$ are at risk to move upward), as well as the bottom effect (only people who are not already at the bottom of the prestige score at risk of moving downwards), and we have corrected for all job exits without a job destination (in this case `presn` is coded -1).

[22]The Statistical Software Components (SSC) archive, maintained by Boston College and accessible via RePEc (www.repec.org), provides user-written Stata commands, often referred to as ado-files. To install the Stata program to estimate CIF, type `ssc install stcompet` in the command line or in your do-file. After the installation, you can type `help stcompet` in the command line for more information on this program.

Chapter 4

Exponential Transition Rate Models

In practical research, analyzing event history data with nonparametric estimation methods has several disadvantages. First, as discussed in the previous chapter, with an increasing number of subgroups, one usually soon reaches a point where the estimation and comparison of survival functions are no longer meaningful due to the small number of cases in the different subgroups. Second, even if it is possible to estimate an increasing number of survivor functions for important subgroups, comparisons of these functions quickly become complex and difficult to interpret. Third, for quantitative characteristics (e.g., income, age), the variables have to be grouped (e.g., "high income group" vs. "low income group"), and this leads to a loss of information. Finally, multiepisode processes can hardly be analyzed with nonparametric methods. Therefore, transition rate models have increasingly taken the place of nonparametric methods in the analysis of event history data in practical research.

Transition rate models are a general statistical technique for analyzing the dependence of the transition rate on a set of covariates. As discussed in section 1.2, viewing the transition rate as a function of change in covariates is naturally linked with a causal approach to the study of social processes. In Chapters 4 to 9, we assume for didactic reasons that there is no unobserved heterogeneity in our models. Therefore, we can interpret the estimated results causally. In general, this modeling approach requires that the covariates are measured on an interval or ratio scale, but nominal and ordinal covariates can be included in the models through the use of "dummies" (i.e., by substituting the original variables by a set of binary variables). If measurement permits, there is also the interesting possibility of controlling for various factors by introducing their metric versions as proxies to the analysis.[1] Well-known examples are the inclusion of social inequality via metric prestige scores (Handl, Mayer, and Müller 1977; Treiman 1977; Wegener 1985; Shavit and Blossfeld 1993) or the approximation of qualification levels by the average number of school years required to reach a certain educational level (Blossfeld 1985; Shavit and Blossfeld 1993). The previous history of the process can also be easily taken into account in transition rate models. For example, in the job duration example, the previous history of the process could be taken into account by a variable "general labor force experience" (measured in number of months worked) or a variable "number of previously held jobs." The use of these metric versions of important factors makes it possible to study or control for a large number of effects without substantially increasing the number of parameters to be estimated in the models. Finally, and most importantly, transition rate models also permit the analysis of the impact of duration dependence and the influence of one or more parallel processes through the use of (qualitative and/or quantitative) time-dependent covariates.

In this chapter, we discuss the application of the basic exponential model. First, we describe the characteristics of the exponential transition rate model and its estimation using the maximum likelihood method. We then estimate exponential models without covariates and with time-constant covariates and give detailed interpretations. We also demonstrate how to deal with multiple destination states and multiple episodes as well as the application of equality constraints. Extensions of the basic exponential model will be discussed in the next two chapters.

4.1 The Basic Exponential Model

The exponential model is the simplest transition rate model. It assumes that the continuous duration variable T can be described by an exponential distribution with density, survivor, and transition rate functions, each of which is described by

$$\begin{aligned} f(t) &= a\exp(-at) \qquad a > 0, \\ G(t) &= \exp(-at), \\ r(t) &= a. \end{aligned}$$

A general definition of the model, for transitions to destination state k, always assuming a given origin state, can be written as

$$r_k(t) \equiv r_k = \exp\left(\alpha_{k_0} + A_{k_1}\,\alpha_{k_1} + \ldots\right) = \exp\left(A_k\,\alpha_k\right)$$

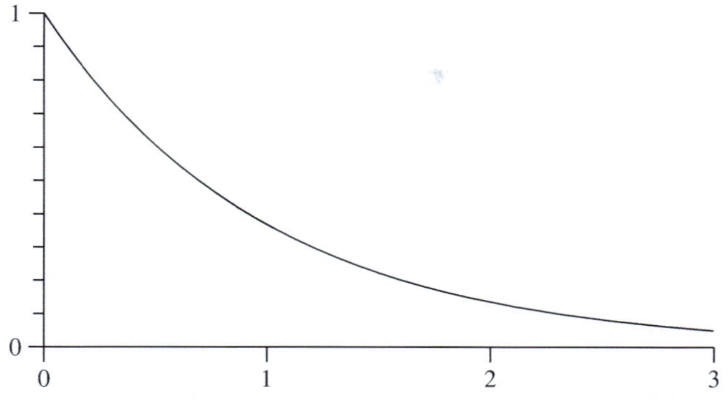

Figure 4.1.1 Survivor function of the standard ($r = 1$) exponential distribution.

where r_k is the time-constant transition rate to destination state k. The exit rate (i.e., the rate of leaving the origin state for any one of the possible destination states) is

$$r = \sum_{k \in \mathcal{D}} r_k$$

with $\mathcal{D}$ denoting the set of all possible destination states. The survivor function for the duration in the origin state can be easily formulated using this exit rate:

$$G(t) = \exp\left(- \int_0^t r \,\mathrm{d}\tau \right) = \exp\big(- rt \big).$$

The exponential model assumes that the transition rate $r_k(t)$ to destination state k can vary with different constellations of covariates but is time-constant: $r_k(t) = r_k$. In other words, it is assumed that the process is not time-dependent. The relationship between the transition rate and the (row) vector of covariates A_k is specified as log-linear to ensure that estimates of the transition rate cannot become negative.[2] The (column) vector of unknown parameters α_k and the vector of observed covariates A_k are specific to each destination state k. The vector of parameters also includes a constant term, α_{k_0}, that can be estimated even if no covariates are included in the model. The covariates in the vector A_k are assumed to be measured at the beginning of each episode and to be time-constant.

Figure 4.1.1 shows the density and survivor functions of a standard exponential distribution in which the transition rate is $r = 1$. In this case, both functions are identical. In general, the density function is the survivor function multiplied by the transition rate (see section 1.2). By definition, the distribution is defined only on a nonnegative time axis.

4.1.1 Maximum Likelihood Estimation

The exponential model, like all parametric transition rate models,[3] is estimated using the maximum likelihood method. To explain the general setup of the likelihood for parametric transition rate models, we proceed in two steps: First, we consider the case of a single transition. Second, then situations with two or more destination states.

The notation to set up the likelihood is as follows: $\mathcal{D}$ is the set of possible destination states; $\mathcal{N}$ and $\mathcal{Z}$ are the set of all episodes and the set of all censored episodes, respectively; and $\mathcal{E}_k$ is the set of all uncensored episodes with destination state k. To simplify the notation, the dependence on parameters is not shown explicitly.

In the case of a single transition to destination state k, the likelihood may be written as

$$\mathcal{L}_k = \prod_{i \in \mathcal{E}_k} f(t_i) \prod_{i \in \mathcal{Z}} G(t_i) = \prod_{i \in \mathcal{E}_k} r(t_i) \prod_{i \in \mathcal{N}} G(t_i)$$

where $f(t)$ is the density function and $G(t)$ is the survivor function for the single transition. The contribution to the likelihood of an episode with an event at t_i is given by the density function, which is evaluated at the ending time t_i along with appropriate covariate values. The contribution of a censored episode is given by the survivor function evaluated at the censored ending time t_i, but this possibly depends on covariates changing their values during the episode. Recall that we currently assume that all episodes have start time zero. The likelihood can then be expressed by using only the transition rate and the survivor function.

The next step is to look at the case with two or more destination states. The destination state space is $\mathcal{D}$. The likelihood may then be written as

$$\mathcal{L} = \prod_{k \in \mathcal{D}} \prod_{i \in \mathcal{E}_k} \tilde{f}_k(t_i) \prod_{i \in \mathcal{Z}} G(t_i).$$

The contribution of an episode with an event at t_i is again given by the density function, but according to the underlying model, it is now transition-specific. It is, in fact, a subdensity function, $\tilde{f}_k(t)$, with the interpretation that $\tilde{f}_k(t)\,dt$ is the probability of leaving the origin state for the destination state k in a small time interval at t. The contribution to the likelihood of a right-censored episode is the value of the survivor function at t_i, denoted by $G(t_i)$; that is, the probability that the episode has no event until t_i.

It is now possible to factorize the likelihood into a product of transition-specific terms. First, the likelihood may be rewritten as

$$\mathcal{L} = \prod_{k \in \mathcal{D}} \prod_{i \in \mathcal{E}_k} r_k(t_i) \prod_{i \in \mathcal{N}} G(t_i).$$

This can be rewritten by using so-called pseudosurvivor functions, defined by

$$\tilde{G}_k(t) = \exp\left(-\int_0^t r_k(\tau)\,\mathrm{d}\tau\right).$$

Using this definition, the survivor function $G(t)$ can be written as a product of pseudosurvivor functions:

$$G(t) = \prod_{k \in \mathcal{D}} \tilde{G}_k(t).$$

Inserting this, the likelihood becomes

$$\mathcal{L} = \prod_{k \in \mathcal{D}} \prod_{i \in \mathcal{E}_k} r_k(t_i) \prod_{k \in \mathcal{D}} \prod_{i \in \mathcal{N}} \tilde{G}_k(t_i),$$

or, with rearranged terms:

$$\mathcal{L} = \prod_{k \in \mathcal{D}} \left\{ \prod_{i \in \mathcal{E}_k} r_k(t_i) \prod_{i \in \mathcal{N}} \tilde{G}_k(t_i) \right\}.$$

All models described in this book are estimated using the logarithm of this likelihood, that is

$$\ell = \sum_{k \in \mathcal{D}} \sum_{i \in \mathcal{E}_k} \log\{r_k(t_i)\} + \sum_{i \in \mathcal{N}} \log\left\{\tilde{G}_k(t_i)\right\}. \tag{4.1}$$

The fact that the total likelihood can be expressed as a product of transition-specific factors implies that, in the case of several transitions, a model for each transition can be estimated separately. The estimation for a transition to destination state k is done using all episodes that start in the given origin state; the episodes ending in destination state k are regarded as having an event, and *all other* episodes are included as censored.

Moreover, the ability to factorize the total likelihood offers an easy way to estimate models with different specifications of transition rates for different transitions. A different model can be assumed for each transition, and then each of these models can be estimated separately.

All parametric transition rate models offered by Stata are estimated by maximizing the log likelihood (4.1) separately for destination states $k \in \mathcal{D}$ that must be specified by the user. Note, however, that log likelihoods are defined only up to an arbitrary additive constant; the values may therefore differ between different computer programs.[4]

4.1.2 Models without Covariates

We begin with a simple exponential model and then gradually introduce complexity. To illustrate the interpretation of the basic exponential model, we first estimate a model without any covariates (i.e., we analyze only the average job change behavior of individuals). For didactic purposes, we further assume that the episodes in the data file `rrdat1` are independent of each other and that there is only one origin state ("being in a job") and one destination state ("having left the job"). Thus, the exponential model to be estimated is

$$r(t) \equiv r = \exp(\alpha_0).$$

Box 4.1.1 Do-file `ehd1.do` (exponential model without covariates)

```
version 19
capture log close
set more off
log using ehd1.log, replace

use rrdat1, clear

gen des = tfin ~= ti            /* destination state */
gen tf  = tfin - tstart + 1     /* ending time */

stset tf, failure(des)    /* define single episode data */

streg, dist(e) nohr    /* fit parametric survival model */

predict hazard, hazard
predict surv,    surv

stcurve, survival scheme(sj)                                    ///
title("Survivor function - based on exponential model")  ///
saving("Figure 4.1.2", replace)

log close
```

Box 4.1.2 Results of using do-file `ehd1.do` (Box 4.1.1)

```
         failure _d:  des
   analysis time _t:  tf

Iteration 0:    log likelihood =  -937.9681
Iteration 1:    log likelihood =  -937.9681

Exponential regression -- log relative-hazard form

No. of subjects =          600    Number of obs    =        600
No. of failures =          458
Time at risk    =        40782
                                  LR chi2(0)       =       0.00
Log likelihood  =    -937.9681    Prob > chi2      =          .

   ------------------------------------------------------------------
    _t |      Coef.   Std. Err.      z    P>|z|     [95% Conf. Interval]
   ---+--------------------------------------------------------------
 _cons |  -4.489127   .0467269   -96.07   0.000     -4.58071   -4.397544
   ------------------------------------------------------------------
```

Box 4.1.1 shows how to estimate this exponential model with Stata. The largest part of the do-file `ehd1.do` is again identical to the previously used do-files and deals with loading the data file (`rrdat1.dta`) and defining the variables as well as the episode data. In addition, we use the command

```
streg, distribution(exponential) nohr
```

to request the estimation of an exponential transition rate model. The option `distribution(exponential)` specifies the exponential model to be fitted. By default, hazard ratios are reported, but you can also instruct Stata to display coefficients by typing `nohr`.

The result of `ehd1.do` is shown in Box 4.1.2. The value of the log likelihood function is -937.9681,[5] and the estimated coefficient is -4.4891. The estimated average rate of job exit is therefore

$$r = \exp(-4.4891) = 0.0112.$$

You can calculate this rate directly within Stata using the `display` command. Type: `display exp(_b[_cons])`.[6] Alternatively, you can use the postestimation command `predict`: `predict hazard, hazard`. To derive the estimated survivor function, type `predict surv, surv`.

Stata also offers a very convenient feature for displaying survivor curves and hazard rates after parametric model estimations. The basic command is `stcurve` followed by a specification of which information should

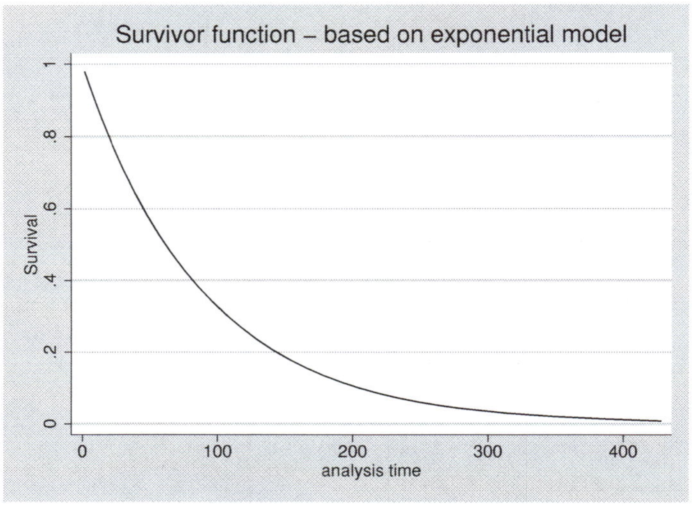

Figure 4.1.2 Plot of survivor function generated with do-file `ehd1.do`.

be put into a figure (see Box 4.1.1). The survivor function produced by `stcurve, surv` is displayed in Figure 4.1.2.

Given an exponential model, the probability of job exit in a given year does not depend on when the year begins. Using equation 1.7 (see Chapter 1), we obtain the estimate

$$\Pr(t \leq T < t + 12 \mid T \geq t) = 1 - exp(-r \cdot 12) = 1 - exp(-0.0112 \cdot 12) = 0.126$$

Since $0.0112 \cdot 12 = 0.134$ is a small value, this value approximates the estimated probability (see equation 1.8 in Chapter 1.2.2).

An estimate of the average duration in a job can be obtained in the exponential model using the following relationship (see Blossfeld, Hamerle, and Mayer 1989)

$$\mathrm{E}(T) = \frac{1}{r} = \frac{1}{0.0112} \approx 89 \text{ months.}$$

This means that, on average, about 7.5 years pass before an individual exits his/her job. If you use `predict` after `streg` and specify the option `mean time`, the predicted mean survival time is calculated within Stata. For our example, type: `predict meantime, mean time`.
Based on the survivor function

$$G(t) = \exp(-rt),$$

you can obtain an estimate of the median duration M, defined by $G(M) = 0.5$. Inserting the estimated rate

$$G(\hat{M}) = \exp(-0.0112\,\hat{M}) = 0.5,$$

our estimate is $\hat{M} \approx 62$ months. Again you can use Stata to calculate the median survival time: `predict mediantime, median time`. Since the exponential distribution is right-skewed (see Figure 4.1.1), the median is smaller than the mean of the job duration. In general, for the exponential model, the median (M) is about 69.3 % of the mean ($\mathrm{E}(T)$):

$$0.5 = \exp\left(-\frac{1}{\mathrm{E}(T)}\,M\right).$$

The probability that an individual is still employed in the same job after 10 years is $G(120) = \exp(-0.0112 \cdot 120) = 0.26$, or 26 %, and the probability that she or he has left the job by this time is $1 - G(120) = 1 - 0.26 = 0.74$, or 74 %. If you have used `predict` after `streg` to obtain each observation's predicted survivor probability, you can simply type `tab surv if _t==120` to get $G(120)$.

4.1.3 Time-Constant Covariates

The simple model without covariates treats the data as a sample of homogeneous individual job episodes, which means that we abstract from all sources of heterogeneity among individuals and their job episodes when estimating such a model. Of course, social scientists are usually more interested in finding differences (i.e., in investigating how the transition rate describing the process under study depends on observable characteristics of the individuals and their environment).

For this purpose, covariates must be included in the model. The simplest option is to include time-constant covariates. This means that the values of these covariates are fixed *at the beginning* of the episodes under study; that is, they do not change their values over the process time. As Yamaguchi (1991) noted, two groups of time-constant covariates can be distinguished: *ascribed statuses* that are (usually) constant throughout individuals' lives such as ethnicity, gender, or social origin; and *statuses attained before to* (or at the time of) entry into the process time and that remain constant thereafter.[7] Examples include highest educational attainment at the time of marriage or age at marriage in the analysis of divorce from first marriage. It should be noted that very often selection bias exists with regard to such states at entry, and this must be considered carefully in event history models.

A special case of time-constant covariates is *information about the history of the process itself* assessed at the beginning of the episodes under study. In the example of employment duration, the dependence between successive job episodes could be controlled by using information on the number of previous jobs (and unemployment episodes) and their durations.[8] The inclusion of such information is important from a substantive point of view, because life-course research has shown that for repeated events, the past course of the process is a crucial factor for understanding the present. But it is also important from a statistical point of view, because model estimation is based on the assumption of conditionally independent episodes (i.e., conditional on the covariates considered in the model specification).[9]

To estimate a model with time-constant covariates, we use the do-file `ehd2.do`, a small modification of `ehd1.do`. The changes are shown in Box 4.1.3. We define the additional required covariates and then include these covariates in the model specification.

First, to include an individual's highest educational attainment at the beginning of each job, we use `edu` as a metric proxy variable. This variable is already inculded in our data file `rrdat1` and measures the average number of school years required to reach a certain educational attainment level in Germany. Second, to distinguish between the three birth cohorts, two dummy variables, `coho2` and `coho3` are defined in the following way (using `coho1` as the reference group):

	`coho2`	`coho3`
Cohort 1929-31	0	0
Cohort 1939-41	1	0
Cohort 1949-51	0	1

Third, a variable `lfx` is constructed to provide an approximate measure of the "general labor force experience" at the beginning of each job episode. This variable is computed as the difference in months between the beginning of the respective job episode (`tstart`) and the time of first entry into the labor force (`te`).[10] Fourth, we define a variable `pnoj` to capture information on the number of previously held jobs. The value of this covariate is simply the serial number of the current job minus one. Finally, we use a variable `pres` to capture the prestige score of the current job. This information is already contained in our data file.

The model specification is again done with the `streg` command. To include the covariates, we simply add the variable names to the command line:

```
streg edu coho2 coho3 lfx pnoj pres, dist(e) nohr.
```

Using a log-linear link function between the transition rate and the vector of explanatory time-constant covariates, the model to be estimated is

$$r(t) \equiv r = \exp(A\alpha)$$

where A is the row vector of covariates and α is a corresponding column vector of coefficients.

Since Stata 18, it has been possible to visually test how well the model fits the data by using the `estat gofplot` command. By default, the Nelson–Aalen estimator is provided (Nelson 1972; Aalen 1978). It is also possible to estimate the minus log of the Kaplan–Meier estimator by adding the option "km" (Kaplan and Meier 1958; see also Chapter 3). Visually, the goodness-of-fit is assessed by plotting the residuals against their estimated cumulative hazard rate. The closer the plotted line is to the 45° line, the better the fit (Cox and Snell 1968).

Estimation results for this model are shown in Box 4.1.4. First, we obtain a value for the log likelihood function: −889.93527. Thus, one can compare this model with the exponential model without covariates (in Box 4.1.2) using a likelihood ratio test. Under the null hypothesis that the additionally included covariates

Box 4.1.3 Do-file `ehd2.do` (exponential model with covariates)

```
version 19
capture log close
set more off
log using ehd2.log, replace

use rrdat1, clear

gen des = tfin ~= ti          /* destination state */
gen tf  = tfin - tstart + 1   /* ending time */

gen coho2 = tb >= 468 & tb <= 504   /* cohort 2 */
gen coho3 = tb >= 588 & tb <= 624   /* cohort 3 */
gen lfx   = tstart - te             /* labor force experience */
gen pnoj  = noj - 1                 /* previous number of jobs */

stset tf, failure(des)   /* define single episode data */

streg edu coho2 coho3 lfx pnoj pres, dist(e) nohr   /* fit parametric
                                                       survival model */

estat gofplot   /* Nelson-Aalen goodness-of-fit visual test */
estat gofplot, km    /* Kaplan-Meier goodness-of-fit visual test */

display 2 * (e(ll) - e(ll_0))    /* likelihood ratio test */


* percentage change in the rate, given that only one covariate
* changes its value
display (exp(_b[edu])-1) * 100        /* education */
display (exp(_b[coho2])-1) * 100      /* cohort 2 */
display (exp(_b[coho3])-1) * 100      /* cohort 3 */
display (exp(_b[lfx])^12-1) * 100     /* labor force experience
                                         (increase by 1 year) */
display (exp(_b[pres])^10-1) * 100    /* prestige scale
                                         (increase by 10 units) */

* simultaneous change in covariates: change in prestige score and lfx
display (exp(_b[pres])^10 * exp(_b[lfx])^12-1) * 100

* selected subgroups
* (edu==13, coho2==0, coho3==0, lfx==0, pnoj==0, pres==60)
display     exp(_b[_cons] + _b[edu] * 13 + _b[pres] * 60)  /* rate */
display 1 / exp(_b[_cons] + _b[edu] * 13 + _b[pres] * 60)  /* average job
                                                               duration */

log close
```

do not significantly improve the model fit, the likelihood ratio test statistic (LR) follows approximately a χ^2 distribution with m degrees of freedom, where m is the number of additionally included covariates. These test statistics can be calculated as two times the difference of the log likelihoods:

$$\text{LR} = 2\,(\text{LogLik(present model)} - \text{LogLik(reference model)})\,.$$

For our example, the test statistic is

$$\text{LR} = 2\,((-889.9353) - (-937.9681)) = 2 \cdot 48.0328 = 96.07$$

with six degrees of freedom (the six additionally included covariates). In Stata, the `display` command can be used to show the test statistic. The command `streg` stores the value of the log likelihood function of the present model and the constant-only model in a matrix called `e(b)`.[11] You can now access the values contained in this matrix. The computation of the test statistic would then look like this:

```
display 2*(e(ll)-e(ll_0)) .
```

Alternatively, you can also use the command `lrtest`, a postestimation tool for `streg`. Before you can execute this command, you must estimate both the current model and the reference model and save the results with

Box 4.1.4 Results of using do-file `ehd2.do` (Box 4.1.3)

```
         failure _d:  des
   analysis time _t:  tf

Iteration 0:   log likelihood = -937.9681
Iteration 1:   log likelihood = -892.94482
Iteration 2:   log likelihood = -889.94245
Iteration 3:   log likelihood = -889.93527
Iteration 4:   log likelihood = -889.93527

Exponential regression -- log relative-hazard form

No. of subjects =          600          Number of obs    =       600
No. of failures =          458
Time at risk    =        40782
                                        LR chi2(6)       =     96.07
Log likelihood  =   -889.93527          Prob > chi2      =    0.0000

------------------------------------------------------------------------
   _t |      Coef.   Std. Err.      z    P>|z|     [95% Conf. Interval]
------+-----------------------------------------------------------------
  edu |   .0773013   .0247033     3.13   0.002     .0288837     .125719
coho2 |   .6080358   .1135509     5.35   0.000     .3854802    .8305914
coho3 |   .6107997    .118542     5.15   0.000     .3784617    .8431377
  lfx |  -.0031793   .0009378    -3.39   0.001    -.0050174   -.0013412
 pnoj |   .0596384   .0441525     1.35   0.177     -.026899    .1461758
 pres |  -.0280065   .0055303    -5.06   0.000    -.0388457   -.0171672
_cons |  -4.489445   .2795003   -16.06   0.000    -5.037255   -3.941634
------------------------------------------------------------------------
```

the `estimates store` command. For our example, simply enter:

```
streg, dist(e) nohr
estimates store ll_0
streg edu coho2 coho3 lfx pnoj pres, dist(e) nohr
lrtest ll_0
```

At a significance level of 0.05, the null hypothesis must be rejected. At least one of the included covariates significantly improves the model fit.

In addition, the maximum likelihood estimation provides standard errors for the estimated coefficients. These standard errors are useful to assess the precision of the estimates of the model parameters. In particular, you can check whether the estimated coefficients differ significantly from zero. If you divide the estimated coefficients (column `Coef.` in Box 4.1.4) by the estimated standard error (column `Std. Err.`) it results in a test statistic (column `z`), which is approximately normally distributed if the model is correct and the sample is large. If this is assumed, one can apply a formal test (see Blossfeld, Hamerle, and Mayer 1989). If one uses, for instance, a 0.05 significance level and a two-sided test, then a covariate A_j has a significant (nonzero) effect, if the following relationship is satisfied:[12]

$$\left|\frac{\hat{\alpha}_j}{\hat{\sigma}(\hat{\alpha}_j)}\right| > 1.96$$

where $\hat{\alpha}_j$ is the estimated coefficient for covariate A_j, and $\hat{\sigma}_j$ is the associated standard error. In Box 4.1.4, all covariates, with the exception of `pnoj` ("previous number of jobs"), have a significant effect.

The effect of a covariate on the transition rate reflects its impact on both the speed of the dependent process and the proportion of people who have experienced an event after a certain time (see Bernardi 2001). The effect of a covariate can be easily interpreted by examining the percentage change in the rate, given that only one covariate changes its value. The formula (see Blossfeld, Hamerle, and Mayer 1989) is

$$\Delta\hat{r} = (\exp(\hat{\alpha}_j)^{\Delta A_j} - 1) \cdot 100\,\%$$

where ΔA_j is the change in variable A_j and $\Delta\hat{r}$ is the resulting percentage change in the estimated rate, $\exp(\alpha_j)$ the antilogarithm of the coefficient α_j, which is referred to in the literature as the "alpha effect" (see Tuma and Hannan 1984). It takes the value 1 if the covariate has no effect ($\alpha_j = 0$); it is smaller than 1 if $\alpha_j < 0$ and greater than 1 if $\alpha_j > 0$. If the value of the covariate is increased by just one unit, the rate changes by

$$\Delta\hat{r} = (\exp(\hat{\alpha}_j) - 1) \cdot 100\,\%.$$

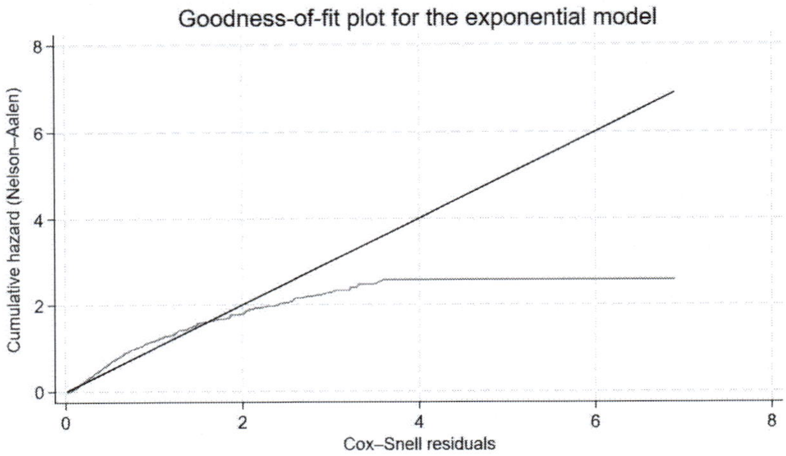

Figure 4.1.3 Nelson-Aalen plot of the goodness-of-fit (GOF) for the exponential model, generated with do-file `ehd1.do`.

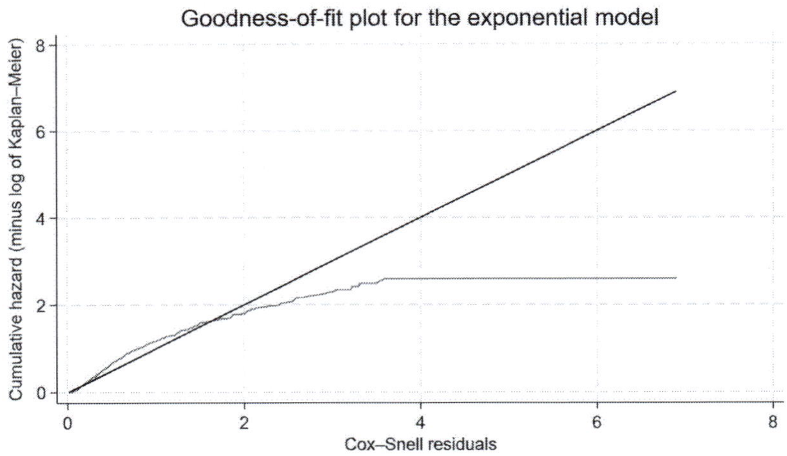

Figure 4.1.4 Kaplan-Meier plot of the goodness-of-fit (GOF) for the exponential model, generated with do-file `ehd1.do`.

In Box 4.1.4, the coefficient of the covariate `edu` ("educational attainment level") has a positive sign. Therefore, each additional school year increases the job-exit rate by about $\exp(0.0773) - 1) \cdot 100\,\% = 8\,\%$. This means that better educated workers are more mobile than less educated ones. As shown in do-file `ehd2.do` in Box 4.1.3, you can calculate this amount directly with Stata using the `display` command. However, this estimate is difficult to interpret in substantive terms, since theory suggests that educational attainment should have a positive effect on upward moves and a negative effect on downward moves, while the effects of educational attainment on lateral moves are theoretically still quite open (Blossfeld and Mayer 1988; Blossfeld 1989). A meaningful substantive interpretation of the effect of variable `edu` can therefore only be achieved if we estimate models with directional moves (see next section).

Each younger birth cohort is more mobile. Compared to the reference group (individuals born 1929–31), the job-exit rate of people born 1939–41 (`coho2`) is about 83.7 % higher: $(\exp(0.6080) - 1) \cdot 100\,\%$; and for the people born in 1949–51 (`coho3`), about 84.2 % higher: $(\exp(0.6108) - 1) \cdot 100\,\%$.[13]

The effect of labor force experience (`lfx`) is negative. Thus in our example, each additional year of labor force experience reduces the job-exit rate by 3.7 %. This is in line with human capital theory, which predicts that with increasing general labor force experience, additional investments into human capital decline, and as a consequence, job exits decrease.[14]

The number of previously held jobs (`pnoj`) also has a positive effect, but this variable is not statistically significant. In our example, this part of the job history therefore has no effect on the job-exit rate.

Finally, the prestige score of the current job (`pres`) influences the job-exit rate negatively. An increase by 10 units on the magnitude prestige scale (Wegener 1985) decreases the job-exit rate by about 24 %.[15]

It is important to note that the effects of the covariates in the model are not independent of each other. They are related *multiplicatively*. For example, a simultaneous change in prestige by 10 units and in labor force experience by one year decreases the job-exit rate by 27.3 %:

$$\big(\exp(-0.0280)^{10} \cdot \exp(-0.0032)^{12} - 1\big) \cdot 100\,\% \approx -27.3\,\%.$$

For selected subgroups, one can also predict the average job duration, the median of the duration, the average number of exits in a given duration interval, and the probability of remaining in the same job up to a certain point in time. For example, for an individual with Abitur (`edu` = 13) belonging to the birth cohort 1929–31 (`coho2` = 0 and `coho3` = 0) who is just entering the labor market (`lfx` = 0 and `pnoj` = 0) into a job with prestige level `pres` = 60 on the Wegener scale, we can calculate the following rate:

$$r = \exp(-4.4894 + 0.0773 \cdot 13 - 0.028 \cdot 60) \approx 0.0057.$$

Consequently, we expect an average job duration of about 175 months, or 15 years ($1/0.0057 \approx 175$), for individuals with these characteristics. Similarly, the median job duration for this group of individuals can be calculated as about 69.3 % of the mean duration, which corresponds to around 10 years ($175 \cdot 0.693 \approx 121$). Finally, the probability that individuals with the assumed characteristics are still employed in the same job after 8 years is about 58 %. This is calculated by using the survivor function $G(96) = \exp(-0.0057 \cdot 96) \approx 0.58$. Similar predictions can also be made for other subgroups, so that a fairly differentiated picture of the job-exit behavior in the sample emerges.

When comparing the estimated line with the 45° line, we see that our exponential model fits the data well only at the beginning of the process. However, the estimated line deviates strongly from the reference line towards the end of the process. In Chapter 9, we estimate this visual goodness-of-fit test for the Cox model, which provides a much better fit.

4.2 Models with Multiple Destinations

So far, we have only looked at job exits. More important are situations in which people can change from a given origin state to any one of several destination states ("competing risks"). Defining an appropriate state space is a matter of substantive consideration. For example, as shown in the previous section, it is very difficult to interpret the effect of educational attainment on the rate of job exit, as theory predicts contradictory effects of educational attainment on upward and downward moves. Furthermore, the effect of educational attainment on lateral moves is still quite open theoretically. In the causal analysis of transitions, the specification of the state space might turn out to be a serious source of misspecification (see chapter 10). Origin and destination states must be carefully selected, and the effects of covariates must be related to the specific transitions in a theoretically meaningful way.

In this section, we extend the exponential model with time-constant covariates to examine transitions from the origin state "being in a job" to a better job (upward exit), a worse job (downward exit), and to a job of about the same reward level (lateral exit). In other words, we estimate a model with one origin state and three destination states or competing risks. The do-file (`ehd3.do`) is shown in Box 4.2.1. The first part, specifying the data file `rrdat1` and providing definitions of the basic variables, is identical with the do-file in Box 4.1.1. In addition, we first specify that we now have three destination states. They are 1 ("better job"), 2 ("same job level"), and 3 ("worse job").[16] Second, as in Box 4.1.3, we define a set of covariates to be used in the model specification. Finally, an exponential model is selected using the `streg, dist(e)` command, and the covariates to be included in the model are specified. Note that we estimate a separate model for each of the possible transitions. In our example, there are three possible transitions: (0–1), (0–2), and (0–3), so we use the `stset` and `streg` commands three times to declare our data to be event history data and estimate an exponential model.[17]

Using do-file `ehd3.do`, Stata's output is shown in Box 4.2.2.[18] The tables report estimates for the rates of upward (`des` = 1), lateral (`des` = 2), and downward moves (`des` = 3). The results for lateral moves are presented but not interpreted.

The *positive effect of education* (`edu`) *on upward moves* is in accordance with status attainment theory (e.g., Blau and Duncan 1967; Sewell and Hauser 1975; Goldthorpe 1987; Erikson and Goldthorpe 1991), human capital theory (Becker 1975), and vacancy competition theory (Sørensen 1979). However, the three theories offer different explanations.[19] Status attainment theory claims that increases in educational attainment always lead to better job opportunities and will have a positive effect on upward moves. Under the condition of an imperfect labor market, human capital theory predicts that upward exits are more likely if employees are underrewarded; and the probability of being underrewarded increases with increasing personal resources (Tuma 1985). Education will therefore have a positive effect on upward moves. Finally, vacancy competition theory explains this positive effect of education in terms of the ability of higher educated persons to obtain better places in the labor queue and therefore be in a better position to take advantage of opportunities.

Box 4.2.1 Do-file `ehd3.do`

```
version 19
capture log close
set more off
log using ehd3.log, replace

use rrdat1, clear

gen     des = 2                            /* lateral moves */
replace des = 1 if (presn/pres -1)>0.2     /* upward moves */
replace des = 3 if (presn/pres -1)<0.0     /* downward moves */
replace des = 0 if presn==-1

gen tf = tfin - tstart + 1    /*ending time*/

gen coho2 = tb >= 468 & tb <= 504    /* cohort 2 */
gen coho3 = tb >= 588 & tb <= 624    /* cohort 3 */
gen lfx   = tstart - te              /* labor force experience */
gen pnoj  = noj - 1                  /* previous number of jobs */

* define single episode data and fit parametric survival model

stset tf, fail(des==1)    /* upward */
streg edu coho2 coho3 lfx pnoj pres if pres <= 65, dist(e) nohr

stset tf, fail(des==2)    /* lateral */
streg edu coho2 coho3 lfx pnoj pres, dist(e) nohr

stset tf, fail(des==3)    /* downward */
streg edu coho2 coho3 lfx pnoj pres, dist(e) nohr

log close
```

The *positive effects of the cohort dummy variables* for the birth cohorts 1939–49 (`coho2`) and 1949–51 (`coho3`) on all directional moves show that each younger cohort is more mobile, independently of whether we focus on upward, downward, or lateral job exits. However, caution is required when interpreting this result in substantive terms. A methodological reason for the cohort effects could be that the younger cohorts with shorter job histories simply remember job shifts better than the older ones with longer job histories. Finally, we also know that job mobility decreases with age. Hence, people with longer histories should have a higher average job duration. We therefore consider only the cohort variables as technical control variables for the different lengths of job histories in the German Life History Study GLHS.

The *negative effect of prestige on upward moves* can also be interpreted in light of two different theories. For Sørensen, opportunities for even better jobs decline as the level of attainment (measured here in terms of prestige scores) already achieved increases. In this theoretical model, employees have control over the decision to leave their job. This control is derived from job-specific skills, collective action, and so on. Hence, an employee will only leave a job if a better job is available; but the higher the attainment level already achieved, the more difficult it will be to find an even better job in a given structure of inequality. It should be noted that, in the context of vacancy competition theory, the size of the coefficient of prestige is also a measure of the opportunity structure of a given society: the larger the absolute magnitude of this coefficient, the fewer the opportunities for gains that are available in a society (Sørensen and Blossfeld 1989). A competing explanation is given by human capital theory, which assumes that there are costs of job searches and that there is imperfect information. In this case, upward moves are more likely if employees are underrewarded, and the probability of being underrewarded decreases with increasing job rewards. In other words, the higher the prestige score of the origin job, the less upward moves are to be expected (Tuma 1985).

Consistent with this modified form of human capital theory (Tuma 1985) is also the *negative effect of education on downward moves* (even if it is not significant in our didactic example due to the small number of events). Downward moves should be more likely if employees are overrewarded, and the probability of being overrewarded rises with decreasing personal resources, for example, educational attainment. Another reason can be seen in the specific way the labour market is organized in Germany. For example, Blossfeld and Mayer (1988) have shown that labor market segmentation in Germany is much more the result of qualification barriers, and that educational certificates tend to protect workers against downward exits to the secondary labor market (see also Blossfeld, Giannelli, and Mayer 1993).

Although the human capital and vacancy competition theories can explain some of the estimated coefficients in Box 4.2.2, there are also results that contradict these theories. Most importantly, the vacancy

Box 4.2.2 Results of using do-file `ehd3.do` (Box 4.2.1)

```
         failure _d:  des == 1
   analysis time _t:  tf

Exponential regression -- log relative-hazard form
No. of subjects =           591               Number of obs   =       591
No. of failures =            84
Time at risk    =         40161
                                              LR chi2(6)      =    131.39
Log likelihood  =    -253.68509               Prob > chi2     =    0.0000

------------------------------------------------------------------------------
      _t |      Coef.   Std. Err.      z    P>|z|     [95% Conf. Interval]
---------+--------------------------------------------------------------------
     edu |   .3020663   .0429622     7.03   0.000     .2178619    .3862708
   coho2 |   .6366232   .2713856     2.35   0.019     .1047172    1.168529
   coho3 |   .7340517   .2766077     2.65   0.008     .1919105    1.276193
     lfx |  -.0022632   .0020781    -1.09   0.276    -.0063363    .0018098
    pnoj |   .1734636   .1003787     1.73   0.084    -.0232751    .3702022
    pres |   -.143771   .0142008   -10.12   0.000     -.171604    -.115938
   _cons |  -5.116249   .6197422    -8.26   0.000    -6.330922   -3.901577

         failure _d:  des == 2
   analysis time _t:  tf

Exponential regression -- log relative-hazard form
No. of subjects =           600               Number of obs   =       600
No. of failures =           219
Time at risk    =         40782
                                              LR chi2(6)      =     39.89
Log likelihood  =      -595.272               Prob > chi2     =    0.0000

------------------------------------------------------------------------------
      _t |      Coef.   Std. Err.      z    P>|z|     [95% Conf. Interval]
---------+--------------------------------------------------------------------
     edu |   .0033448    .037983     0.09   0.930    -.0711004      .07779
   coho2 |   .6722249   .1642522     4.09   0.000     .3502966    .9941533
   coho3 |   .6843349   .1732453     3.95   0.000     .3447803    1.023889
     lfx |   -.003085   .0013664    -2.26   0.024     -.005763    -.000407
    pnoj |   .0322934   .0644585     0.50   0.616     -.094043    .1586298
    pres |   .0081357     .00806     1.01   0.313    -.0076616    .0239329
   _cons |  -5.804317   .4054322   -14.32   0.000    -6.598949   -5.009684

         failure _d:  des == 3
   analysis time _t:  tf

Exponential regression -- log relative-hazard form
No. of subjects =           600               Number of obs   =       600
No. of failures =            96
Time at risk    =         40782
                                              LR chi2(6)      =     17.34
Log likelihood  =    -345.80692               Prob > chi2     =    0.0081

------------------------------------------------------------------------------
      _t |      Coef.   Std. Err.      z    P>|z|     [95% Conf. Interval]
---------+--------------------------------------------------------------------
     edu |  -.0542781   .0640498    -0.85   0.397    -.1798134    .0712572
   coho2 |   .5946145   .2431647     2.45   0.014     .1180204    1.071209
   coho3 |   .5510851   .2620141     2.10   0.035      .037547    1.064623
     lfx |  -.0040521   .0021377    -1.90   0.058    -.0082419    .0001377
    pnoj |   .0388227   .0990888     0.39   0.695    -.1553878    .2330332
    pres |   .0034328   .0123049     0.28   0.780    -.0206844      .02755
   _cons |  -5.699672   .6715966    -8.49   0.000    -7.015977   -4.383367
```

competition theory regards downward moves as an exception. However, this is not the case. The number of downward moves ($n_{\text{down}} = 96$) is greater than the number of upward moves ($n_{\text{up}} = 84$).[20] Furthermore, according to the vacancy competition model, the effect of time spent in the labor force (or labor force experience) on upward moves should not be significant once education and prestige are controlled for. Otherwise, this variable is no adequate proxy of the discrepancy between resources and current job rewards.[21] However, an explanation for the negative effect of the covariate "general labor force experience" on upward moves is given by Mincer (1974). He argues that people invest in their resources as long as their expected returns exceed their expected costs. Therefore, training is mainly concentrated in the earlier phases of employment, where there is more time left to recover training costs. In this way, the job mobility process is time-dependent, as time spent in the labor force (or general labor force experience) reduces the probability of new training and resulting gains in attainment.

It should also be noted that human capital theory is only interesting for sociological mobility research to the extent that an imperfect labor market is assumed. As shown by Tuma (1985), human capital theory only leads to specific hypotheses about mobility if imperfect information and search costs are assumed. Otherwise

Box 4.2.3 Do-file ehd4.do

```
version 19
capture log close
set more off
log using ehd4.log, replace

use rrdat1, clear

gen     des = 2                             /* lateral moves */
replace des = 1 if (presn/pres -1) > 0.2    /* upward moves */
replace des = 3 if (presn/pres -1) < 0.0    /* downward moves */
replace des = 0 if  pres n== -1

gen tf = tfin - tstart + 1    /* ending time */

gen coho2 = tb >= 468 & tb <= 504  /* cohort 2 */
gen coho3 = tb >= 588 & tb <= 624  /* cohort 3 */
gen lfx   = tstart - te            /* labor force experience */
gen pnoj  = noj - 1                /* previous number of jobs */

set seed 339487731  /*specify the initial value of the random-number seed*/

gen r = uniform()   /* uniformly distributed pseudorandom numbers
                       on the interval [0,1) */

stset tf, fail(des==1)    /* upward */
streg edu coho2 coho3 lfx pnoj pres if pres <= 65, ///
 dist(e) nohr                  /* uwpard */
estimates store upward         /* save most recent estimation results
                                  under a name (upward) */

stset tf, fail(des==2)         /* lateral */
streg edu coho2 coho3 lfx pnoj pres, ///
 dist(e) nohr                  /* unstandardized lateral */
estimates store unstd_lat
streg edu coho2 coho3 lfx pnoj pres if r <=84/219, ///
 dist(e) nohr                  /* standardized lateral */
estimates store std_lat

stset tf, fail(des==3)         /* downward */
streg edu coho2 coho3 lfx pnoj pres, ///
 dist(e) nohr                  /* unstandardized downward */
estimates store unstd_down
streg edu coho2 coho3 lfx pnoj pres if r <=84/96, ///
 dist(e) nohr                  /* standardized downward */
estimates store std_down

estimates table upward unstd_lat std_lat unstd_down std_down, ///
 b(%4.3f) se(%4.3f) p(%4.2f) varwidth(8) modelwidth(10) ///
 stats(N N_fail ll chi2 df_m p)    /*write stored estimates in table*/

log close
```

the labor market would be in equilibrium, and in equilibrium there are only random exits because no one can improve their current situation.

Comparison of Covariate Effects Across Destinations

When comparing the statistical significance of covariate effects across various destinations in competing risks models, care is needed, because statistical significance tests of parameters (see section 4.1.3) are normally affected by a varying number of competing events. For example, in Box 4.2.2, these tests are based on 84 upward moves, 219 lateral moves, and 96 downward moves.[22] Given a fixed size of an effect for all directional moves, the results of the significance tests are therefore dependent on the number of events. Thus, it is more likely that the statistical test will produce a significant result for lateral moves than for downward moves, and for downward moves than for upward moves.

To demonstrate the impact of the various number of events, we standardize the number of events across the three directional moves. Since the number of upward moves is smallest, we use this number as the baseline and standardize the number of events by drawing probability samples from the input data for lateral and downward job moves (Box 4.2.3).

Box 4.2.4 Comparison of estimates on upward moves with unstandardized (`unstd`) and standardized (`std`) estimates on lateral (`lat`) and downward (`down`) moves (results of using do-file `ehd4.do`)

Variable	upward	unstd_lat	std_lat	unstd_down	std_down
edu	0.302	0.003	-0.016	-0.054	-0.030
	0.043	0.038	0.062	0.064	0.068
	0.00	0.93	0.79	0.40	0.66
coho2	0.637	0.672	0.706	0.595	0.748
	0.271	0.164	0.260	0.243	0.277
	0.02	0.00	0.01	0.01	0.01
coho3	0.734	0.684	0.339	0.551	0.640
	0.277	0.173	0.295	0.262	0.293
	0.01	0.00	0.25	0.04	0.03
lfx	-0.002	-0.003	-0.004	-0.004	-0.006
	0.002	0.001	0.002	0.002	0.003
	0.28	0.02	0.11	0.06	0.02
pnoj	0.173	0.032	0.046	0.039	0.096
	0.100	0.064	0.114	0.099	0.111
	0.08	0.62	0.68	0.70	0.39
pres	-0.144	0.008	0.024	0.003	0.002
	0.014	0.008	0.014	0.012	0.014
	0.00	0.31	0.08	0.78	0.88
_cons	-5.116	-5.804	-5.989	-5.700	-6.006
	0.620	0.405	0.588	0.672	0.736
	0.00	0.00	0.00	0.00	0.00
N	591	600	224	600	507
N_fail	84.000	219.000	84.000	96.000	77.000
ll	-253.685	-595.272	-229.403	-345.807	-280.862
chi2	131.388	39.890	15.064	17.340	20.120
df_m	6.000	6.000	6.000	6.000	6.000
p	0.000	0.000	0.020	0.008	0.003

legend: b/se/p

First, we specify the initial value of the random-number seed to ensure that we obtain the same results at all times (as long as we use the same Stata version): `set seed 33948773`.[23] Next, we use the command `gen r = uniform()` to generate uniformly distributed pseudorandom numbers on the interval [0,1). In Box 4.2.3, we estimate upward moves in the same way as in the previous do-file (see Box 4.2.1). What is new is that we store the estimation result (`estimates store`). Then we also re-estimate the model on lateral moves and store the estimates. Now an episode is randomly selected if a random number is less than or equal to the value 84/219 (the number of upward moves divided by the number of lateral moves). In the next steps, we reestimate the model on downward moves, store it and then restrict the estimation of the exponential model to those observations in which the random number is less than or equal to the value 84/96 (the number of upward moves divided by the number of downward moves).

Finally we instruct Stata to display results of all five estimations in one single table (`estimates table`) and to report the coefficents, the standard errors, and significance level. In addition, we instruct Stata to include in this table the case numbers, the number of failures, and the final values of the log likelihood function as well as the likelihood ratio test statistics (LR), their degrees of freedom, and significance levels.

The unstandardized and standardized estimation results for upward, lateral, and downward moves are summarized in Box 4.2.4 to make comparison easier. We start with the lower part of the table: The drawing a random sample for lateral moves results in 219 episodes with 84 events; drawing for downward moves results in 507 episodes with 77 events. Stata reports the coefficient followed by its standard error and significance level. In the case of lateral moves, the effect of the covariate `lfx` is no longer significant in the standardized estimate. It is therefore possible that one could come to completely different substantive conclusions if one were to take into account the number of events when comparing the effects across different transitions. If you want to compare the statistical significance of effects of covariates across competing transitions, you should always be aware that different event numbers might have an effect on statistical tests.

Continuous Time, Continuous State Space Models

Our job-exit example with directional moves has the disadvantage that a continuous state space of inequality (measured by job prestige values) must be decomposed into a set of ordered destination states ("better job," "job of same level," and "worse job"). Thus for job moves, the exact size of the increases in prestige scores could not be evaluated. To solve this problem, Petersen (1988a) proposed a method for analyzing what he called continuous time, continuous state space failure time processes. In these processes, the dependent variable is continuous but, in contrast to diffusion processes, it remains constant for finite periods of time. He shows that the probability for this type of process can be written as the product of the transition rate of a change in the state variable $Y(t)$ and the density for the new value of $Y(t)$, given that a change in $Y(t)$ has occurred. For the first term, as already shown, transition rate models for the analysis of duration data apply. For the second term, we can use transition rate models for the outcomes of continuous dependent variables. Introducing covariates into both model equations is straightforward. Petersen (1988a) illustrated this method by first analyzing the rate of upward exits in socioeconomic status and then the new value of socioeconomic status, given that an upward exit has occurred.

Two other methodological solutions for the analysis of continuous time, continuous outcome processes have been suggested: (1) growth curve analysis (see e.g. Herle et al. 2020), and (2) stochastic differential equation models (Hannan et al. 1990). Herle et al. (2020) provide a review of the three most used methods for the identification of latent trajectories (growth mixture models, latent class growth analysis, and longitudinal latent class analysis). For example, they used longitudinal data on childhood body mass index. Hannan et al. (1990) analyzed wage trajectories across job histories of men and women and estimated (1) the wage rate at time of first entry into the labor market, (2) wage changes when workers changed jobs, and (3) wage growth within jobs. Especially for estimating the wage growth within jobs, they applied a stochastic differential equation model.

In recent years, in the study of continuous-time, continuous-state models, latent growth curve models have also been increasingly used (Miller et al. 2024).

4.3 Models with Multiple Episodes

As demonstrated in this chapter, most studies of social mobility research (e.g., Sørensen and Tuma 1981; Sørensen 1984; Tuma 1985; Carroll and Mayer 1986; Sørensen and Blossfeld 1989) model transitions to a destination job as a function of the time since entry into an origin job. The career trajectory of an individual is therefore divided into a series of job spells (single episodes), each starting with time equal to 0. The process time is job-specific labor force experience. The effects of causal variables are usually considered to be independent of the different job episodes and are therefore estimated as constant across episodes.[24] Blossfeld and Hamerle (1989b; Hamerle 1989, 1991) refer to this type of job-exit model *uni-episode analysis with episode-constant coefficients*.[25]

Another way to model job mobility is to regard job transitions as being dependent on the time a person has spent in the labor force since entry into the labor market. The process time in this case is the amount of *general labor force experience*.[26] Starting and ending times of the job spells in a person's career are then given as the time since entry into the labor market.[27] The coefficients of covariates can be estimated as constant or as changing across episodes. This type of job exit analysis can be referred to as *multiepisode models with episode-constant or episode-changing coefficients* (Blossfeld and Hamerle 1989b).

In the case of an exponential model, where we assume that general labor force experience does not affect job-exit rates, an interesting application of a multiepisode model could be to study how covariate effects change across episodes. Such a model, estimated with do-file `ehd5.do`, is shown in Box 4.3.1.

However, before we begin to describe this do-file, we should make some more general comments. It is important to point out that, compared to American workers, German workers have considerably fewer jobs and change jobs less frequently. The average duration of a given job for a man in Germany is about 6 years (Carroll and Mayer 1986), while in the United States it is only 2.2 years (Tuma 1985). The stable nature of job trajectories in Germany implies that a job change is more likely to be substantively meaningful than a job exit in the United States. But, conversely, it also means that the distribution of job episodes is clustered more to the left of the mean, with extreme values to the right (see Box 2.2.8). This has implications for the analysis of career mobility as a multiepisode process: the smaller the number of job transitions for a given serial job number, the more likely it is to yield statistically nonsignificant results. Hence, a comparison of episode-specific parameters across spell numbers based on significance tests may be misleading; and the number of serial job episodes that can be used to study career mobility as a multiepisode process is limited. In such a model, it would be unwise to include episodes of serial numbers with a very small number of job transitions. Therefore, based on the distribution of job transitions, as displayed at the end of Box 2.2.8, only the first four job episodes are used for our estimation of the multiepisode model. This ensures that

Box 4.3.1 Do-file `ehd5.do` (multiepisode exponential model)

```
version 19
capture log close
set more off
log using ehd5.log, replace

use rrdat1, clear

keep if noj <= 4    /*keep only episodes with serial number <= 4*/

gen des = tfin ~= ti                /*destination state*/
gen tf  = tfin - tstart + 1         /*ending time*/

gen coho2 = tb >= 468 & tb <= 504   /*cohort 2*/
gen coho3 = tb >= 588 & tb <= 624   /*cohort 3*/
gen lfx   = tstart - te             /*labor force experience*/
gen pnoj  = noj - 1                 /*previous number of jobs*/

stset tf, failure(des)              /*define single episode data*/

table noj _d, c(n _t mean _t) f(%5.2f) /*descriptive statistics for
the individuals' first four jobs; or type: stci, by (noj des) emean */

by noj, sort: streg edu coho2 coho3 lfx pres, dist(e) nohr
/*fit parametric survival models*/

log close
```

Box 4.3.2a Result (first part) of using do-file `ehd5.do` (Box 4.3.1)

```
. table noj _d, c(n _t mean _t) f(%5.2f)

---------------------------
Serial     | 1 if failure;
number of  | 0  if censored
the job    |      0       1
-----------+---------------
         1 |     16     185
           | 233.44   52.73
           |
         2 |     36     126
           | 150.17   52.30
           |
         3 |     38      69
           | 121.74   50.36
           |
         4 |     24      38
           |  85.33   37.71
---------------------------
```

the number of job spells and transitions in each episode is large enough to allow meaningful comparison of estimates between episodes. It also ensures that only a small number of job spells are lost in the analysis. On the basis of this decision, about 89 % (532/600) of all job spells and 91 % (418/458) of all job moves (or events) are included in this multiepisode model.

The do-file `ehd5.do`, which is used to estimate this multi-episode model, is shown in Box 4.3.1. First, we select episodes with the `keep` command. Using `keep if noj <= 4` instructs Stata to keep only job episodes with a serial number (`noj`) less than or equal to 4. The episodes (jobs) belonging to each individual are distinguished by their serial number given by the `noj` variable. As discussed in chapter 2, the serial numbers must be positive integers and should be contiguous. We display descriptive statistics for the individuals' first four jobs using the command `table noj _d, c(n _t mean _t) f(%5.2f)`. Results are shown in Box 4.3.2a.

We sort the data by `noj` and use the `by` prefix to repeat the `streg` command for the first four job episodes. Note that the covariate `lfx`, the general labor force experience measured at the beginning of each new job episode, cannot be included as a covariate for the *first* job episode because this variable is zero for all first episodes (or workers). With the exception of this variable, all other covariates are time-constant, as they do not change their value after the beginning of the process time, which is the *time of entry into the first job*. Variable `lfx` ("general labor force experience at the beginning of each new job") is a special form of a (process-related) time-dependent covariate, as it changes its value at the beginning of each job. Using this covariate in a multiepisode exponential model whose process time is defined by the clock "general labor

Box 4.3.2b Result (second part) of using do-file `ehd5.do` (Box 4.3.1)

```
-> noj = 1

note: lfx dropped due to collinearity
Exponential regression -- log relative-hazard form
No. of subjects =          201          Number of obs    =        201
No. of failures =          185
Time at risk    =        13490
LR chi2(4)      =        21.90
Log likelihood  =   -320.73616          Prob > chi2      =     0.0002

---------------------------------------------------------------------
_t |      Coef.   Std. Err.      z    P>|z|     [95% Conf. Interval]
-------+-------------------------------------------------------------
edu |  .0622383   .0326678    1.91   0.057    -.0017894    .126266
coho2 |  .7220939   .1888079    3.82   0.000     .3520372    1.09215
coho3 |  .6637726   .1791567    3.70   0.000     .3126319   1.014913
pres |  -.009535   .0091225   -1.05   0.296    -.0274147   .0083448
_cons |  -5.017885   .4173747  -12.02   0.000    -5.835924  -4.199846
---------------------------------------------------------------------

-> noj = 2

Exponential regression -- log relative-hazard form
No. of subjects =          162          Number of obs    =        162
No. of failures =          126
Time at risk    =        11996
LR chi2(5)      =        42.88
Log likelihood  =   -237.89354          Prob > chi2      =     0.0000

---------------------------------------------------------------------
_t     |      Coef.   Std. Err.      z    P>|z|     [95% Conf. Interval]
------+--------------------------------------------------------------
edu    |   .1261256   .0516029    2.44   0.015     .0249857    .2272655
coho2 |   .3422823   .2157199    1.59   0.113    -.0805208    .7650855
coho3 |   .2703772   .2334466    1.16   0.247    -.1871696    .7279241
lfx    |  -.0061697   .0018403   -3.35   0.001    -.0097767   -.0025627
pres   |  -.0510282   .0110363   -4.62   0.000     -.072659   -.0293974
_cons |  -3.735184   .5778047   -6.46   0.000    -4.867661   -2.602708
---------------------------------------------------------------------

-> noj = 3

Exponential regression -- log relative-hazard form
No. of subjects =          107          Number of obs    =        107
No. of failures =           69
Time at risk    =         8101
LR chi2(5)      =        18.00
Log likelihood  =   -152.73377          Prob > chi2      =     0.0029

---------------------------------------------------------------------
_t     |      Coef.   Std. Err.      z    P>|z|     [95% Conf. Interval]
------+--------------------------------------------------------------
edu    |   .0522256   .0748627    0.70   0.485    -.0945026    .1989538
coho2 |   .1975193    .297243    0.66   0.506    -.3850663    .7801048
coho3 |   .7143032   .3288906    2.17   0.030     .0696895    1.358917
lfx    |  -.0039478   .0019895   -1.98   0.047    -.0078472   -.0000483
pres   |  -.0223009   .0139315   -1.60   0.109    -.0496061    .0050044
_cons |  -4.262784   .7396699   -5.76   0.000    -5.712511   -2.813058
---------------------------------------------------------------------

-> noj = 4

Exponential regression -- log relative-hazard form
No. of subjects =           62          Number of obs    =         62
No. of failures =           38
Time at risk    =         3481
LR chi2(5)      =        21.98
Log likelihood  =   -73.004575          Prob > chi2      =     0.0005

---------------------------------------------------------------------
_t     |      Coef.   Std. Err.      z    P>|z|     [95% Conf. Interval]
------+--------------------------------------------------------------
edu    |   .0437997   .1329081    0.33   0.742    -.2166953    .3042947
coho2 |   1.135352   .3830044    2.96   0.003      .384677    1.886027
coho3 |   .3163115   .4492212    0.70   0.481    -.5641459    1.196769
lfx    |  -.0051987   .0027173   -1.91   0.056    -.0105246    .0001272
pres   |  -.0654294   .0204556   -3.20   0.001    -.1055215   -.0253372
_cons |   -2.22315    1.34894   -1.65   0.099    -4.867025    .4207244
```

force experience" would be meaningless. On the one hand, if the assumption of the exponential model is correct, then this covariate cannot have a significant effect, because the process does not depend on history, and the rate of job change is constant across the process time (i.e., general labor force experience). If, on the other hand, the covariate "general labor force experience" has a significant effect, then the assumption of an exponential distribution would be wrong, and one should use a model with a time-dependent transition rate. The transition rate would then become a function of time, so that events become either less likely (negative duration dependence) or more likely (positive duration dependence) as the clock "general labor force experience" progresses. Thus, this multiepisode model can serve as a test for the assumption that the job history process is *time-stationary* or *time-homogeneous* (i.e., transitions do not depend on time).

Box 4.3.2b provides the job-number-specific estimates for the covariates using do-file ehd5.do (Box 4.3.1). It is easy to see that the covariate "general labor force experience" (`lfx`) has significant effects on the rate of job change for episode numbers 2 and 3, and is almost significant for the fourth job. This means that the job history is not time-stationary or time-homogeneous, but that the job movement rate becomes less likely as general labor force experience increases.

It is interesting to see how the effects of the covariates change across the number of episodes. For mobility out of the first job, only the cohort dummy variables are statistically significant. The parameters for `coho2` and `coho3` are quite similar in size. Hence, the greatest difference is between the reference group (birth cohort 1929–31) and the two younger birth cohorts (1939–41 and 1949–51). After having left the first job and entered the second one, the educational attainment level has a positive impact and prestige of the job has a negative impact on job mobility. In the third job, only the birth cohort 1949–51 (`coho3`) behaves differently. Members of this cohort move to a fourth job significantly more often. Finally, in the fourth job, the birth cohort 1939–41 (`coho2`) is more mobile, and the prestige level (`pres`) reduces job mobility. Of course, all these estimates are based on a small number of events, and the interpretation given here serves didactic purposes only.

Notes

[1]In this case it is assumed that qualitative states reflect points (or intervals) on an underlying metric scale. If the states are ordered, it can be argued that the sequence of states corresponds to segments of an underlying continuous variable.

[2]We generally adopt the convention of writing covariates as row vectors and the associated coefficients as column vectors.

[3]See chapters 5, 6, and 7.

[4]The log-likelihood value reported by Stata is the unmodified value ℓ plus $\Sigma_{i \in \mathcal{E}_k} \log(t_i)$.

[5]In this example, $\Sigma_{i \in \mathcal{E}_k} \log(t_i) = 1576.0520$. The unmodified log likelihood is therefore $-937.9681 - 1576.0520 = -2514.0201$.

[6]After `streg`, all coefficients are stored in a matrix called `e(b)`. For further calculations, you can access the values contained in this matrix. With `_b[varname]`, you can refer to the coefficient for a specific variable; with `_b[_cons]`, you can refer to the constant of the model.

[7]In the case of multiepisode processes, the status must be attained prior to the *first* entry into the process.

[8]Heckman and Borjas (1980) called the effect of the number of previous occupancy of labor force states (e.g., jobs) "occurrence dependence" and the effect of the durations in previous states (e.g., general labor force experience) "lagged duration dependence."

[9]Maximum likelihood estimation assumes independent episodes. The better the dependencies between the episodes are controlled for in the case of repeated events, the less biased are the estimated parameters. However, in practical research it is very difficult to achieve this goal completely. Therefore it is always a source of possible misspecification; see chapter 10.

[10]Of course, this variable is only a very rough proxy, as it assumes that individuals were employed continuously. In a "real" job-exit analysis, one would construct the variable "general labor force experience" for each individual on the basis of the number of months she or he had actually worked.

[11]Stata stores various other results in this matrix. If you type `ereturn list`, you will get the names and contents of all saved results.

[12]It is important to distinguish between statistical and substantive significance in empirical analysis. Given a fixed significance level, statistical significance depends on the ratio of the effect size (column `Coef.`) to the estimated standard error (column `Std. Err.`), and the standard error is inversely related to the number of events. Thus, if the number of events in a sample increases, the standard error decreases and the test statistic (column `z`) increases, which means that these tests have an increasing probability to produce a statistically significant finding (given a fixed significance level). In other words, statistical significance is highly dependent on the number of events, whereas substantive significance (an evaluation based on a substantive theory) is not. Therefore, some variables may be substantively significant, but with a small sample size, the results of the test may not be statistically significant; and in many cases, some variables may be without any substantive significance, but with a large sample size, the test leads to a statistically significant result. Taken together, these arguments caution against any mechanical application of statistical tests, stress the importance of substantive theory in the interpretation of test results, and highlight that interpretations are often quite ambiguous.

[13]At this stage of our analysis, it is unclear whether younger birth cohorts actually have a higher mobility rate (e.g., due to modernization processes) or whether this is just a methodological artefact (younger birth cohorts are still in a life phase with higher mobility rates compared to the 1929–31 cohort).

[14]In this case, the effect on directional moves can also be better interpreted in substantive terms (see next section).

[15]Again, in substantive terms this result is easier to interpret for directional moves. For example, vacancy competition theory (Sørensen 1977, 1979) states that upward moves are increasingly less likely the higher the job is located in the pyramid of inequality (see next section).

[16]This is consistent with section 3.3 (see Box 3.3.3) in which we defined upward exits as job mobility leading to an increase in the prestige score of 20 % or more; downward exits as job mobility connected with a decrease in the prestige score; and lateral exits as having no change in the prestige score or experiencing an increase in the prestige score of up to 20 %.

[17]In this example, the same list of covariates is linked to each of the three transitions. In general, it would be possible to use different sets of covariates for each transition.

[18]These results differ from the results in Blossfeld and Rohwer (2002), because we have taken into account the ceiling effect for upward moves here (only people who can experience a 20 % or more increase in the prestige score from job n to job $n + 1$

are at risk to move upward) along with the bottom effect (only people who are not already at the bottom of the prestige score are at risk to move downward) and we have corrected for all job-exits without a job destination (in this case `presn` is coded -1).

[19]This is a good example of the well-known situation in the social sciences in which a causal effect can be explained by different theories. Comparing the relative merits of these competing theories is very difficult, simply because it is difficult to find a "crucial test" (see Lieberson 1985).

[20]These numbers are also a function of the technical definition of upward, downward, and lateral moves. In our example, however, the number of upward moves remains more or less the same if we lower the threshold from a 20 % increase to a 10 % or even 5 % increase in the prestige score. However, there are good theoretical reasons to classify job moves as upward moves only if they are connected with a significant step upward (Blossfeld 1986).

[21]Similar problems exist with regard to the effect of labor market experience on downward exits.

[22]These results differ from the results in Blossfeld and Rohwer (2002), because we have taken into account the ceiling and bottom effects here, and we have corrected for all job-exits without a job destination.

[23]Stata has changed its random-number generator. From version 14 onward, it uses a Mersenne Twister; prior to that, the "32-bit KISS" generator (for more information type `help set rng`). For this reason, the results differ from those presented in Blossfeld et al. (2007: 108-110). To reproduce the results published earlier, you have to set the random-number generator to "32-bit KISS" (`set rng kiss32`) before specifying the initial value (`set seed 339487731`) and ask for random numbers (`gen r2 = uniform()`).

[24]In exponential models, for example, interaction terms between the serial number of the job (or a set of dummy variables for it) and the other time-constant covariates could be used to study changes in such covariate effects across spells.

[25]In the literature, these models usually also include job duration dependence (see chapters 5, 6, and 7). But, in this case, time dependence means that the event of a job exit depends primarily on the time spent in each of these jobs (or job-specific labor force experience) regardless of the specific location of the job in an individual's job career (which is general labor force experience).

[26]Again, definitions of different clocks or process times are particularly important in the case of models with time-dependence (chapter 7) and semiparametric (Cox) models (chapter 9).

[27]In time-dependent models, the event of a job exit depends on the specific location of the spell in a person's life course. Consequently, the job spells of a person's career are not treated as autonomous units.

Chapter 5

Piecewise Constant Exponential Models

In most applications of transition rate models, there is no theoretical justification for assuming that the forces of change are constant over time. For appropriate modeling of social processes, it is therefore important to be able to include time-dependent covariates in transition rate models. Before looking at this issue in more detail in chapter 6, we provide an overview of the *piecewise constant exponential model*. This is a simple generalization of the standard exponential model (chapter 4), but it is extremely useful in many practical research situations. It is particularly helpful when researchers are unable to explicitly measure and include important time-dependent covariates explicitly, or when they do not have a clear idea of the shape of the time-dependence of the process. In both cases, a small modification of the exponential model leads to a very flexible analytical tool. The basic idea is to split the time axis into time periods and to assume that transition rates are constant in each of these intervals but can change between them.

5.1 The Basic Model

If there are L time periods, the piecewise constant transition rate is defined by L parameters. Stata offers two different options for the inclusion of covariates, both of which are presented in this section. The first option assumes that only a baseline rate, given by period-specific constants, can vary across time periods, but that the covariates have the same (proportional) effects in each period.[1] In the second option, period-specific effects of covariates are possible. We start with the demonstration of the first option.

Stata 18 does not have a built-in command for piecewise constant exponential models. But Stata offers a solution: you need to split the episodes into two or more episodes and estimate an exponential model using the command `streg`. There is also an even smarter solution: you can also install an ado-file that automatically splits the episodes and estimate the piecewise constant exponential model.[2] Both procedures lead to exactly the same estimation results. We discuss both methods in this chapter.

The commands require a definition of time periods. This is done on the basis of split points on the time axis

$$0 = \tau_1 < \tau_2 < \tau_3 < \ldots < \tau_L .$$

With $\tau_{L+1} = \infty$, one gets L time periods

$$I_l = \{t \mid \tau_l < t \leq \tau_{l+1}\} \qquad l = 1, \ldots, L.$$

Given these time periods, the transition rate from a given origin state to destination state k is

$$r_k(t) = \exp\left\{\bar{\alpha}_l^{(k)} + A^{(k)}\alpha^{(k)}\right\} \quad \text{if} \quad t \in I_l. \tag{5.1}$$

For each transition to destination state k, $\bar{\alpha}_l^{(k)}$ is a constant coefficient associated with the lth time period, $A^{(k)}$ is a (row) vector of covariates, and $\alpha^{(k)}$ is an associated vector of coefficients that are assumed not to vary across time periods. Note that in this model, the vector of covariates cannot contain an additional constant.

Maximum Likelihood Estimation

The maximum likelihood estimation of this model is performed following the outline given in section 4.1.1. To simplify notation, we omit the indices for the transitions and define $l[t]$ as the index of the time period in which t lies (i.e. always $t \in I_{l[t]}$). The following notation is also helpful:

$$\delta[t, l] = \begin{cases} 1 & \text{if} \quad t \in I_l \\ 0 & \text{otherwise} \end{cases}$$

$$\Delta[s, t, l] = \begin{cases} t - \tau_l & \text{if} \quad s \leq \tau_l,\ \tau_l < t < \tau_{l+1} \\ \tau_{l+1} - \tau_l & \text{if} \quad s \leq \tau_l,\ t \geq \tau_{l+1} \\ \tau_{l+1} - s & \text{if} \quad t \geq \tau_{l+1},\ \tau_l < s < \tau_{l+1} \\ 0 & \text{otherwise.} \end{cases}$$

Box 5.2.1 Do-file `ehe1.do` (piecewise constant exponential model)

```
version 19
set scheme sj
capture log close
set more off
log using ehe1.log, replace

use rrdat1, clear

* 1st option: use stsplit

gen des = tfin ~= ti              /*destination state*/
gen tf  = tfin - tstart + 1       /*ending time*/

gen newid = _n
stset tf, failure(des) id(newid) /*define single episode data*/

stsplit time, at(0 (12) 96)       /*split records*/
tab time, ge(t)                   /*generate time dummies*/

streg t1 t2 t3 t4 t5 t6 t7 t8 t9, ///
dist(exp) nohr noconstant         /*fit parametric survival model*/

predict hazard, hazard            /*obtain predictions*/

line hazard _t, sort title("Piecewise Constant Exponential Rate") ///
xtitle("analysis time") saving("Figure 5_2_1",replace)

predict csurv, csurv
sts graph, addplot(line csurv _t, sort) ///
title ("Non-parametric and parametric survivor function") ///
xtitle("analysis time") saving("Figure 5_2_2",replace)


drop time t1-t9 hazard csurv
stjoin                            /*optional*/

* 2nd option: use stpiece

ssc install stpiece
/*install wrapper to estimate piecewise-constant hazard rate models*/

stset tf, failure(des) id(newid) /*define single episode data*/

stpiece, tp(0(12)96) nohr     /*fit piecewise constant exponential model*/

log close
```

The conditional survivor function may then be written as

$$G(t \mid s) = \exp\left\{ -\sum_{l=1}^{L} \Delta[s,t,l] \exp(\bar{\alpha}_l + A\alpha) \right\}.$$

Using this expression, the log likelihood can be written as

$$\ell = \sum_{i \in \mathcal{E}} (\bar{\alpha}_{l[t_i]} + A_i\alpha) - \sum_{i \in \mathcal{N}} \sum_{l=1}^{L} \Delta[s_i, t_i, l] \exp(\bar{\alpha}_l + A_i\alpha).$$

5.2 Models without Covariates

To illustrate the piecewise constant exponential model, we continue with the example in Box 4.1.1 (section 4.1), but now consider the possibility that the transition rate varies over time periods.

In general, the time periods can be defined arbitrarily, but there is a certain trade-off. If one chooses a large number of time periods, one will get a better approximation of the unknown baseline rate, but this implies a large number of coefficients to be estimated. If, on the other hand, a small number of time periods are chosen, there are fewer estimation problems, but a poorer approximation to the baseline rate is likely to be obtained. Therefore, a compromise is required in most cases.

Box 5.2.2 Part of estimation results of do-file `ehe1.do` (Box 5.2.1)

```
         failure _d:  des
   analysis time _t:  tf
                 id:  newid

Iteration 0:   log likelihood = -1002.8964
Iteration 1:   log likelihood =   -892.003
Iteration 2:   log likelihood = -888.99801
Iteration 3:   log likelihood = -888.99048
Iteration 4:   log likelihood = -888.99048

Exponential regression -- log relative-hazard form

No. of subjects =          600          Number of obs    =       2715
No. of failures =          458
Time at risk    =        40782
                                        Wald chi2(9)     =    8882.37
Log likelihood  =   -888.99048          Prob > chi2      =     0.0000

------------------------------------------------------------------------
   _t |      Coef.   Std. Err.      z    P>|z|     [95% Conf. Interval]
-----+------------------------------------------------------------------
   t1 |  -4.360092   .1072113   -40.67   0.000    -4.570223   -4.149962
   t2 |  -4.063008   .1031421   -39.39   0.000    -4.265162   -3.860853
   t3 |  -3.920421   .1097643   -35.72   0.000    -4.135555   -3.705287
   t4 |  -4.413101   .1581139   -27.91   0.000    -4.722999   -4.103204
   t5 |  -4.219508   .1561738   -27.02   0.000    -4.525603   -3.913413
   t6 |  -4.558253   .2041241   -22.33   0.000    -4.958329   -4.158177
   t7 |  -4.995652   .2773501   -18.01   0.000    -5.539248   -4.452056
   t8 |  -5.175585   .3162278   -16.37   0.000     -5.79538    -4.55579
   t9 |  -5.223871   .1230915   -42.44   0.000    -5.465126   -4.982616
```

Another important requirement is that there should be some episodes with ending event times within the interval for all time periods. Otherwise it is generally not possible to attain meaningful estimates.

We already have some information about the length of episodes in our example data set (see Box 2.2.6, Box 3.1.1, Figure 3.2.1). Therefore, it seems appropriate to use eight time periods, each having a length of one year, plus an additional open-ended interval.

First, as shown in do-file `ehe1.do` (Box 5.2.1), we generate a new ID variable and declare the data to be event history data.[3] Next, we have to split the records into two or more episodes by using the command

```
stsplit time, at(0 (12) 96).
```

Stata splits the records at specified times and creates 2,115 new subepisodes.[4] In total, there are 2,715 subepisodes. In the next step, we generate time dummies named `t1` to `t9`: `tab time, ge(t)`. We then fit an exponential model that includes these time dummies as covariates. Alternatively, you can use a wrapper that does all these steps automatically for you (see the second part of do-file `ehe1.do`). Before explaining this procedure, we introduce another command that is very helpful after episode splitting. The command `stjoin` joins episodes together if this can be achieved without losing information. This means that two records can be joined if they are next to each other and contain the same data. In our example, we must first remove all variables created by `stsplit` from the data set before we enter `stjoin`. This is achieved by using the command `drop`. Then you can enter the new Stata command `stpiece`. The split points are defined by specifying the option `tp(0(12)96)`.

The results of both approaches are exactly the same, which is why we only show part of the estimation results in Box 5.2.2. The estimated parameters for the baseline transition rate at first increase, from −4.36 to −3.92, and then decrease. In our application example, this means that with increasing duration in a specific job, the force of job exit (or the job-exit rate) is nonmonotonic. This can be seen more easily when the estimation result in Box 5.2.2 is plotted in Figure 5.2.1.[5] The commands that plot the estimated rates of the piecewise constant exponential model are

```
predict hazard, hazard
line hazard _t, sort .
```

In substantive terms, this bell-shaped rate pattern could be interpreted as an interplay of two opposing causal forces (increases in job-specific investments and decreases in the need to resolve mismatches) that are

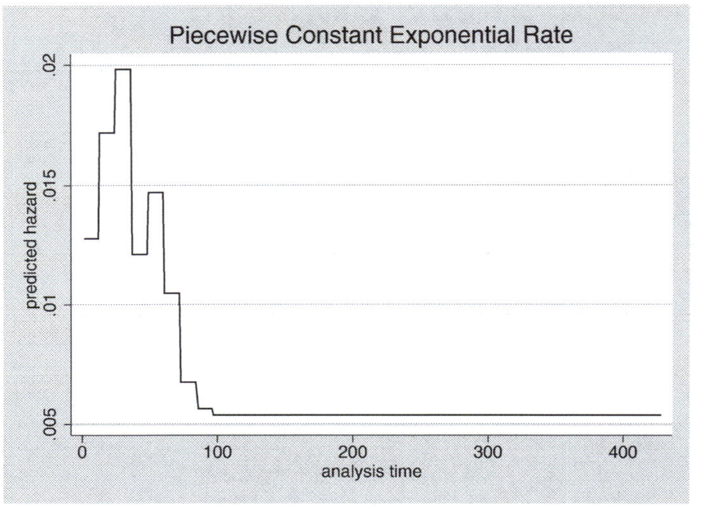

Figure 5.2.1 Piecewise constant transition rate, estimated with do-file `ehe1.do`.

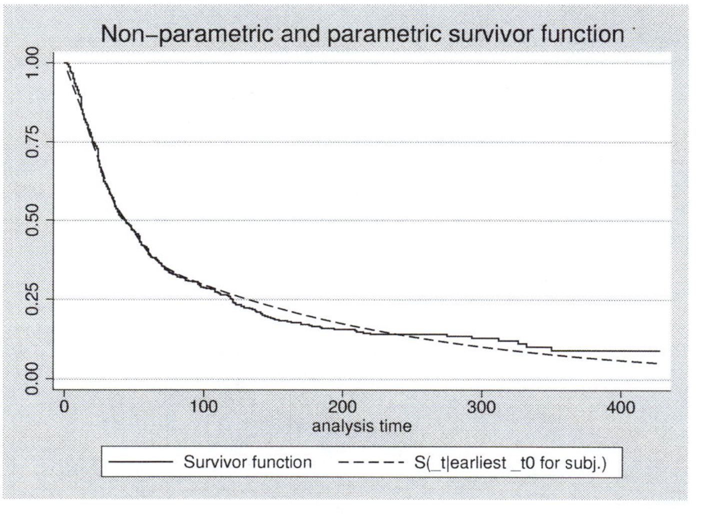

Figure 5.2.2 Plot of non-parametric and parametric survivor function, estimated with do-file `ehe1.do`.

not easily measured, so that the duration in a job has to serve as a proxy variable for them. The argument is that mismatches can occur when people are matched to jobs under conditions of imperfect information and high search costs. Particularly during the first months of any new employment, an intensive adjustment process takes place in which the respective expectations of the employer and the employee are confronted with reality. German labor law, for example, allows for a probationary period of 6 months during which it is relatively easy for employers to fire somebody if they are not satisfied with them. But employees are also more likely to quit their jobs at the beginning of each new employment, because their job-specific investments are still low. Consequently, one would expect that the rate of moving out of a job increases at the beginning of each new employment. However, as the mismatches are increasingly resolved and investments in job-specific human capital rise continously, a point is reached where these two forces will become equally strong. At this point, the peak of the transition rate of moving out of the job is reached (Figure 5.2.1). In our example, this point is reached after about 2 years of employment in a new job. After this job duration, further increases in job-specific investments increasingly outweigh the force of resolving mismatches, so that the job-exit rate declines with increasing duration.

In chapter 4, we also introduced the possibility of calculating survivor functions after model estimation using the command `predict` and specifying the option `surv` (see Box 4.1.1). In contrast to the `hazard`-option, which we used both in chapter 4 and in `ehe1.do` (Box 5.2.1), the option `surv` is not useful here. Since we are splitting the process time, the survivor function would restart at the beginning of each time-period (i.e., after 12, 24, 36, ... months in a job). Instead, we need a conditional survivor function that makes survival in time period I_l conditional on having survived the previous time period I_{l-1}. This is calculated in Stata with the option `csurv`. You can simply plot this predicted survivor function with `line csurv _t, sort`. In do-file `ehe1.do` (Box 5.2.1), we present a convenient way to put two survivor functions into one graph. The first is the non-parametric survivor function estimated with the Kaplan–Meier method (see chapter 3.2); the second is the survivor function based on the piecewise constant exponential model. Both survivor functions are hardly distinguishable in the first 120 months in a job (Figure 5.2.2).

5.3 Models with Proportional Covariate Effects

The problem with the analysis without covariates in the previous section is that important heterogeneity between the individuals was not taken into account. Thus, in the next step, we study whether the nonmonotonic pattern of the job-exit rate is still present when time-constant covariates are included. This can be done by a simple modification of the do-file `ehe1.do` (Box 5.2.1). The new do-file, `ehe2.do`, is shown in Box 5.3.1. The modification simply consists in constructing the covariates and adding the names of covariates to be included into the model to the `stpiece` command.

The estimation results of using do-file `ehe2.do` are shown in Box 5.3.2. Based on these parameter estimates, we can draw at least two conclusions: First, the nonmonotonic rate pattern of the piecewise constant exponential model without covariates (Box 5.2.2) did not change. Thus, this pattern of time dependence cannot be explained by the additionally included time-constant covariates.[6] Second, the parameter estimates for the time-constant covariates resemble the estimates for the basic exponential model in Box 4.1.5. Hence, the parameter estimates for the time-constant covariates in the exponential model seem to be quite robust.[7]

5.4 Models with Period-Specific Effects

In this section, we further generalize the piecewise constant exponential model by also allowing the effects of the time-constant covariates (i.e., their associated parameters) to vary across time periods. This model was first proposed by Tuma (1980).

To request estimation of an exponential model with period-specific effects, you can use the `streg` command. Yet it is important to note here that you need to generate period-specific dummies that will then be included in the `streg` command. One way to do this is shown in Box 5.4.1.

The time periods are defined by split points on the time axis in exactly the same way as in the standard exponential model with time periods. For time periods, which are defined by split points $\tau_1, \tau_2, \ldots, \tau_L$, the transition rate to destination state k is

$$r_k(t) = \exp\left\{\bar{\alpha}_l^{(k)} + A^{(k)}\alpha_l^{(k)}\right\} \quad \text{if} \quad \tau_l < t \leq \tau_{l+1}. \tag{5.2}$$

For each transition to destination state k, $\bar{\alpha}_l^{(k)}$ is a constant coefficient associated with the lth time period, the (row) vector of covariates is $A^{(k)}$, and $\alpha_l^{(k)}$ is an associated vector of coefficients showing the effects of these covariates in the lth time period.

It is obvious that the standard exponential model with time periods defined in (5.1) (see section 5.1) is a special case of the model defined in (5.2). Estimating the latter model with constraints that require the $\alpha_l^{(k)}$ parameters to be equal across time periods would yield identical results as the standard exponential model with time periods.

Maximum likelihood estimation. The maximum likelihood estimation of this model is similar to the standard piecewise constant exponential model described earlier. Using the same abbreviations, $l[t]$ and $\Delta[s,t,l]$, the conditional survivor function is

$$G(t \mid s) = \exp\left\{-\sum_{l=1}^{L} \Delta[s,t,l] \exp(\bar{\alpha}_l + A\alpha_l)\right\}$$

and using this expression, the log likelihood can be written as

$$\ell = \sum_{i\in\mathcal{E}} (\bar{\alpha}_{l[t_i]} + A_i\alpha_{l[t_i]}) - \sum_{i\in\mathcal{N}}\sum_{l=1}^{L} \Delta[s_i,t_i,l] \exp(\bar{\alpha}_l + A_i\alpha_l).$$

The initial estimates that start the iterative process of maximizing the log likelihood are calculated in the same way as for the standard piecewise constant exponential model.

An application. In the context of our job-exit example, the piecewise constant exponential model with period-specific effects is particularly interesting because it offers the opportunity to assess hypotheses based on *filter* or *signaling theory* (Arrow 1973; Spence 1973, 1974). This labor market theory states that, in the

Box 5.3.1 Do-file `ehe2.do`

```
version 19
capture log close
set more off
log using ehe2.log, replace

use rrdat1, clear

gen des = tfin ~= ti                 /*destination state*/
gen tf  = tfin - tstart + 1          /*ending time*/

gen newid = _n
stset tf, failure(des) id(newid) /*define single episode data*/

gen coho2 = tb>=468 & tb<= 504       /*cohort 2*/
gen coho3 = tb>=588 & tb<= 624       /*cohort 3*/
gen lfx   = tstart - te              /*labor force experience*/
gen pnoj  = noj - 1                  /*previous number of jobs*/

stpiece edu coho2 coho3 lfx pnoj pres, ///
tp(0(12)96) nohr  /*fit piecewise constant exponential model*/

log close
```

Box 5.3.2 Estimation results of do-file `ehe2.do` (Box 5.3.1)

```
          failure _d:  des
    analysis time _t:  tf
                  id:  newid

Iteration 0:   log likelihood = -1002.8964
Iteration 1:   log likelihood = -854.90874
Iteration 2:   log likelihood = -851.88418
Iteration 3:   log likelihood = -851.86828
Iteration 4:   log likelihood = -851.86828

Exponential regression -- log relative-hazard form

No. of subjects =          600     Number of obs    =      2715
No. of failures =          458
Time at risk    =        40782
                                   Wald chi2(15)    =   8626.69
Log likelihood  =   -851.86828     Prob > chi2      =    0.0000

----------------------------------------------------------------------
    _t |      Coef.   Std. Err.       z    P>|z|     [95% Conf. Interval]
------+---------------------------------------------------------------
   tp1 |   -4.25273    .294021   -14.46   0.000    -4.829001   -3.676459
   tp2 |  -3.916678   .2920046   -13.41   0.000    -4.488996   -3.344359
   tp3 |  -3.738957   .2925247   -12.78   0.000    -4.312295   -3.165619
   tp4 |   -4.22092   .3133731   -13.47   0.000     -4.83512    -3.60672
   tp5 |  -4.009912   .3122496   -12.84   0.000     -4.62191   -3.397914
   tp6 |  -4.323992   .3395438   -12.73   0.000    -4.989486   -3.658498
   tp7 |  -4.772435   .3848382   -12.40   0.000    -5.526704   -4.018166
   tp8 |  -4.940688   .4134257   -11.95   0.000    -5.750988   -4.130389
   tp9 |   -4.95165   .2898812   -17.08   0.000    -5.519807   -4.383493
   edu |   .0635991   .0248868     2.56   0.011     .0148219    .1123763
 coho2 |   .4572261   .1153366     3.96   0.000     .2311704    .6832817
 coho3 |   .3521179    .122506     2.87   0.004     .1120105    .5922254
   lfx |  -.0037174   .0009285    -4.00   0.000    -.0055373   -.0018976
  pnoj |   .0595917   .0442055     1.35   0.178    -.0270495    .1462329
  pres |  -.0252979   .0054807    -4.62   0.000    -.0360398   -.0145559
----------------------------------------------------------------------
```

hiring process, easily observable characteristics (such as educational qualifications, number of previously held jobs, or years of labor force experience) serve as signals for employers. In particular, employers use these characteristics to infer differences in productivity between people applying for a particular job. They are therefore very important for employment decisions and in the initial phase of any new job. With increasing job duration, however, an employer is able to gain more direct information on the employee and is increasingly

Box 5.4.1 Do-file `ehe3.do`

```
version 19
capture log close
set more off
log using ehe3.log, replace

use rrdat1, clear

gen des = tfin ~= ti            /*destination state*/
gen tf  = tfin - tstart + 1     /*ending time*/

gen newid = _n
stset tf, failure(des) id(newid) /*define single episode data*/

gen cohc2 = tb>=468 & tb<= 504   /*cohort 2*/
gen cohc3 = tb>=588 & tb<= 624   /*cohort 3*/
gen lfx   = tstart - te          /*labor force experience*/
gen pnoj  = noj - 1              /*previous number of jobs*/

stsplit time, at(0,24,60)        /*split records*/
tab time, ge(t)                  /*generate time dummies*/

*generate period-specific dummies
for any edu edu edu       \ any t1 t2 t3: gen eduY  = X*Y
for any coho2 coho2 coho2 \ any t1 t2 t3: gen coho2Y= X*Y
for any coho3 coho3 coho3 \ any t1 t2 t3: gen coho3Y= X*Y
for any lfx lfx lfx       \ any t1 t2 t3: gen lfxY  = X*Y
for any pnoj pnoj pnoj    \ any t1 t2 t3: gen pnojY = X*Y
for any pres pres pres    \ any t1 t2 t3: gen presY = X*Y

stset tf, failure(des) id(newid)  /*define single episode data*/

streg t1 t2 t3 edut* coho2t* coho3t* lfxt* pnojt* prest*, ///
dist(exp) nohr noconstant /*fit piecewise constant exponential model*/

log close
```

less forced to rely on such signals to evaluate the competence or productivity of employees. It can therefore be expected that the importance of such signals for job decisions will decrease with increasing length of employment. On the other hand, one can also assume that job decisions of an employee, especially at the beginning of each new job, depend strongly on the image (or prestige) of the job position and that the relevance of these characteristics for her or his job decisions declines with increasing experience in the job. In summary, both arguments lead to the hypothesis that the effects of the time-constant covariates, which serve as signals, decrease with increasing job duration.

In the following example, we therefore divide the duration in a job into periods and estimate period-specific effects of covariates with a piecewise constant exponential model. We use a new do-file, `ehe3.do`, as shown in Box 5.4.1. It is basically the same as do-file `ehe2.do` that we already used for the standard piecewise constant exponential model. Note, however, the new specification of time periods. Since our model estimates the parameters for all covariates separately in each time period, we have to use a reasonably small number of periods to obtain meaningful estimates for the 600 job episodes in our example data set.[8] Thus, we define only three periods (`t1`, `t2`, `t3`) by using three split points, namely, 0, 24, and 60 months. There are several ways to generate period-specific dummies. Here we use the `for` command, which repeats the `generate` command for each variable in the `forlist` where `X` refers to the first list of variables and `Y` to the second. To save space, we use abbreviations for variable names (*) when specifying `streg`.

Part of the estimation result is shown in Box 5.4.2. It supports our hypothesis that the effects of the signaling or filter variables, such as educational attainment (`edu`), general labor force experience at the beginning of each job (`lfx`), number of previously held jobs (`pnoj`), or prestige of the job (`pres`), are strong in the first phase of each job (period up to 2 years), and then become less important across the later periods. In the third period (job duration greater than 5 years), none of these signaling or filter variables are still significant.

Box 5.4.2 Estimation results of do-file `ehe3.do` (Box 5.4.1)

```
         failure _d:  des
   analysis time _t:  tf
                 id:  newid

Iteration 0:   log likelihood = -1002.8964
Iteration 1:   log likelihood =  -859.0758
Iteration 2:   log likelihood = -853.15786
Iteration 3:   log likelihood = -853.13814
Iteration 4:   log likelihood = -853.13814

Exponential regression -- log relative-hazard form

No. of subjects =          600             Number of obs   =       1202
No. of failures =          458
Time at risk    =        40782
                                           Wald chi2(21)   =    8642.88
Log likelihood  =   -853.13814             Prob > chi2     =     0.0000

------------------------------------------------------------------------
      _t |      Coef.   Std. Err.      z    P>|z|     [95% Conf. Interval]
---------+--------------------------------------------------------------
      t1 |   -4.35138   .4224051   -10.30   0.000    -5.179279   -3.523481
      t2 |   -3.32689   .4831846    -6.89   0.000    -4.273915   -2.379866
      t3 |  -5.615024   .5755181    -9.76   0.000    -6.743019   -4.487029
   edut1 |   .1097628   .0357066     3.07   0.002     .0397792    .1797464
   edut2 |   .0011625   .0481818     0.02   0.981     -.093272    .0955971
   edut3 |   .0682161   .0492593     1.38   0.166    -.0283304    .1647626
  coho2t1 |   .5809915   .1917358     3.03   0.002     .2051963    .9567867
  coho2t2 |   .4070205   .1892725     2.15   0.032     .0360532    .7779879
  coho2t3 |   .4125671   .2304085     1.79   0.073    -.0390254    .8641595
  coho3t1 |   .5206901   .1931976     2.70   0.007     .1420298    .8993504
  coho3t2 |   .0866651   .2018906     0.43   0.668    -.3090332    .4823634
  coho3t3 |   .7097271   .2600232     2.73   0.006     .2000909    1.219363
   lfxt1 |   -.004681   .0015347    -3.05   0.002     -.007689   -.001673
   lfxt2 |  -.0041873    .001523    -2.75   0.006    -.0071723   -.0012023
   lfxt3 |  -.0011644   .0018192    -0.64   0.522      -.00473    .0024011
  pnojt1 |   .1455211   .0681359     2.14   0.033     .0119773     .279065
  pnojt2 |   .0229359   .0750371     0.31   0.760    -.1241341    .1700059
  pnojt3 |  -.0477606   .0940906    -0.51   0.612    -.2321749    .1366536
  prest1 |  -.0371613   .0082883    -4.48   0.000    -.0534062   -.0209165
  prest2 |  -.0189583   .0094382    -2.01   0.045    -.0374569   -.0004597
  prest3 |  -.0075705   .0116297    -0.65   0.515    -.0303643    .0152232
```

Notes

[1]The exponential model and the piecewise constant exponential model, in which the covariates have the same effects across the time periods, are both special cases of the more general proportional transition rate models (see chapter 9).

[2]The Statistical Software Components (SSC) archive maintained by Boston College (`http://www.repec.org`) provides user-written Stata commands. To install the Stata program to estimate piecewise constant exponential models, use the `ssc install` command: `ssc install stpiece`. This command downloads the ado-file and implements the new Stata command `stpiece`.

[3]Note that `_n` is a Stata system variable that indicates the position of an observation in the data set.

[4]Note that it is necessary to define an ID variable before using the command `stsplit`. We use the option `at(numlist)` to define time periods. Type `help numlist` to learn more about number lists in Stata.

[5]Postestimation commands such as `predict` are not possible after `stpiece`.

[6]In chapter 10, we give some examples of how unobserved time-constant heterogeneity can lead to time-dependent rates.

[7]In the next chapters, we will see that the parameter estimates for the time-constant covariates are surprisingly stable across a broad range of different models.

[8]Clearly, a model that attempts to estimate the changing effects of covariates across a number of time periods generally requires a large data set, even in the case of moderate numbers of periods and covariates.

Chapter 6

Exponential Models with Time-Dependent Covariates

In our view, the most important advance in using event history analysis for the empirical study of social change has been the explicit measurement and inclusion of time-dependent covariates in transition rate models. In such models, covariates can change their values over process time. Time-dependent covariates can be qualitative or quantitative, and can remain constant for finite periods of time or change continuously.

Three basic approaches for including time-dependent covariates in transition rate models can be distinguished: (1) using a piecewise constant exponential model as shown in the previous chapter, (2) applying the method of episode splitting in parametric and semiparametric transition rate models, and (3) specifying the distributional form of the time dependence and estimating its parameters directly using the maximum likelihood method (see chapter 7). In this chapter, we begin with a general discussion of time-dependent covariates in the context of parallel and interdependent processes, and then focus on the method of episode splitting.[1] Parametric models of time-dependence are presented in chapter 7.

6.1 Parallel and Interdependent Processes

By using time-dependent covariates, we can study the effects of change in one phenomenon over time on change in another phenomenon over time (Tuma and Hannan 1984).[2] From a substantive point of view, it is therefore useful to conceptualize time-dependent covariates as observations of the sample path of parallel processes (Blossfeld, Hamerle, and Mayer 1989).[3] These processes can occur on different levels. For example:

1. *At the level of the individual, parallel processes can occur in different areas of life* (e.g., we can ask how upward and downward moves in an individual's job career influence her or his family trajectory; see Blossfeld and Huinink 1991; Blossfeld et al. 1999; Mills 2000b; Blossfeld and Timm 2003; Blossfeld 2022).
2. There may be *parallel processes on the level of some few individuals interacting with each other* (e.g., we could study the effect of the husband's career on his wife's labor force participation; see Bernasco 1994; Blossfeld and Drobnič 2001).
3. There may be *parallel processes on the intermediate level* (e.g., we can analyze how organizational growth influences career advancement or how changing household structure determines women's, or how institutional factors and the organization of PhD programs influence PhD completion. labor force participation; see Blossfeld and Hakim 1997; Blossfeld and Drobnič 2001; Skopek et al. 2022).
4. There may be *parallel processes at the macrolevel* (e.g., we may be interested in the effect of changes in the business cycle on family formation or career advancement; see Blossfeld 1987a; Huinink 1989, 1992, 1993; Blossfeld and Becker 2017; Becker et al. 2024).
5. There can be *any combination of Processes 1 to* 4. For example, studies of life-course, cohort, and period effects, must simultaneously include time-dependent covariates at different levels (Blossfeld 1986; Mayer and Huinink 1990; Becker et al. 2024). Such an analysis combines individual-level processes (life-course change) with two types of macro-level processes: variations in structural conditions across successive (birth, marriage, entry, etc.) cohorts, and changes in historical conditions affecting all cohorts in the same way.

When dealing with such systems of parallel processes, the methodological literature usually addresses the issue of reverse causation (see, e.g., Kalbfleisch and Prentice 1980; Tuma and Hannan 1984; Blossfeld, Hamerle, and Mayer 1989; Yamaguchi 1991; Courgeau and Lelièvre 1992). Reverse causation refers to the (direct or indirect) effect of the dependent process on the independent covariate process. Reverse causation is considered a problem, because the effect of a time-dependent covariate on the transition rate is confounded with a feedback effect of the dependent process on the values of the time-dependent covariate.[4] Nonetheless, the following two types of time-dependent covariates are described in the literature as not being subject to reverse causation (Kalbfleisch and Prentice 1980):

1. *Defined time-dependent covariates* whose overall time path (or functional form of change over time) is *determined in advance in the same way for all subjects* under study. For example, process time, such as age or duration in a state (e.g., job-specific labor force experience), is a defined time-dependent covariate, because its values are fixed in advance for all subjects. Thus, by definition, the values of these time-dependent covariates cannot be affected by the dependent process under study.

2. *Ancillary time-dependent covariates* whose time path is the outcome of a stochastic process that is *external* to the units under study. Again, by definition, the values of these time-dependent covariates are not influenced by the dependent process itself. Examples of time-dependent covariates that are *approximately external* in the analysis of individual life courses are variables that reflect *changes on the macrolevel* of society (unemployment rates, occupational structure, etc.) or at the *population level* (composition of the population by age, sex, race, etc.), provided that the contribution of each unit is small and does not really affect the structure in the population (Yamaguchi 1991). For example, consider changes in the occupational structure. Although an individual's job change may contribute to a change in the occupational structure, its effect on the job structure as a whole is negligible.[5]

In contrast to defined or ancillary time-dependent covariates, *internal time-dependent covariates* are considered problematic for causal analysis with event history models (e.g., Kalbfleisch and Prentice 1980; Tuma and Hannan 1984; Blossfeld, Hamerle, and Mayer 1989; Yamaguchi 1991; Courgeau and Lelièvre 1992). An internal time-dependent covariate X_t describes a stochastic process that is regarded as a cause in a causal model and which in turn is influenced by another stochastic process Y_t that is regarded as an effect in the causal model. Thus, there are direct effects in which the processes affect each other autonomously (X_t affects Y_t, and Y_t affects X_t), and there are "feedback" effects in which these processes are affected by themselves via the respective other process (Y_t affects Y_t via X_t, and X_t affects X_t via Y_t). In other words, such processes are interdependent and form a so-called dynamic system (Tuma and Hannan 1984). Interdependence is typical at the individual level for processes in different domains of life and at the level of a few individuals who interact with each other (e.g., career trajectories of partners). For example, the empirical literature suggests that the employment trajectory of an individual is influenced by her or his marital history, and that marital history is dependent on the employment trajectory.

At least two main approaches have been proposed for dealing with dynamic systems. We consider both and call them the "system approach" and the "causal approach."

6.2 Interdependent Processes: The System Approach

The system approach to the analysis of interdependent processes suggested in the literature (see, e.g., Tuma and Hannan 1984; Courgeau and Lelièvre 1992) defines change in the system of interdependent processes as a new "dependent variable." Thus, instead of analyzing one of the interdependent processes with respect to its dependence on the respective others, the focus of modeling is on a system of state variables.[6] In other words, the interdependence between the different processes is only implicitly considered.

We will first demonstrate the logic of this approach for a system of *qualitative* time-dependent variables and give some examples, before discussing its limitations, and finally describing what we consider more appropriate for the analysis of coupled processes from an analytical point of view: the causal approach.

Suppose there are J interrelated qualitative time-dependent variables (processes): $Y_t^A, Y_t^B, Y_t^C, \ldots, Y_t^J$. A new time-dependent variable (process) Y_t, which represents the system of these J variables, is then defined by associating each discrete state of the ordered J-tuple with a particular discrete state of Y_t. As shown by Tuma and Hannan (1984), as long as change in the whole system depends only on the various states of the J qualitative variables and on exogenous variables, this model is identical to modeling change in a single qualitative variable.[7] Thus, the idea of this approach is to simply define a new *joint* state space based on the different state spaces of the coupled qualitative processes, and then to proceed as in the case of a single dependent process.

For example, suppose we have repeated episodes of two interdependent processes, "employment trajectory" and "marital history," represented by two dichotomous variables Y_t^A and Y_t^B, where Y_t^A takes the values 1 ("not employed") or 2 ("employed") and Y_t^B takes the values 1 ("not married") or 2 ("married"). Then, as shown in Figure 6.2.1, a new variable Y_t, which represents the bivariate system, has $L = 4$ different states.[8] These are:

$1 \equiv (1,1)$	"not employed and not married"
$2 \equiv (1,2)$	"not employed and married"
$3 \equiv (2,1)$	"employed and not married"
$4 \equiv (2,2)$	"employed and married"

In general, with L different states in the combined process, there are $L(L-1)$ possible transitions. However, if the possibility of simultaneous changes in two or more of the processes is excluded,[9] the number of possible transitions is reduced by the number of simultaneous transitions. In our example of two dichotomous variables, eight transition rates describe the joint process completely, as can be seen in Figure 6.2.2. Each of these origin- and destination-specific rates can be estimated in a model without covariates or with respect to their dependence on exogenous covariates.

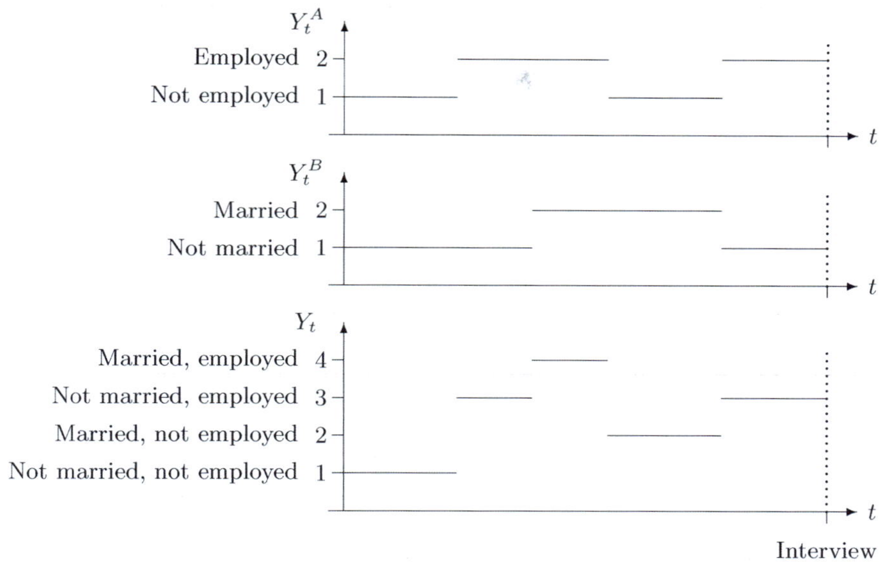

Figure 6.2.1 Hypothetical sample paths of two coupled processes, Y_t^A and Y_t^B, and the sample path of the combined process, Y_t.

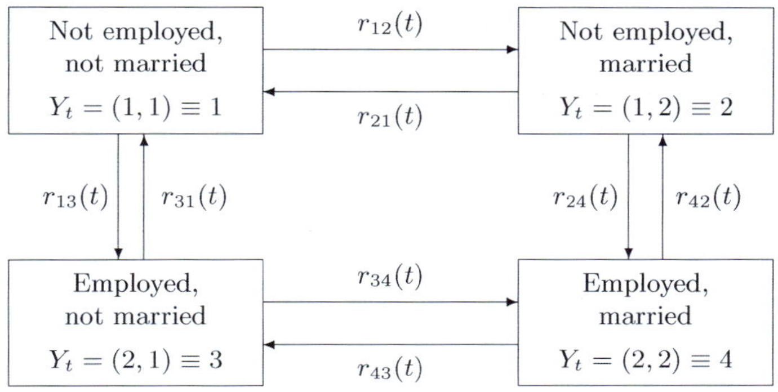

Figure 6.2.2 States and transitions for the system Y_t consisting of employment and marital histories.

In the case of coupled processes, and if we consider only irreversible events such as first marriage and first pregnancy, the diagram of possible transitions can be further simplified to four possible transitions, as shown in Figure 6.2.3. As demonstrated by Courgeau and Lelièvre (1992), we can use equality tests that compare origin- and destination-specific transition rates to determine whether (see Figure 6.2.3):[10]

1. The two processes are independent: $r_{12} = r_{34}$ and $r_{13} = r_{24}$.
2. One of the two processes is exogenous and the other endogenous:[11]
   a) $r_{12} = r_{34}$ and $r_{13} \neq r_{24}$: pregnancy affects marriage positively; $r_{13} < r_{24}$, pregnancy affects marriage negatively: $r_{13} > r_{24}$, or
   b) $r_{12} \neq r_{34}$ and $r_{13} = r_{24}$: marriage affects pregnancy positively; $r_{12} < r_{34}$, marriage affects pregnancy negatively: $r_{12} > r_{34}$.
3. The processes are interdependent (or endogenous) and affect each other: $r_{12} \neq r_{34}$ and $r_{13} \neq r_{24}$.

These equality tests are easy to perform as long as no covariates are involved and only the baseline transition rates for specific origin and destination states have to be estimated. However, if the episodes are heterogeneous and a greater number of covariates has to be taken into account to make the episodes in each origin state independent of each other, the number of possible equality tests increases rapidly, which poses practical problems for comparisons (Courgeau and Lelièvre 1992).

Thus, although the system approach provides some insight into the behavior of the dynamic system as a whole, it has several drawbacks: (1) From a causal-analytical point of view, *it does not provide direct estimates of effects of coupled processes* on a process under study. In other words, when using the system approach, we usually do not know to what extent one or more of the other coupled processes affect the process of interest, even though we control for other exogenous variables and the history of the dependent process. It is only possible to compare transition rates for general models without covariates, as shown in the previously

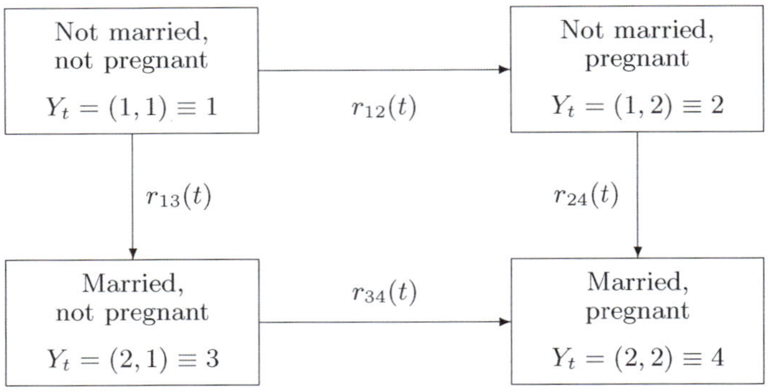

Figure 6.2.3 States and transitions for the system Y_t consisting of first marriage and first pregnancy.

mentioned pregnancy-marriage example. (2) In particular, a *mixture of qualitative and quantitative processes*, where the transition rate of a qualitative process depends on the levels of one or more metric variables, turns out to be a problem in this approach.[12] Tuma and Hannan (1984) suggested that in such situations, each quantitative variable should be collapsed into a set of ordered states. But in many situations, this is not very useful. (3) This approach is also unable to deal with *interdependencies* between coupled processes that *occur only in specific phases of the process* (e.g., processes may be interdependent only in specific phases of the life course) or interdependencies that are dynamic over time (e.g., an interdependence may reverse at later stages of life; see Courgeau and Lelièvre 1992). (4) Finally, the number of origin and destination states of the combined process, Y_t, which represents the system of J variables, can lead to practical problems. Even if the number of variables and their distinct values is small, the *state space of the system variable is large*. Therefore, event history data sets must contain a great number of events, even if only the most general models of change (i.e., models without covariates) are to be estimated. In summary, the systems approach for analyzing interdependent processes has many limitations. We therefore suggest using a different perspective in modeling dynamic systems, which we call the "causal approach."

6.3 Interdependent Processes: The Causal Approach

The basic idea of the causal approach to analyzing interdependent processes can be outlined as follows (Blossfeld 1994): Based on theoretical reasons, the researcher focuses on one of the interdependent processes and considers it as the dependent one. Future changes in this process are linked to the present state and history of the whole dynamic system as well as to other exogenous variables (see Blossfeld 1986; Gardner and Griffin 1986; Blossfeld and Huinink 1991). Thus, in this approach, the variable Y_t, which represents the system of joint processes at time t, is not used as a multivariate dependent variable. Instead, the history and the current state of the system are considered as a condition of the change in (any) of its processes.

The question remains as to how to find a more precise formulation for the causal approach. Two ideas may be helpful: First, it seems somewhat misleading to regard processes as causes. As argued in section 1.2, only events, or changes in a state variable, can sensibly be viewed as possible causes. Consequently, we would not say that a process Y_t^B is a cause of a process Y_t^A, but that *a change* in Y_t^B could be a cause (or represents a new condition) of a change in Y_t^A. This leads directly to a second idea: that every event needs a certain amount of time to become the cause of an effect, because effects follow their causes in time (see section 1.2). This time span may be very short, but it, nonetheless, does not seem reasonable to assume an instantaneous reaction—at least not in the social world where most effects are mediated by decision-making agents.[13]

Of course, we are considering only interdependent processes here that are not just an expression of another underlying process. Hence, it makes sense to assess the properties of the two processes without regarding the underlying one. This means, for instance, that what happens next to Y_t^A should not be related directly to what happens to Y_t^B at the same point in time, and vice versa. This condition, which we call local autonomy (see Pötter and Blossfeld 2001), can be formulated in terms of the uncorrelatedness of the prediction errors of both processes, Y_t^A and Y_t^B, and excludes stochastic processes that are functionally related.

By combining these ideas, taking a causal view on parallel and interdependent processes becomes easy—at least in principle. For two parallel processes, Y_t^A and Y_t^B, a change in Y_t^A at any (specific) point in time t' may depend on the history of both processes *up to, but not including,* t'. Or to put it another way: What happens to Y_t^A at any point in time t' is *conditionally* independent of what happens with Y_t^B at t', conditional on the history of the joint process $Y_t = (Y_t^A, Y_t^B)$ up to, but not including, t'. Of course, the same reasoning can be applied if one focuses on Y_t^B instead of Y_t^A as the "dependent variable." We call this the *principle of conditional independence* for parallel and interdependent processes.[14]

Conditional independence is a regulative idea in the modeling of causal relationships. It is not an empirical concept and therefore cannot be demonstrated from observation alone (see Pötter and Blossfeld 2001). In particular, it depends on the number of processes considered in a particular analysis. If a process is added or removed from the analysis, the conditional independence may change. Although the theoretical background on which an analysis is based always determines to some extent the histories and processes to be excluded and included, it will rarely be specific enough to determine exactly which processes are to be considered (see Pötter and Blossfeld 2001). In this sense, there may be several valid causal analyses based on different sets of stochastic processes (see Giere 1999).

The same idea can be formally developed further. Starting from a transition rate model for the joint process, $Y_t = (Y_t^A, Y_t^B)$, and assuming the principle of conditional independence, the likelihood for this model can be factorized into a product of the likelihoods for two separate models: a transition rate model for Y_t^A, which is dependent on Y_t^B as a time-dependent covariate; and a transition rate model for Y_t^B, which is dependent on Y_t^A as a time-dependent covariate.[15]

This result has an important implication for modeling event histories. From a technical point of view, it is not necessary to distinguish between defined, ancillary, and internal covariates, as all of these time-dependent covariate types can be treated equally in the estimation procedure. A distinction between defined and ancillary covariates on the one hand and internal covariates on the other is, however, useful from a theoretical perspective, because it is only in the case of internal covariates that it makes sense to examine whether parallel processes are independent, whether one of the parallel processes is endogenous and the other ones are exogenous, or whether parallel processes form an interdependent system (i.e., they are all endogenous).[16]

In the next section, we show how qualitative time-dependent covariates, whose sample paths follow a step function, can be included in transition rate models on the basis of the episode-splitting method. This procedure leads to direct estimates of how parallel qualitative processes affect the rate of change in another qualitative process and allow us to conduct significance tests of their parameters (see section 4.1.3). Then, in section 6.5, we demonstrate that a generalization of the episode-splitting technique can also be used to include quantitative time-dependent covariates. In particular, this method offers an efficient strategy for approximating (1) the effects of any type of duration dependence in a state, (2) the effects of any sort of parallel quantitative process, as well as (3) complex temporal shapes of effect patterns of covariates over time.

6.4 Episode Splitting with Qualitative Covariates

The effects of time-dependent processes on the transition rate can be easily estimated using the method of episode splitting. The idea of this method can be described as follows: Time-dependent qualitative covariates change their values only at discrete points in time. At all points in time at which (at least) one of the covariates changes its value, the original episode is split into pieces—called *splits* (of an episode) or *subepisodes*. For each subepisode, a new record is created containing the following:

1. Information about the origin state of the original episode.
2. The values of all the covariates at the beginning of the subepisode.
3. The starting and ending times of the subepisode (information on duration would only be sufficient in the case of an exponential model).
4. Information indicating whether the subepisode ends with the destination state of the original episode or is censored. All subepisodes, apart from the last one, are regarded as right-censored. Only the last subepisode is given the same destination state as the original episode.

Let us consider one of the original episodes (j, k, s, t) with single origin and destination states j and k and with starting and ending times s and t, respectively. It is assumed that the episode is defined on a process time axis such that $s = 0$. Now assume that this episode is split into L subepisodes[17]

$$(j, k, s, t) \equiv \{(j_l, k_l, s_l, t_l) \mid l = 1, \ldots, L\}. \tag{6.1}$$

The likelihood of this episode can be written as the product of a transition rate $r(t)$, and a survivor function $G(t)$. Obviously, only $G(t)$ is influenced by the process of episode splitting. However, $G(t)$ can be written as the product of the conditional survivor functions for each split of the episode:

$$G(t) = \prod_{l=1}^{L} G(t_l \mid s_l)$$

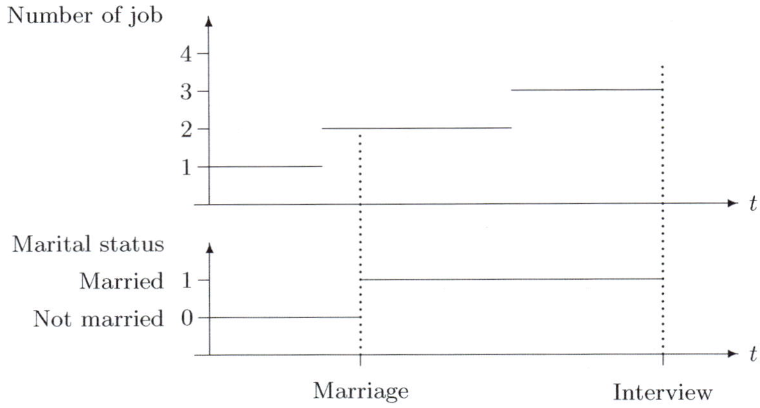

Figure 6.4.1 Modeling the effect of marriage on job mobility as a time-dependent covariate.

with conditional survivor functions defined by

$$G(t_l \mid s_l) = \frac{G(t_l)}{G(s_l)} = \exp\left\{ - \int_{s_l}^{t_l} r(\tau)\, d\tau \right\}.$$

On the right-hand side, the transition rate $r(\tau)$ could be specific for each split (j_l, k_l, s_l, t_l) and depend on covariate values for this split.[18]

It would now be possible to rewrite the general likelihood function for transition rate models given in section 4.1.1 by inserting the product of conditional pseudosurvivor functions. However, one does not really need to do this. One can write a general likelihood function for transition rate models, assuming a given origin state, as

$$\mathcal{L} = \prod_{k \in \mathcal{D}} \prod_{i \in \mathcal{E}_k} r_k(t_i) \prod_{i \in \mathcal{N}} \tilde{G}_k(t_i \mid s_i). \tag{6.2}$$

Written in this way, the likelihood can also be used with a sample of original (unsplit) episodes where all starting times are zero, instead of a sample of splits.[19] Of course, it is the user's responsibility to split each episode in such a way that the splits add up to a sample of meaningful episodes.

The maximum likelihood estimation of transition rate models is always performed in Stata using the likelihood function (6.2).[20] Therefore, the program is easy to use with the episode-splitting method.

Episode Splitting with Stata

Let us assume that we are studying job mobility and want to examine whether first marriage has an effect on the job-exit rate. In this case, we first need to create a time-dependent dummy variable that changes its value from 0 to 1 at the time of first marriage. We must add this information to the original job episodes as a new variable by splitting the original job episodes into two subepisodes in which the marriage date lies between the starting and ending times of a job spell (see the second job in Figure 6.4.1).

Using Stata, we can split the episodes and estimate the coefficients of the time-dependent covariates in one run. First, we have to create a data file containing the split episodes. As an example, we use do-file `ehf1.do` shown in Box 6.4.1. Most of the commands in this file are already known from previous do-files. For didactic purposes, we duplicate the variable on job duration and name the duplicate tf2. What is new and necessary is to add the lines

```
gen     mdate = tmar - tstart if tmar >0
replace mdate = 9999          if tmar==0
```

to the do-file to define a new variable, `mdate`, which is used for episode splitting. This variable contains the marriage date (in number of months) relative to the beginning of each job episode. `mdate` is negative or zero if the individual was already married at the beginning of each job episode, because then `tmar` is less than or equal to `tstart`. `mdate` is positive, but less than the duration of the job (`tf2`) if the individual marries during that job episode. Finally, `mdate` is greater than the duration of the job episode (`tf2`) when the individual marries after the end of the job or has not married until the interview. For the last case, we have arbitrarily chosen a high value, namely 9999, which far exceeds the longest job duration in our data

Box 6.4.1 Do-file `ehf1.do`

```
version 19
capture log close
set more off
log using ehf1.log, replace

use rrdat1, clear

gen des = tfin ~= ti              /*destination state*/
gen ts  = 0                       /*starting time*/
gen tf  = tfin - tstart + 1       /*ending time*/
gen tf2 = tf                      /*duplicate tf (optional)*/

* variable for episode splitting
gen     mdate = tmar - tstart if tmar >0
replace mdate = 9999          if tmar==0

gen newid = _n
stset tf, failure(des) id(newid)
stsplit ma_d , after(mdate) at(0)

generate marr = 1 if ma_d ==  0  /*time-dependent dummy variable*/
replace  marr = 0 if ma_d == -1  /*0 for single, 1 for married*/

list id tstart tfin tmar ts tf tf2 mdate des _t0 _t _d ma_d marr ///
if id<3 | id==4 | id==5 | id==45, noobs sepby(id)

streg marr, dist(exp) nohr        /*fit parametric survival model*/

* percentage change in the rate: entry into marriage
display (exp(_b[marr])-1)*100

log close
```

Box 6.4.2 Selected records of `rrdat1.dta` after episode splitting (generated with `ehf1.do` in Box 6.4.1)

id	tstart	tfin	tmar	ts	tf	tf2	mdate	des	_t0	_t	_d	ma_d	marr
1	555	982	679	0	124	428	124	.	0	124	0	-1	0
1	555	982	679	0	428	428	124	0	124	428	0	0	1
2	593	638	762	0	46	46	169	1	0	46	1	-1	0
2	639	672	762	0	34	34	123	1	0	34	1	-1	0
2	673	892	762	0	89	220	89	.	0	89	0	-1	0
2	673	892	762	0	220	220	89	1	89	220	1	0	1
4	872	926	872	0	55	55	0	1	0	55	1	0	1
5	583	650	701	0	68	68	118	1	0	68	1	-1	0
5	651	787	701	0	50	137	50	.	0	50	0	-1	0
5	651	787	701	0	137	137	50	1	50	137	1	0	1
5	788	982	701	0	195	195	-87	0	0	195	0	0	1
45	896	913	0	0	18	18	9999	1	0	18	1	-1	0
45	914	982	0	0	69	69	9999	0	0	69	0	-1	0

set. To actually perform the episode splitting, we create a new ID, declare the data to be survival time data (`stset`), and use the command `split` already introduced in chapter 5.2, but here with an additional option.

```
stsplit ma_d , after(mdate) at(0)
```

For our data set, Stata creates 161 new episodes.

To demonstrate how episode splitting with covariates works in Stata, we present relevant variables from selected records in the data set (see Box 6.4.2). The first person in the data set has only one job and got married during the time on this job. Thus, the original job episode is divided into a first subepisode from 0 to 124 and a second subepisode from 124 to 428 (see `_t0` and `_t`). Note that Stata updates the variable on job duration `tf` (but not `tf2`), because `tf` was used to declare the data as survival time. In contrast, `ts`

Box 6.4.3 Result of do-file `ehf1.do` (Box 6.4.1)

```
         failure _d:  des
   analysis time _t:  tf
                 id:  newid

Iteration 0:   log likelihood =  -937.9681
Iteration 1:   log likelihood = -922.78222
Iteration 2:   log likelihood =  -922.4817
Iteration 3:   log likelihood = -922.48159
Iteration 4:   log likelihood = -922.48159

Exponential regression -- log relative-hazard form

No. of subjects =          600        Number of obs   =        761
No. of failures =          458
Time at risk    =        40782
                                      LR chi2(1)      =      30.97
Log likelihood  =   -922.48159        Prob > chi2     =     0.0000

------------------------------------------------------------------------
     _t |      Coef.   Std. Err.      z    P>|z|     [95% Conf. Interval]
-------+----------------------------------------------------------------
   marr |  -.5211523   .0936828    -5.56   0.000    -.7047673   -.3375374
  _cons |  -4.212944   .0638877   -65.94   0.000    -4.338161   -4.087726
```

was not updated. If you want to have updated values in `ts`, you have to add the option `time0(ts)` after `stset`. The `stsplit` command is used to request a new variable. This new variable called `ma_d` contains only two values: -1 and 0. If the marriage has already taken place, `ma_d` is 0, otherwise -1. The first subepisode is right-censored (`_d = 0`). It has the same origin and destination states; that is, 0 ("being in a job").[21] Only the second split is given the same destination state as in the original episode. In this case, the original episode was also censored.

The person with `id = 2` married many years after the end of her first two jobs. Therefore these job episodes remain unchanged and the variable `ma_d` has the value -1. She married during her third job, and this job episode is split in the same way as already described for the first case in the data set. The person with `id = 4` started a job and married in the same month. Consequently, the job episode is not subdivided into parts, but the variable `ma_d` is now 0 from the beginning. A further interesting case is the person with `id = 45`. She has had two jobs, but has remained single up to the time of the interview. The high value we have assigned here guarantees that there is no episode splitting.[22]

In a second step, we create the time-dependent dummy variable `marr` (marriage) by adding the instruction

```
generate marr = 1 if ma_d ==  0
replace  marr = 0 if ma_d == -1
```

to create a dummy variable with the value 0 until the marriage date and the value 1 if the individual has married (see Box 6.4.2).

Using do-file `ehf1.do`, the estimation results are shown in Box 6.4.3. Obviously, the coefficient for the time-dependent covariate `marr` is statistically significant and has a negative sign. This means that entry into marriage reduces the job-exit rate by about 41 %, calculated as

$$\big(\exp(-0.5212) - 1\big) \cdot 100\,\% = -40.6\,\%.$$

In other words, marriage is associated with a lower rate of individual job-exit rates. If the goal of the analysis is *descriptive* or *correlating*, we do not need to adjust for further covariates. In this case, we only want to determine whether married people have a different rate of job mobility than unmarried people. We are simply interested in the association between being married and the rate of job-exit rate in a certain population. Therefore, no adjustments for other factors need to be made in the data analysis. We will generally suspect that this association measure is confounded and therefore not interpret it as a causal effect measure. Indeed, "there is no such thing as a 'spurious association' "(Hernán 2018, p. 618). However, when we ask *causal questions*, we are interested in determining *whether marrying lowers the rate of job-exit rate.* In a *causal analysis*, we then have to think carefully about selection processes and other covariates that might be confounders, so that the estimates for exposure to marriage can be interpreted causally as changing the job-exit rate. The ability to use time-dependent covariates in event history analysis is a big step forward in causal analysis, as these covariates allow us to determine the timing of exposure to a causally presumed variable much more exactly. However, this advantage comes at a price, as we need to carefully consider not only which time-constant variables might be confounders, but also which *other time-dependent processes* might be important and need to be adjusted for in the causal analysis. The problem is that in addition to the causal process under consideration, there are many other parallel and interdependent processes that are changing their states over time. For example, if the qualitative time-dependent covariate `marr` changes its state at the time of marriage in our analysis above, a distinction is also made for other processes between a period "before exposure" (to marriage) and a period "during exposure" (to marriage). If other parallel and interdependent processes are changing over time and they are not controlled for in the analysis, the coefficient `marr` will then absorb these effect contributions like a sponge and assign them to the coefficient of `marr`. For example, in Figure 6.4.2, the estimated coefficient of `marr` captures the average difference of being a parent or not and having a low or high general labor force experience. Since, in the birth cohorts studied in Germany, the job-exit rate of women is likely to increase after the birth of a child (due to interruptions related to the traditional family role of wives) and the job-exit rate of men is likely to decrease (due to additional responsibilities imposed on the traditional breadwinners), it is also likely that these gender-specific differences influence the effect of `marr` on the job-exit rate, when gender is not controlled for in the analysis. Since the marriage and the parenthood events are highly correlated (pregnancy might increase the probability of marriage, and marriage might increase the probability of pregnancy), the coefficient of `marr` in the rate equation for the job-exit rate is likely to be confounded. Today, one would expect that if we additionally include a time-dependent covariate "entry into parenthood" in the rate equation, it is parenthood (and not marriage) that "causes" the job-exit rate. Due to (traditional) gender roles, we would also expect the effect of entering parenthood (and entering marriage) to be positive for women and negative for men. Thus, gender and its interaction with the time-dependent covariate `marr` could also be important variables in a causal analysis. In Figure 6.4.2, general labor force experience could also increase the `marr` coefficient. It is well known that individuals are much more mobile at the beginning of their careers and gradually become more stable with increasing labor force experience (see the coefficient of the variable `lfx` in the analyses in Chapter 4). Thus, we expect the average job-exit rate to generally be higher before individuals are married then afterwards. The difference in these averages is likely to be picked up by the coefficient `marr`, when general labor force experience is not adjusted in the analysis. In other words, the (negative) coefficient of `marr` should decline, when general labor force experience is controlled for in the model. There are many other processes that may influence the effect of the time-dependent covariate `marr` on the job-exit rate. For example, an important macro variable might be the business cycle that influences marriage, parenthood, and the job-exit rate at the same time (Blossfeld and Becker 2017; Becker et al. 2024). In general, the estimated effects of time-dependent covariates in event history analysis are more vulnerable than the effects of time-constant covariates. The examples in our book demonstrate that the estimated effects of time-constant covariates are quite robust in terms of signs and significances (and therefore in terms of substantive conclusions) across a broad range of nested models and assumptions regarding time dependence (see Chapter 7). However, we will see in Chapter 10 that unobserved heterogeneity can lead to apparent time dependence even when there is no time dependence in the process data. Thus, the application of event history models in causal analysis in general and of time-dependent covariates in particular requires *profound subject-matter knowledge*.

Of course, readers might rightly point out that with observational data, there is no guarantee that a causal model incorporates all substantively important (time-constant and time-dependent) variables; and, therefore, there is no guarantee that the parameter estimates of variables in event history analysis can be interpreted causally, even approximately (Hernán 2018, p. 618). This is certainly true, and we do not believe that this problem can be solved satisfactorily. There seems to be no solution to this fundamental problem, even if we apply more sophisticated statistical models (see, e.g., Chapter 10). Therefore we follow Xie (2007, p. 155) who concludes that we should "be savvy researchers who are interested in advancing our empirical understanding about substantive topics of sociological import but unwilling to make sweeping claims."

The question how the effect of the time-varying variable `marr` changes when other relevant covariates are included in the model is demonstrated in example do-file `ehf2.do` in Box 6.4.4.

The output of `ehf2.do` is shown in Box 6.4.5. The effect of the time-dependent covariate is still highly significant and has a negative sign. However, its absolute size is smaller than in Box 6.4.3. When controlling for the time-constant covariables, entry into marriage only reduces the job-exit rate by about 29 %.[23] Thus, part of the time-constant heterogeneity between the individuals was captured by the time-dependent covariate "marriage." The estimated effects of the time-constant covariates, however, are very similar to the results in Box 4.1.4 in which the time-dependent covariate "marriage" was not included.

Interaction Effects with Time-Dependent Covariates

Although we obtain a significant effect for the time-dependent covariate "marriage" in Box 6.4.5, it is not easy to interpret this theoretically. For example, it is well known from the empirical literature that family

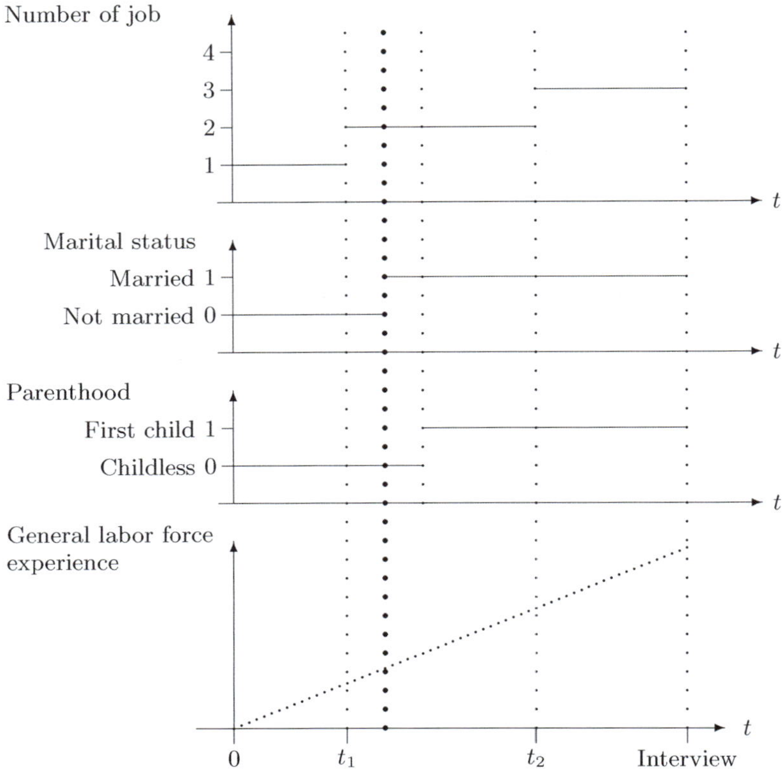

Figure 6.4.2 Parallel and interdependent processes and causal inference

events have an opposite influence on the job-exit rates for men and women. Marriage increases the job exit rate for women, at least in the older cohorts (because they normally take care of the household and children), while marriage lowers the job exit rate for men (because they normally bear additional economic responsibility for their wives and children).

In order to investigate whether these gender-specific relationships between marriage and employment career apply, we create an interaction variable for the joint effect of the time-dependent covariate "marriage" and the time-constant variable "sex." The command is shown in the lower part of Box 6.4.4. The interaction variable `marrmen` has the value 1 as soon as a man marries.

Part of the estimation results of the model with the `marrmen` interaction effect are shown in Box 6.4.6. The estimated parameters are in line with our expectations. The effect of marriage on the rate of moving out of a job is *positive* for women. Marriage *increases* the job-exit rate for women by about 31 %.[24] Of course, the time-dependent covariate marriage also serves here as a proxy variable for other time-dependent processes that are connected with the marriage event such as childbirth and so forth. In a "real" analysis, one would therefore also include time-dependent covariates for childbirth and additional possible interaction effects with, for example, the birth cohort.[25]

For men, on the other hand, the effect of marriage on the job change rate is *negative*. Marriage *decreases* the rate of moving out of a job by about 53 %.[26] In other words, marriage makes men less mobile.

This didactic example shows again how important theory is for modeling event history data. The only reason for the negative effect of the time-dependent covariate in Box 6.4.6 is that the negative effect for men is stronger than the positive effect for women. In general, it is very difficult to evaluate whether an estimated model is appropriate without a strong theory. Although there are a few technical criteria and approaches that can be applied to assess models (see chapters 8 and 10), our experience has shown that they are of limited use in choosing between competing models. Thus, event history modeling (like any causal modeling) must be guided by theoretical ideas.

Box 6.4.4 Do-file `ehf2.do`

```
version 19
capture log close
set more off
log using ehf2.log, replace

use rrdat1, clear

gen des = tfin ~= ti            /*destination state*/
gen ts  = 0                     /*starting time*/
gen tf  = tfin - tstart + 1     /*ending time*/
gen tf2 = tf                    /*duplicate tf (optional)*/

gen coho2 = tb>=468 & tb<=504   /*cohort 2*/
gen coho3 = tb>=588 & tb<=624   /*cohort 3*/
gen lfx   = tstart - te         /*labor force experience*/
gen pnoj  = noj - 1             /*previous number of jobs*/

* variable for episode splitting
gen     mdate = tmar - tstart if tmar >0
replace mdate = 9999          if tmar==0

gen newid = _n
stset tf, failure(des) id(newid)
stsplit ma_d , after(mdate) at(0)

generate marr = 1 if ma_d ==  0  /*time-dependent dummy variable*/
replace  marr = 0 if ma_d == -1  /*0 for single, 1 for married*/

streg edu coho2 coho3 lfx pnoj pres marr, dist(exp) nohr
/*fit parametric survival model*/

* percentage change in the rate: entry into marriage
display (exp(_b[marr])-1)*100

gen marrmen = sex==1 & marr==1   /*interaction effect*/

streg edu coho2 coho3 lfx pnoj pres marr marrmen, ///
dist(exp) nohr /*fit parametric survival model*/

log close
```

Box 6.4.5 Result of do-file **ehf2.do** (Box 6.4.4)

```
       failure _d:  des
 analysis time _t:  tf
               id:  newid

Iteration 0:   log likelihood =  -937.9681
Iteration 1:   log likelihood =   -887.864
Iteration 2:   log likelihood = -884.20472
Iteration 3:   log likelihood = -884.19606
Iteration 4:   log likelihood = -884.19606

Exponential regression -- log relative-hazard form

No. of subjects =          600        Number of obs   =        761
No. of failures =          458
Time at risk    =        40782
                                      LR chi2(7)      =     107.54
Log likelihood  =   -884.19606        Prob > chi2     =     0.0000

-----------------------------------------------------------------------
   _t |      Coef.   Std. Err.      z    P>|z|     [95% Conf. Interval]
------+----------------------------------------------------------------
  edu |   .0770113   .0245584     3.14   0.002     .0288778    .1251448
coho2 |   .5970898   .1136579     5.25   0.000     .3743244    .8198553
coho3 |   .6137878   .1188341     5.17   0.000     .3808773    .8466984
  lfx |  -.0022972    .000968    -2.37   0.018    -.0041945      -.0004
 pnoj |   .0645368   .0439811     1.47   0.142    -.0216645    .1507381
 pres |   -.027352    .005522    -4.95   0.000    -.0381748   -.0165291
 marr |  -.3447105   .1022392    -3.37   0.001    -.5450956   -.1443254
_cons |  -4.387819   .2755794   -15.92   0.000    -4.927944   -3.847693
```

Box 6.4.6 Result of do-file **ehf2.do** (Box 6.4.4)

```
failure _d:  des
analysis time _t:  tf
id:  newid

Iteration 0:   log likelihood =  -937.9681
Iteration 1:   log likelihood = -866.10127
Iteration 2:   log likelihood =  -858.1433
Iteration 3:   log likelihood = -858.12342
Iteration 4:   log likelihood = -858.12342

Exponential regression -- log relative-hazard form

No. of subjects =          600    Number of obs    =        761
No. of failures =          458
Time at risk    =        40782
LR chi2(8)      =     159.69
Log likelihood  =   -858.12342    Prob > chi2      =     0.0000

--------+--------------------------------------------------------------
_t      |     Coef.   Std. Err.       z    P>|z|     [95% Conf. Interval]
--------+--------------------------------------------------------------
edu     | .0943202   .0234471     4.02   0.000     .0483646    .1402757
coho2   | .5422956   .1131758     4.79   0.000     .3204752     .764116
coho3   | .5303917   .1192044     4.45   0.000     .2967555     .764028
lfx     |-.0030792    .000976    -3.16   0.002    -.0049921   -.0011664
pnoj    | .1029071   .0444981     2.31   0.021     .0156924    .1901218
pres    | -.026412   .0053505    -4.94   0.000    -.0368988   -.0159252
marr    | .2734695   .1231325     2.22   0.026     .0321343    .5148047
marrmen |-1.023141   .1401201    -7.30   0.000    -1.297771   -.7485103
_cons   |-4.584152   .2693982   -17.02   0.000    -5.112163   -4.056141
```

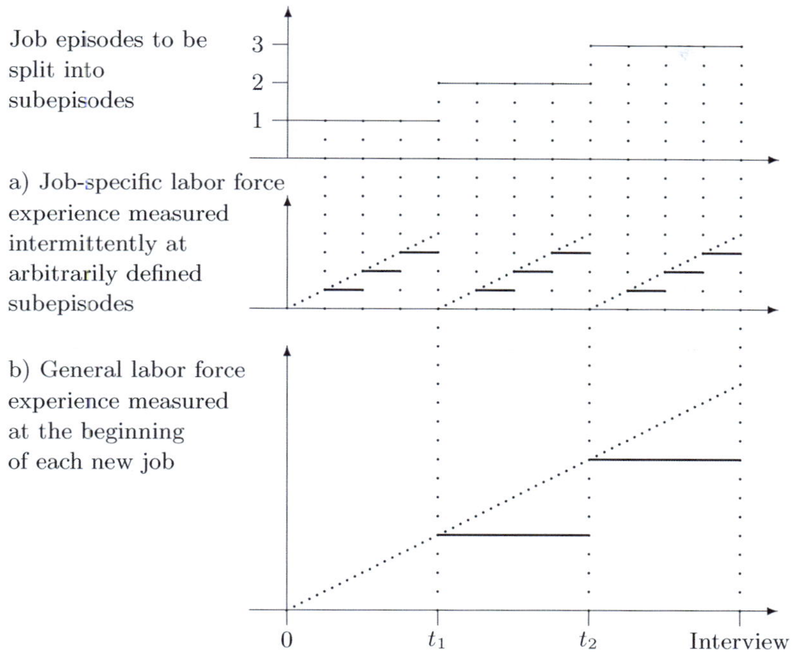

Figure 6.5.1 Modeling labor force experience as (a) time-dependent covariate *job-specific labor force experience* and (b) time-constant covariate *general labor force experience* illustrated with three consecutive job episodes arbitrarily split into subepisodes.

6.5 Episode Splitting with Quantitative Covariates

Many of the theoretical models of interest to sociologists combine quantitative processes with qualitative outcomes over time. For example, "job-specific labor force experience" can be considered as a metric latent (or unobserved) variable. There are good reasons to measure it with the proxy variable "duration in a job." Other examples for metric causal processes are continuously changing "investments in specific marriages" in divorce studies (Blossfeld et al. 1995); "career success of the partner" in analyses of parallel careers of husbands and wives (Bernasco 1994); or measures of the continuously changing "business cycle" or "modernization process" in job mobility studies (see Blossfeld 1986; Blossfeld and Becker 2017; Becker et al. 2024). As long as the effects of such quantitative variables can be considered a specific function of process time (e.g., based on a Weibull, Gompertz (-Makeham), a log-logistic, a lognormal, or a sickle model), it is possible to select and estimate available parametric models of time dependence (see chapter 7).[27]

However, if these parametric models are not appropriate, or if the sample path of the quantitative covariate is so irregular or complex that it is impossible to specify its shape with a parametric distribution, then the method of episode splitting can be generalized and used as an effective tool to *approximate the sample path of a metric covariate* (Blossfeld, Hamerle, and Mayer 1989). In this approach, the original episodes are arbitrarily divided into small subepisodes, and the quantitative time-dependent covariate is measured intermittently at the beginning of each of these subepisodes (see panel a) of Figure 6.5.1).[28] The result of this procedure is a step function approximating the changes in the metric causal variable. The smaller the selected sub-episodes and the more frequently the metric time-dependent causal variable is measured intermittently, the better the approximation (Blossfeld 2025).

We use the variable "job-specific labor force experience" as an example. We assume that this (unobserved) variable increases linearly with the time a person spends in a job. In section 4.1.3, labor force experience was considered only as "general labor force experience" measured at the beginning of each new job episode and treated as constant over the whole duration in a job. Panel b) in Figure 6.5.1 shows that this usually leads to a very bad approximation of what might be called labor force experience—particularly for employees who do not change jobs very often. Now, we also include the variable "job-specific labor force experience" which is measured intermittently at the beginning of arbitrarily defined subepisodes within each job (see panel a) in Figure 6.5.1). We start our example by defining subepisodes with a maximal length of 60 months (or 5 years) in the Stata do-file `ehf3.do` shown in Box 6.5.1. Since the longest duration in a job is 428 months (see Box 2.2.6), we have to define eight intervals which are performed with the `stsplit` command:

```
stsplit lfx60, at(60 (60) 420)
```

Once we submit this command, Stata creates 421 new episodes. Box 6.5.2 shows the first records of the data

Box 6.5.1 Do-file `ehf3.do`

```
version 19
capture log close
set more off
log using ehf3.log, replace

use rrdat1, clear

gen des = tfin ~= ti                /*destination state*/
gen ts  = 0                         /*starting time*/
gen tf  = tfin - tstart + 1         /*ending time*/

gen coho2 = tb >= 468 & tb <= 504   /*cohort 2*/
gen coho3 = tb >= 588 & tb <= 624   /*cohort 3*/
gen lfx   = tstart - te             /*labor force experience*/
gen pnoj  = noj - 1                 /*previous number of jobs*/
gen newid = _n

stset tf, failure(des) id(newid) time0(ts) /*define single episode data*/

stsplit lfx60, at(60 (60) 420)     /*split time-span records*/
list newid id ts tf des _t0 _t _d lfx lfx60 in 1/20, sepby(newid) noobs
/*first records of data file*/
stset          /* verification of data declaration - optional*/
streg edu coho2 coho3 lfx pnoj pres lfx60, dist(exp) nohr
/*fit parametric survival model*/
estimates store lfx60
/*save most recent estimation results under name lfx60*/

* different episode splits: lfx24
drop ts lfx60  /*drop variables with different values across split obs.*/
stjoin         /*combine records*/
gen ts = 0
stsplit lfx24, at(24 (24) 420)
list newid id ts tf des _t0 _t _d lfx lfx24 in 1/20, sepby(newid) noobs
stset          /*optional*/
streg edu coho2 coho3 lfx pnoj pres lfx24, dist(exp) nohr
estimates store lfx24

* different episode splits: lfx12
stsplit lfx12, at(12 (12) 420)
list newid id ts tf des _t0 _t _d lfx lfx24 lfx12 in 1/40, ///
 sepby(newid) noobs
streg edu coho2 coho3 lfx pnoj pres lfx12, dist(exp) nohr
estimates store lfx12

estimates table lfx60 lfx24 lfx12, b(%4.3f) se(%4.3f) p(%4.2f) ///
 modelwidth(10) stats(N N_fail ll chi2 df_m p)
 /*write stored estimates in table*/

log close
```

set after episode splitting.

For example, for the individual with `id = 1`, the original job episode with a duration of 428 months is split into seven subepisodes of 60 months and an eighth subepisode of 8 months. Note again that the first seven subepisodes are right-censored, that is, they have the same origin and destination states; that is, 0 ("being in a job"). Only the eighth split is given the same destination state as in the original episode. In this case, the original episode was also censored. There is no need to re-declare the data as event history data. But you might type `stset`, and the output will show you how the total number of observations has changed while the number of individuals and events has remained the same.

We use the `streg` command to estimate an exponential model with our standard set of covariates. The estimation results are shown in Box 6.5.3. This box provides estimates for two different forms of labor force experience. The parameter for the variable `lfx` is an estimate for the effect of "*general* labor force experience" and the parameter for variable `lfx60` is an estimate for the effect of "*job-specific* labor force experience." As expected, both variables have a significantly negative effect. In other words, an increase in general and job-specific labor force experience reduces the rate of job mobility. However, as can be seen from the absolute size of the coefficients, job-specific labor force experience reduces job mobility more than general labor force experience. This is in line with the hypotheses suggested in the literature (see chapter 4).

Box 6.5.2 First records of data file `rrdat1.dta` with episode splits

```
+-----------------------------------------------------------------+
| newid   id    ts    tf   des    _t0    _t   _d   lfx    lfx60 |
|-----------------------------------------------------------------|
|     1    1     0    60     .      0    60    0     0        0 |
|     1    1    60   120     .     60   120    0     0       60 |
|     1    1   120   180     .    120   180    0     0      120 |
|     1    1   180   240     .    180   240    0     0      180 |
|     1    1   240   300     .    240   300    0     0      240 |
|     1    1   300   360     .    300   360    0     0      300 |
|     1    1   360   420     .    360   420    0     0      360 |
|     1    1   420   428     0    420   428    0     0      420 |
|-----------------------------------------------------------------|
|     2    2     0    46     1      0    46    1     0        0 |
|-----------------------------------------------------------------|
|     3    2     0    34     1      0    34    1    46        0 |
|-----------------------------------------------------------------|
|     4    2     0    60     .      0    60    0    80        0 |
|     4    2    60   120     .     60   120    0    80       60 |
|     4    2   120   180     .    120   180    0    80      120 |
|     4    2   180   220     1    180   220    1    80      180 |
|-----------------------------------------------------------------|
|     5    3     0    12     1      0    12    1     0        0 |
|-----------------------------------------------------------------|
|     6    3     0    30     1      0    30    1    12        0 |
|-----------------------------------------------------------------|
|     7    3     0    12     1      0    12    1    42        0 |
|-----------------------------------------------------------------|
|     8    3     0    60     .      0    60    0    54        0 |
|     8    3    60    75     1     60    75    1    54       60 |
|-----------------------------------------------------------------|
|     9    3     0    12     1      0    12    1   129        0 |
```

Box 6.5.3 Result of do-file `ehf5.do` (Box 6.5.1)

```
        failure _d:  des
  analysis time _t:  tf
                id:  newid

Iteration 0:   log likelihood =  -937.9681
Iteration 1:   log likelihood = -865.09263
Iteration 2:   log likelihood = -856.34496
Iteration 3:   log likelihood = -856.25564
Iteration 4:   log likelihood = -856.25557

Exponential regression -- log relative-hazard form

No. of subjects =          600     Number of obs   =      1021
No. of failures =          458
Time at risk    =        40782
                                   LR chi2(7)      =    163.43
Log likelihood  =   -856.25557     Prob > chi2     =    0.0000

------------------------------------------------------------------------
   _t |      Coef.   Std. Err.       z    P>|z|     [95% Conf. Interval]
------+-----------------------------------------------------------------
  edu |   .0646117   .0248022     2.61   0.009     .0160002    .1132231
coho2 |   .4073219   .1148612     3.55   0.000     .1821981    .6324456
coho3 |   .2946223   .1216231     2.42   0.015     .0562455    .5329992
  lfx |  -.0039428   .0009293    -4.24   0.000    -.0057641   -.0021214
 pnoj |   .0647431    .044029     1.47   0.141    -.0215522    .1510384
 pres |  -.0253234    .005449    -4.65   0.000    -.0360032   -.0146435
lfx60 |  -.0070767   .0009954    -7.11   0.000    -.0090276   -.0051259
_cons |   -4.01378    .276443   -14.52   0.000    -4.555598   -3.471962
```

The approximation of job-specific labor force experience in each of the jobs is still relatively rough. Therefore, we reduce the maximum interval length of subepisodes from 60 months to 24 months and finally to 12 months, and then examine how the estimated coefficients for job-specific labor force experience change. For this purpose, we first use `stjoin` to reverse episode splitting, adjust the `stsplit` command, and refit

Box 6.5.4 Exponential model with episode splits (60, 24, and 12 months)

Variable	lfx60	lfx24	lfx12
edu	0.065	0.063	0.063
	0.025	0.025	0.025
	0.01	0.01	0.01
coho2	0.407	0.406	0.405
	0.115	0.115	0.115
	0.00	0.00	0.00
coho3	0.295	0.301	0.300
	0.122	0.122	0.122
	0.02	0.01	0.01
lfx	-0.004	-0.004	-0.004
	0.001	0.001	0.001
	0.00	0.00	0.00
pnoj	0.065	0.063	0.063
	0.044	0.044	0.044
	0.14	0.15	0.15
pres	-0.025	-0.025	-0.025
	0.005	0.005	0.005
	0.00	0.00	0.00
lfx60	-0.007		
	0.001		
	0.00		
lfx24		-0.006	
		0.001	
		0.00	
lfx12			-0.006
			0.001
			0.00
_cons	-4.014	-3.950	-3.919
	0.276	0.279	0.280
	0.00	0.00	0.00
N	1021	2004	3668
N_fail	458.000	458.000	458.000
ll	-856.256	-859.029	-858.988
chi2	163.425	157.877	157.961
df_m	7.000	7.000	7.000
p	0.000	0.000	0.000

legend: b/se/p

the exponential model (see Box 6.5.1). The data could be split a second or third time without prior removal of previous episode splitting. Since 12 months are fully nested in 24 months, we immediately proceed with episode splitting in the last case. The results of these estimations are shown in Box 6.5.4. In each cell of the table, Stata first reports coefficients followed by their standard errors and significance levels. It is easy to see that the estimates of the three models are quite similar. The estimates of the metric time-dependent covariate change a little when we reduce the maximum length of the subepisodes from 60 to 24 months, but they remain virtually unchanged when we reduce the maximum length even further from 24 to 12 months. Thus, we can conclude that, given a maximum interval length of 24 months, we achieve a relatively good approximation of the linearly changing time-dependent covariate "job-specific labor force experience." Furthermore, we can see that the coefficients of the time-constant covariates are essentially the same as for the exponential model in 4.1.4, at least in substantive terms. And finally, all the coefficients (including "job-specific labor force experience") are very similar compared with the equivalent Gompertz model with covariates (see section 7.2).[29]

6.6 Application Examples

We will illustrate the usefulness of the episode-splitting method in empirical research using four studies that apply this method in different ways. We will concentrate on the modeling techniques used in these studies and summarize their main findings.

Example 1: A Dynamic Approach to the Study of Life-Course, Cohort, and Period Effects in Career Mobility

Blossfeld (1986) examined the intragenerational mobility of German men and proposed including the changing labor market structure into event history models of career mobility to adequately address the time-dependent nature of the attainment process (see also the papers by Blossfeld and Becker 2017; Becker et al. 2024). The changing labor market affects career opportunities in two ways: First, people start their careers in different structural labor market contexts (when this influence is more or less stable over people's careers, it is normally called a cohort effect; see also Blossfeld 1986); second, the labor market structure influences the opportunities of *all* people within the labor market at any point in time (this is commonly referred to as a period effect).

Any model that uses education (in terms of school years), labor market experience (as a life-course measure), the year of entry into the labor market (as a cohort measure), and the chronological year of measurement (as a period measure) poses an identification problem (Mason and Fienberg 1985), as these variables are linearly interdependent. Blossfeld took a more substantive approach and tried to find more explicit measures for the theoretically important macro effects. He suggested using 14 time series from official statistics indicating the long-term development of the labor market structure in West Germany.[30] However, time series often measure similar characteristics of a process and are highly correlated. Therefore a further identification problem always arises when several time series are included in a model simultaneously. One strategy to solve this problem is to use only one series or to choose only uncorrelated series. The problem is then that the selected time series may only capture specific features of the labor market's structural development. If time series represent aspects of an underlying regularity, it is more appropriate to look for these latent dimensions. A statistical method for this is factor analysis. Blossfeld's factor analysis with principal factoring and equimax rotation extracted two orthogonal structural factors explaining 96.4 % of the variance in the 14 time series.

The first factor could be interpreted as representing the changing "Level of modernization"; the second one, as a measure of the changes in the labor market with regard to the business cycle, so it was called "Labor market conditions." As can be seen from the plots of the scores of the two factors in Figure 6.6.1, the factor Level of modernization shows a monotonic trend with a slightly increasing slope, while the factor Labor market conditions shows a cyclical development with downturns around 1967 and 1973. This is in accordance with the historical development of the business cycle in Germany. As both factors are constructed orthogonally, it is possible to include both measures simultaneously in a model estimation.

To represent the changing conditions under which cohorts enter the labor market, Blossfeld used the factor scores on Level of modernization and Labor market conditions in the year persons entered the labor force. To introduce period as a time-dependent covariate, he used the previously described method of episode splitting. In accordance with this method, the original job episodes were split into subepisodes every year so that the factor scores for Level of modernization and Labor market conditions could be updated in each job episode for all the employees each year. In the terminology of Kalbfleisch and Prentice (1980), the period measures are ancillary time-dependent covariates.

Modeling cohort and period effects in this way not only provides more direct and valid measures of the explanatory concepts but also results in the disappearance of the problem of nonestimable parameters. It is therefore not necessary to impose (in substantive terms normally unjustifiable) constraints on the parameters to identify life-course, cohort, and period effects (see, e.g., Mason et al. 1973). The result of the estimation of life-course, cohort, and period effects on upward moves in an exponential model is shown in Table 6.6.1. In this table, we focus our attention on Model 5 and refrain from substantive interpretations of the effects of the variables "time in the labor force," "education," or "prestige," as these would be basically the same as those already given in section 4.1.3.

We concentrate mainly on interpreting the effects of changes in the labor market structure. In Model 5, the level of modernization at time of entry into the labor market has a negative effect on upward moves. Thus, the higher the level of modernization, the better the first job of beginners and the less likely they are to experience further upward moves. The same applies to the negative effect of labor market conditions at entry into the labor market. The more favorable the business cycle at entry into the labor market is for a particular cohort, the easier it is for its members to find "good jobs" at the beginning of their careers, and the more difficult it will be to find even better ones.

Conversely, the period effect of level of modernization is positive on upward moves. Thus the continuous restructuring of the economy in the process of technological and organizational modernization leads to increasing opportunities for advancement for all people on the labour market. The same applies for the period effect of labor market conditions. It is positive for upward mobility, suggesting that the better the labor market conditions, the more opportunities the economy offers.

If we take into account the effect of labor force experience (life-course effect), then this analysis supports the thesis that the career process is strongly affected by cohort, period, and life-course effects. The attainment process is time-dependent in a threefold sense: It depends on the time spent in the labor force, the historical time of entry into the labor market, and the actual historical time. Thus, analyses of standard mobility tables (e.g., Handl 1988; Haller 1989; Erikson and Goldthorpe 1991), which only distinguish between structural mobility and exchange mobility on a cross-sectional basis, therefore necessarily provide a misleading picture of the mechanisms of attainment (see also Sørensen 1986; Becker et al. 2024). The creation of vacancies and the loss of positions in the course of structural change count as the central mechanisms of career mobility and affect people's mobility changes in different ways.

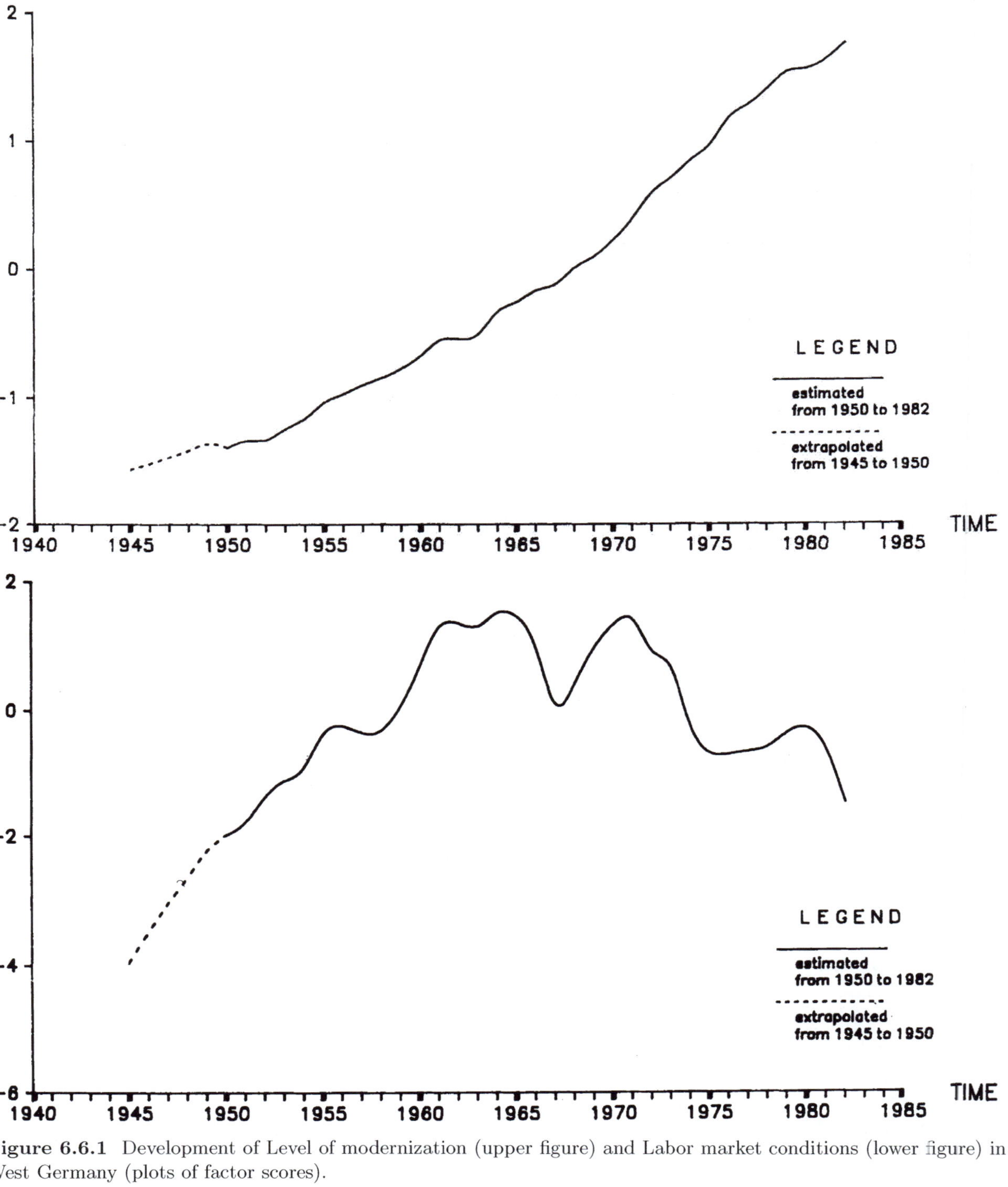

Figure 6.6.1 Development of Level of modernization (upper figure) and Labor market conditions (lower figure) in West Germany (plots of factor scores).

Example 2: Changes in the Process of Family Formation and the New Role of Women

The second example is based on an investigation by Blossfeld and Huinink (1991). They assessed the hypothesis of "new home economics" (e.g., Becker 1981) that women's rising educational and career investments will lead to a decline in marriage and motherhood (see also Oppenheimer 1988). Since the accumulation of educational and career resources is a lifetime process, it must be modeled as a dynamic process over the life course. In West Germany in particular, it is necessary to distinguish between the accumulation of general and vocational qualifications within the educational system on the one hand, and the accumulation of workplace-related labor force experience on the other.

To model the accumulation of general and vocational qualifications in the school system, the vocational training system, and the university system of the Federal Republic of Germany, Blossfeld and Huinink (1991)

Table 6.6.1 Estimates of models for transition rates to better jobs (upward moves) for German men born 1929–31, 1939–41, and 1949–51.

Estimates for model	Upward moves 1	2	3	4	5
Log of mean rate	-6.135				
Constant		-4.241*	-4.288*	-4.943*	-5.585*
Time in labor force (life-course effect)		-0.012*	-0.015*	-0.012*	-0.082*
Education			0.187*	0.224*	0.268*
Prestige			-0.042*	-0.041*	-0.042*
Level of modernization at at entry into labor market (cohort effect)				-0.294*	-9.664*
Labor market conditions at entry into labor market (cohort effect)				0.009	-1.394*
Level of modernization (period effect)					9.066*
Labor market conditions (period effect)					1.203*
Number of exits	590	590	590	590	590
Subepisodes	22843	22843	22843	22843	22843
χ^2		842.21	1007.46	1024.02	2165.29
df		1	3	5	7

* Statistically significant at 0.001 level. Rates are measured with months as units.

used the average number of years required to obtain such qualifications (see variable `edu` in Box 2.2.1). However, this variable was not treated as a time-constant variable but as a time-dependent covariate. To model changes in the accumulation of qualifications over the life course, they updated the level of education at the age at which a woman obtained a particular educational rank in the hierarchy. For example, for a woman who attains a lower school qualification at age 14, reaches the intermediate school qualification at age 16, leaves school with an *Abitur* (university entrance qualification) at age 19, and finally finishes university studies at age 25, one would obtain a step function for the highest level of education over the life course as shown in the upper panel of Figure 6.6.2. The hypothesis of new home economics is that such increasing levels of education raise a woman's labor-market attachment and thus lead to a greater delay in marriage and childbirth.

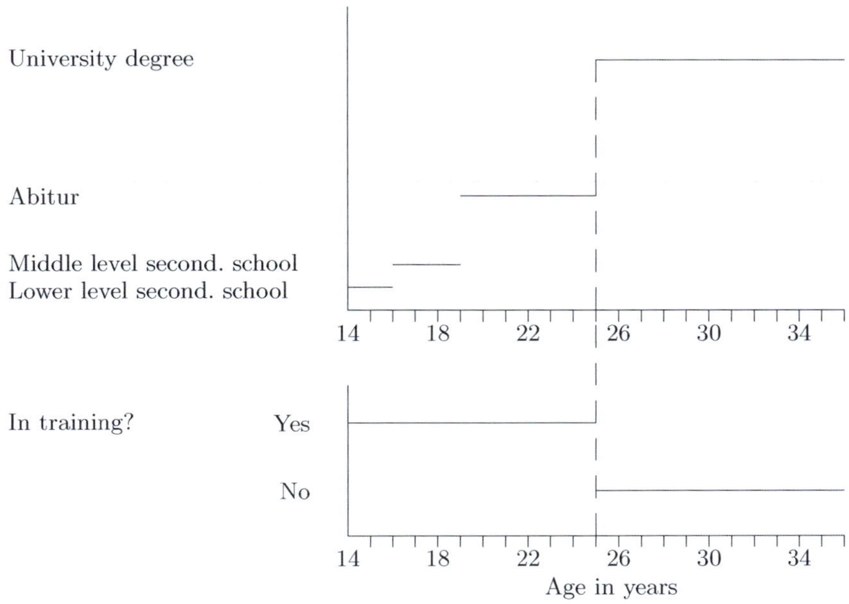

Figure 6.6.2 Educational careers over the life course in West Germany.

However, from a sociological point of view, one could also expect an effect from the fact that women are enrolled in school. When a woman is attending school, university, or a vocational training program, she is normally economically dependent on her parents. Furthermore, there is a normative expectation in modern societies that young people who attend school are "not at risk" of entering marriage and parenthood. Moreover, the roles of students and mothers are sufficiently demanding, so that most women delay childbearing until they have left school. Finishing one's education therefore counts as one of the important prerequisites for entering into marriage and parenthood. To include this influence in their model, Blossfeld and Huinink generated a time-dependent dummy variable indicating whether or not a woman is attending the education system at a specific age (see the lower panel of Figure 6.6.2).

After leaving the education system and entering employment, women accumulate labor force experience at their workplaces. As shown earlier, economists (Mincer 1974; Becker 1975) and sociologists (Sørensen 1977; Sørensen and Tuma 1981) have often used time in the labor force as a proxy for the accumulation of labor force experience. However, this procedure can be criticized on the basis of research into labor-market segmentation (see Blossfeld and Mayer 1988). First, there is a so-called secondary labor market in the economy which offers relatively low-paying and unstable employment with poor chances of accumulating any labor force experience at all (see, e.g., Doeringer and Piore 1971; Blossfeld and Mayer 1988). Second, in some positions within so-called internal labor markets, opportunities to accumulate labor force experience are distributed very unequally (e.g., Doeringer 1967; Piore 1968; Doeringer and Piore 1971; Blossfeld and Mayer 1988). Likewise, differences in the opportunities for acquiring labor force experience may also exist among the self-employed and people working in different kinds of professions. This means that the *speed and levels of the accumulation of labor force experience* must be modeled as a function of the type of employment. For the dynamic modeling of job-specific investments in human capital over the life course, Blossfeld and Huinink have therefore made the following three conjectures:

Development of Career Resources After Entry Into First Employment. Women who have left the educational system and entered their first jobs will accumulate labor force experience with decreasing increments. As on-the-job training is concentrated mainly in the earlier phases of employment, increases are large at the beginning and level off as the duration of employment increases. This means that increments and final levels of labor force experience should be modeled as a function of a measure of job quality, for example, the prestige score, P, of a job. A possible mathematical model of the growth rate $r(P,t)$ of career resources at age t, assuming that the first job was entered at age t_0, is therefore ($t \geq t_0$):

$$r(P,t) = \exp\big(-\alpha\,(t - t_0)\big)$$

where

$$\alpha = \frac{P_{\text{max}} - P_{\text{min}}}{2}\frac{1}{P} = \frac{83.4}{P}.$$

Here P is Wegener's (1985) prestige score, which is used as a proxy measure for the job quality and for the opportunity to accumulate labor force experience within a job. Given this model, the level of career resources, $K(P,t)$, within a job episode at age t is then defined as

$$K(P,t) = \exp\left\{\int_{t_0}^{t} r(P,u)\,\mathrm{d}u\right\} - 1.$$

Until entry into the first job, the level of career resources, $K(P,t)$, is equal to zero. The maximum level of career resources, $\max(K(P,t))$, within a job with the lowest prestige score (a helper with a prestige score of 20.0 on the Wegener scale), for example, is reached after 9 months and has the value 0.27. In a job with the highest prestige score on the Wegener scale (a medical doctor), the maximum level of career resources is reached after about 9–10 years and has a value of 8.15.

Change of Jobs. If a woman changes from a job with prestige level P_0 to a job with prestige level $P_h > P_0$ at time t_1, which is an *upward move*, her career resources will increase until the maximum career level of the new job is reached. In this case the career function for $t > t_1$ is[31]

$$K(P_h,t) = \min\big\{K(P_0,t_1) + K(P_h,t-t_1), \max(K(P_h,t))\big\}.$$

If a woman moves from a job with prestige level P_0 to a job with a prestige level $P_n < P_0$ at time t_2 (a *downward move*), the career resources of the previous job decrease linearly over time and the career resources of the subsequent job increase over time. However, the maximum career level of the successive job is considered to be the lower limit. Thus, with increasing time, the level of career resources decreases and

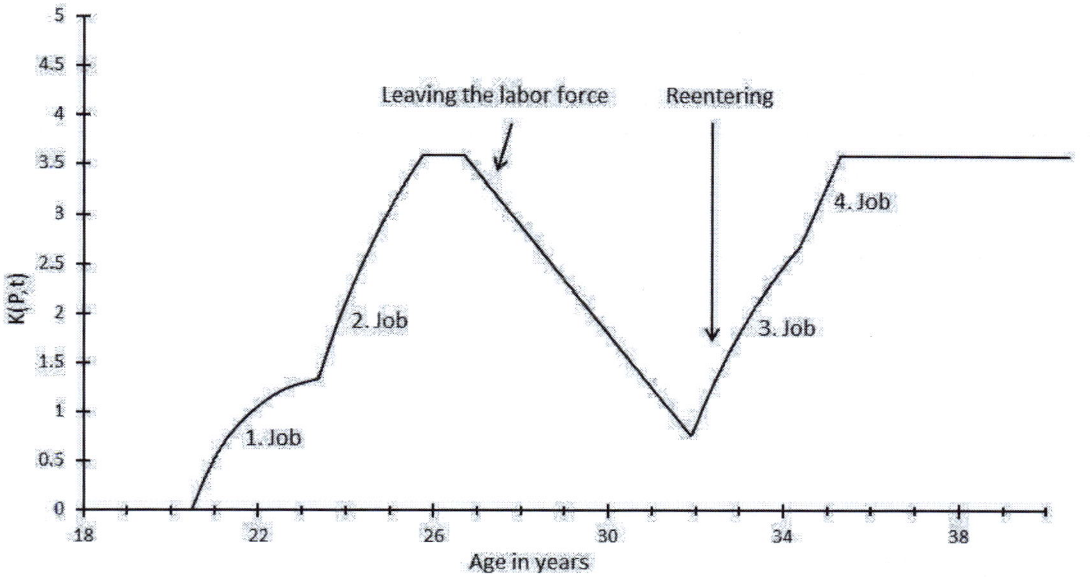

Figure 6.6.3 Career resources over the life course: an occupational career with a phase of nonemployment.

approaches the maximum career level of the successive job. For $t > t_2$, the level of career resources is as follows:

$$K(P_n, t) \;=\; \begin{cases} \min\left\{(1-(1.5/P_0)\,(t-t_2))\,K(P_0,t_2)+K(P_n,t-t_2),\right. \\ \qquad \left. K(P_0,t_2)\right\} \quad \text{if} \quad t-t_2 < P_0/1.5 \\ K(P_n, t-t_2) \quad \text{otherwise.} \end{cases}$$

Discontinuity of Work Experience. Besides continuous changes of the level of career resources as a result of upward and downward moves, it must also be taken into account that women tend to enter and exit the labor market several times after leaving school due to family events (marriage, birth of children, etc.; Mincer and Polachek 1974). Given this discontinuity of work experience, the usual assumption for the career process of men in labor-market research (Sørensen 1977) that career resources increase monotonically with decreasing increments over the work life is no longer valid. When women interrupt their employment careers, they lose career resources that have to be accumulated again after reentry into the labor force. In order to model the path of labor force experience of women, Blossfeld and Huinink assumed that career resources decline when women interrupt (I) their employment career at age t_3 as long as women's career resources are still positive. The speed of the decrease depends on the prestige level (P_0) of the job held immediately before the interruption of the career. For $t > t_3$ one gets:

$$K(I, t) = \max\left\{0, 1 - \frac{1.5}{P_0}\,(t-t_3)\,K(P_0, t_3)\right\}.$$

Figure 6.6.3 shows an example for a trajectory of career resources including an upward move (from Job 1 to Job 2), a work interruption, and a reentry into a third job.

The goal of the transition rate analysis in this research was to specify the rates of entry into marriage or motherhood $r(t)$ as a function of time-constant (X_1) and time-dependent covariates ($X_2(t)$) in an exponential model:

$$r(t) = \exp\left(X_1\,\beta_1 + X_2(t)\,\beta_2\right).$$

Table 6.6.2 Estimates for models of the rate of entry into marriage (women of cohorts 1929–31, 1939–41, and 1949–51).

	Model						
Variables	1	2	3	4	5	6	7
Intercept	-4.69*	-17.62*	-17.58*	-17.68*	- 16.28*	- 16.28*	-16.04*
Log (current age - 15)		1.76*	1.80*	1.73*	1.46*	1.47*	1.45*
Log (45 - current age)		3.20*	3.27*	3.37*	3.09*	3.09*	2.93*
Number of siblings			-0.00	0.01	0.00	0.00	-0.01
Father's social class 2			-0.13	-0.14	-0.08	-0.08	-0.04
Father's social class 3			-0.31*	-0.31*	-0.14	-0.14	-0.06
Father's social class 4			-0.61*	-0.61*	-0.33*	-0.32*	-0.25
Urban residence at age 15			-0.08	-0.11	-0.05	-0.05	-0.07
Cohort 1939–41				-0.04	-0.04	-0.03	-0.05
Cohort 1949–51				-0.09	-0.01	-0.00	0.00
Economic development				0.19*	0.21*	0.21*	0.18*
In training (dynamic measure)					-0.97*	-0.99*	-0.80*
Level of education (dynamic measure)					-0.01	-0.00	0.00
Level of career resources (dynamic measure)						-0.03	0.04
Is pregnant (dynamic measure)							2.84*
Subepisodes	85404	85404	85404	85404	85404	85404	85404
χ^2		457.80	485.55	525.01	598.39	598.47	1085.47
df		2	7	10	12	13	14

* Statistically significant at 0.05 level.

In this model, observation begins at age 15 and ends with the event of first marriage or the birth of the first child; or, for right-censored cases, with the date of the interview or age 45—whichever is earlier.

A combination of two variables was used to control for the well-known nonmonotonic age dependence of the marriage rate and the rate of the first birth (Coale 1971; Bloom 1982). This approach considers women at risk of entering first marriage and having a first child between the ages of 15 and 45 (i is an index for the ith one-month interval):

$$\begin{aligned} \log(D_i) &= \log(\text{current age} - 15) \\ \log(R_i) &= \log(45 - \text{current age}). \end{aligned}$$

Including these variables in the exponential model as time-dependent covariates,

$$\exp\big(\log(D_i)\,\beta' + \log(R_i)\,\beta''\big) = D_i^{\beta'}\,R_i^{\beta''},$$

models the typical bell-shaped curve of the rates of entry into first marriage and first motherhood. This curve is symmetric around the age of 30 for $\beta' = \beta''$, left-skewed for $\beta' < \beta''$, and right-skewed for $\beta' > \beta''$.

The first marriage and the first childbirth are interdependent processes and form a dynamic system. Premarital conception increases women's willingness to marry, and marriage increases the risk of childbirth. Therefore, Blossfeld and Huinink included time-dependent dummy variables for being pregnant in the marriage model and for being married in the first-birth model.

To control for cohort and period effects of historical and economic developments on family decisions, Blossfeld and Huinink introduced two different types of variables. First, they used two dummy variables for the three birth cohorts (reference group = cohort 1929–31) to measure differences among cohorts. Second, they used a variable that reflects the continuous development of labor market conditions as a period measure (see Figure 6.6.1 in the previous example).

To include all these various time-dependent covariates in the rate equation, Blossfeld and Huinink (1991) applied the method of episode splitting, as described previously. As time-constant background variables, they incorporated father's social class, place of residence at age 15 (urban vs. rural, with rural as the reference category), number of siblings, and the educational level of the partner. We cannot go into the details of this sophisticated dynamic analysis here. We merely demonstrate the strength of using time-dependent covariates by reporting the main findings of this study.

Table 6.6.3 Estimates for models of the rate of entry into motherhood (women of cohorts 1929–31, 1939–41, and 1949–51).

Variables	Model 2	3	4	5	6	7	8
Intercept	-19.33*	-18.82*	-18.59*	-17.64*	-16.03*	-16.72*	-14.21*
Log (curr. age - 15)	2.17*	2.19*	2.08*	2.11*	1.68*	1.76*	0.08
Log (45 - curr. age)	3.36*	3.30*	3.33*	3.28*	2.84*	2.95*	2.24*
Number of siblings		0.04*	0.05*	0.04*	0.04*	0.04*	0.09*
Father's social class 2		-0.10	-0.10	-0.03	-0.04	-0.01	0.12
Father's social class 3		-0.34*	-0.34*	-0.21	-0.19	-0.18	-0.03
Father's social class 4		-0.45*	-0.46*	-0.13	-0.08	-0.05	0.16
Urban residence at age 15		-0.23*	-0.47*	-0.18*	-0.17*	-0.18*	-0.23*
Cohort 1939–41			-0.11	-0.03	-0.07	-0.03	-0.22
Cohort 1949–51			-0.16	-0.02	-0.05	0.03	-0.57*
Economic development			0.20*	0.19*	0.20*	0.20*	0.07
In training (dynamic measure)					-1.98*	-2.24*	-1.32*
Level of education (dynamic measure)					0.05	0.08*	0.08*
Level of career resources (dynamic measure)						-0.39*	-0.18*
Married (dynamic measure)							3.82*
Subepisodes	99506	99506	99506	99506	99506	99506	99506
χ^2	480.04	518.86	547.14	579.51	674.56	695.86	1744.10
df	2	7	10	11	13	14	15

* Statistically significant at 0.05 level.

The first interesting aspect of this analysis is that the effects of the dummy variables for father's social class on entry into marriage and motherhood in Model 3 show that women from lower social classes marry earlier and have their babies earlier than women from higher social classes (see Tables 6.6.2 and 6.6.3). However, if the various time-dependent covariates for women's educational and career investments over the life course are included, this effect disappears completely. Thus, by extending education and improving career opportunities, families of higher social classes *indirectly* delay the rate of marriage and having children. In general, event history analysis provides the opportunity to study the importance of indirect effects that operate either by influencing the timing of entry into or exit from specific states as well as with regard to variations of time dependence within states (see also Yamaguchi 1991).

Model 4 (Tables 6.6.2 and 6.6.3) incorporates measures of changes in the historical background such as cohort membership and economic development. There are no significant cohort effects, but one observes a significant positive effect of economic conditions. This means that women marry earlier when the economic situation is favorable (see also similar results by Blossfeld 2022). In a period of economic expansion, the life-cycle prospects of young people are more predictable, making it easier for women to make long-term decisions such as entering into marriage and having a baby.

After controlling for age dependence (Model 2), social background (Model 3), and changes in the historical setting (Model 4), we can give an answer to the question of how important the improvements in educational and career opportunities were for women's marriage timing (Table 6.6.2). We first look at the *dynamic effects of education* in Model 5. This model shows that attending school, vocational training programs, or university actually has a strong negative effect on the marriage rate. What is very interesting, however, is that the effect of the *level of education* is not significant. Women's timing of marriage is therefore independent of the quantity of human capital investments. In assessing the consequences of educational expansion on family formation, one can conclude that marriage is postponed because women delay their transition from youth to adulthood and not because women acquire more human capital and thereby increase their labor force attachment.

In Model 6 of Table 6.6.2, the effect of women's improved career opportunities on the timing of their marriage can be assessed. This variable, which is of great theoretical interest, also proves to be insignificant. Women's entry into marriage appears to be independent of their career opportunities.

Finally, Model 7 (Table 6.6.2) includes a time-dependent pregnancy indicator. It does not change the substantive findings cited earlier, but its effect is positive and very strong. This indicates that the marriage rate rises sharply for women who experience a premarital pregnancy.

Let us now consider the estimates for first motherhood in Table 6.6.3. Again, age dependence, social background, historical period, cohort effects, and partner's educational attainment are controlled for in the first five models. In Model 6, women's continuously changing level of education and an indicator for their participation in school are included to explain the rate of entry into motherhood. As in the marriage model, the level of education, which measures women's general human capital investments, has no significant effect on the timing of the first birth. Only school attendance has a negative effect on women's rate of having a first child. This means that there are temporal conflicts between the role of women as students and mothers (Rindfuss and John 1983) and that there are normative expectations that young women who attend school should not enter motherhood.

When the changes in career resources of women over the life course are introduced in Model 7 of Table 6.6.3, the effect of educational level proves to be significantly positive. This is contrary to the expectations of the economic approach to the family, and means that the process of attaining successively higher levels of qualification has a reinforcing rather than a diminishing effect on the rate of having a first child. The reason for this is that the attainment of increasing levels of education takes time and is connected with an increasing age of women (Figure 6.6.2). Women who remain in school longer and attain high qualifications are subject not only to pressure from potential medical problems associated with late childbearing, but also from societal age norms ("women should have their first child at least by age 30"; Menken 1985). Thus, it is not human capital investments, as claimed by the new home economics, but increasing social pressure that may be at work if the level of education has an impact on the timing of the first birth.

These relationships are also illustrated in Figure 6.6.4.[32] This figure reports the estimates of the age-specific cumulative proportion of childless women (survivor function) for different levels of education.[33] The longer women are in school, the more their first birth is delayed; this is why there is a high proportion of childless women among the highly educated. After leaving the educational system, women who have delayed childbearing "catch up" with their contemporaries who are less educated and started earlier. However, they do not just catch up. The positive effect of the educational level leads to an increase in the proportion of childless women with an upper-secondary school qualification (at about age 20) and even those with university degrees (at about age 27) below the proportion of childless women with lower school qualifications.

However, confirmation of the economic approach to the family can be seen in the negative effect of the *level of career resources* on the rate of entry into motherhood (Model 7, Table 6.6.3). The accumulation of women's career resources is at odds with societal expectations of women's role as mothers. Women still bear the primary responsibility for child care and are still disadvantaged in their careers if they have to interrupt their working life to have a child. Therefore, women who have accumulated a high stock of career resources try to postpone or even avoid the birth of a first child.

Figure 6.6.5 displays examples of age-specific, cumulated proportions of childless women (survivor function) for ideal-typical career lines.[34] This example shows that there is a conflict between career and motherhood in Germany. An increase in career opportunities augments the proportion of childless women at any age.

Finally, in Model 8 (Table 6.6.3), Blossfeld and Huinink introduced a time-dependent dummy variable that changes its value at marriage and thus shows whether or not a woman was married before the birth of her first child. This variable significantly increases the rate of entry into motherhood. When this variable is introduced, we can observe that the effects of "in training" and "level of career resources" become weaker. Part of their influence is therefore mediated by the marriage process (see also Blossfeld and De Rose 1992; Blossfeld and Jaenichen 1992; Blossfeld 2022 and, for an international comparison of these models, Blossfeld 1995).

In summarizing these results, it is interesting to note that empirical support for the new home economics has typically been claimed on the basis of cross-sectional and aggregated time-series data. However, these data do not permit a differentiation between the effect of accumulation of human capital over the life course and the effect of participation in the educational system in keeping women out of the marriage market. Therefore, it seems that such empirical support for the new home economics is merely the result of inappropriate methods and the type of data used.

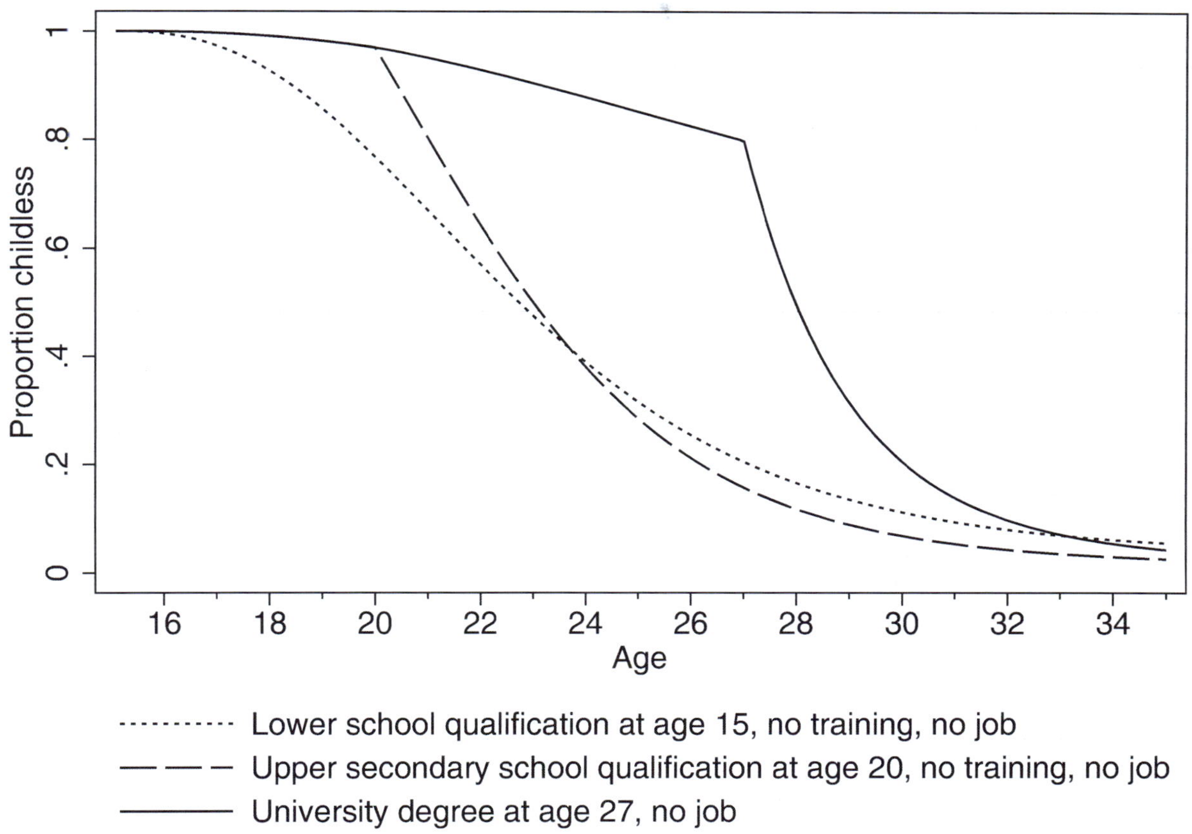

Figure 6.6.4 Estimated cumulative proportion of childless women by education (survivor function).

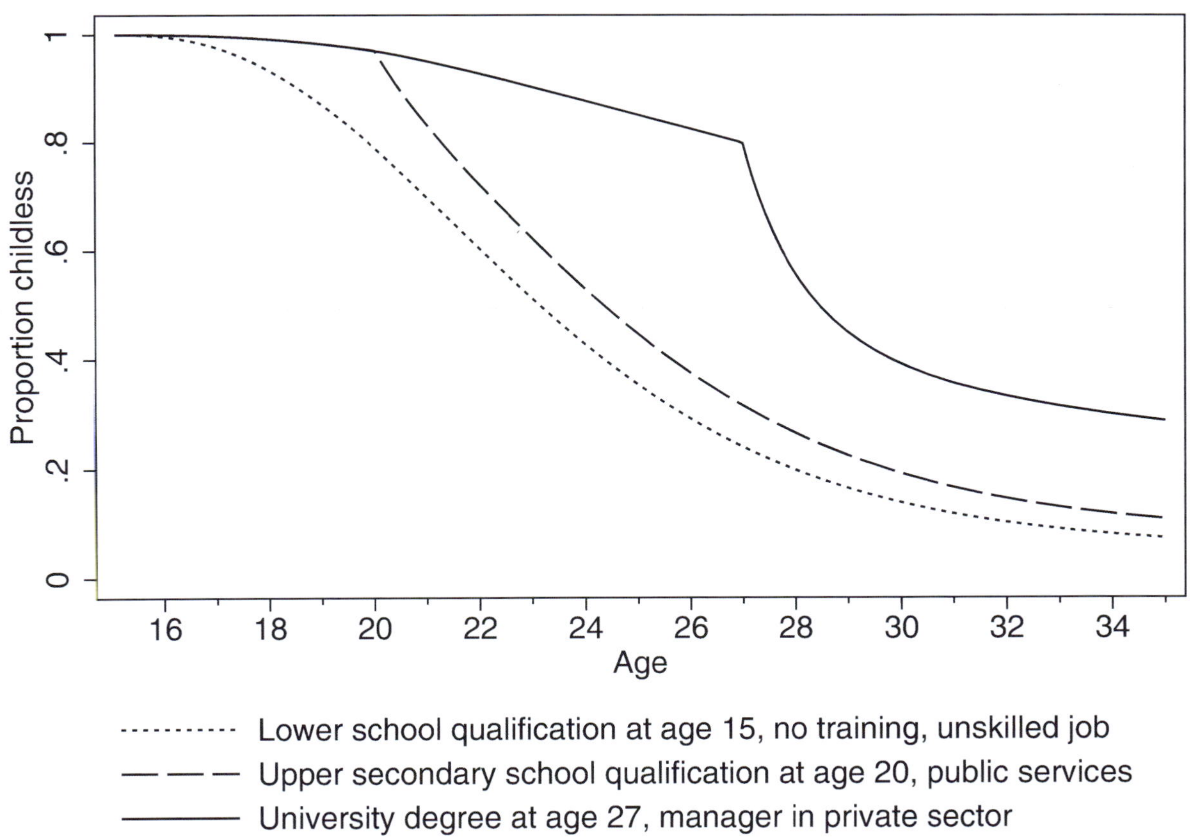

Figure 6.6.5 Estimated cumulative proportion of childless women for ideal-typical career lines (survivor function).

Example 3: The Effects of Pregnancy on Entry into Marriage for Couples Living in Consensual Unions

We finally report on two investigations by Blossfeld, Manting, and Rohwer (1993) and Blossfeld et al. (1999). These studies are instructive because they demonstrate how sensitive the effects of time-dependent covariates can be with respect to the points in time at which they are supposed to change their values.

In substantive terms, the purpose of the first study was to gain an insight into the relationship between consensual unions and marriage in West Germany and the Netherlands. The study focused on the effect of pregnancy on the rate of entry into marriage, controlling for other important covariates in a transition rate model.

Historically, marriage usually preceded the birth of a child, but in recent decades, the interplay between marriage and childbirth has become more complex. Some cohabiting couples wait until the woman gets pregnant and then marry. For other couples, an unplanned pregnancy can lead to a marriage that might not have happened otherwise. Pregnancy can also lead to a dissolution of the consensual union if there is no agreement as to whether a child is desired. Other couples, who wish to have children in the near future, may decide to marry before the woman becomes pregnant. Nationally representative longitudinal data were used to investigate these complex relationships. The first analysis by Blossfeld, Manting, and Rohwer (1993) used the German Socio-Economic Panel for West Germany and the Netherlands Fertility Survey. Both data sets provide important information about the dynamics of consensual unions in the 1980s. In each country, attention was limited to members of the cohorts born between 1950 and 1969 who started a consensual union between 1984 and 1989 in West Germany and between 1980 and 1988 in the Netherlands.[35]

However, we will only discuss the effects of pregnancy here. Pregnancy was included in the transition rate model as a series of time-dependent dummy variables (coded as centered effects)[36] with the following states: "not pregnant," "pregnant," "first childbirth," and "6 months after birth." As shown in Table 6.6.4, the effects of the pregnancy dummy variables are significant for both countries, and they basically work in the same direction. As long as women are not pregnant, a comparatively low rate of entry into marriage for people living in a consensual union is observed. But as soon as a woman becomes pregnant (and in West Germany, also around the time a woman has her child), the rate of entry into marriage increases strongly. Thus, there still seems to be a great desire among the young generations in West Germany and the Netherlands to avoid out-of-wedlock births and to legalize the union when a child is born. However, if the couple does not marry within a certain time (about 6 months) after the birth of the child, the marriage rate in West Germany falls back to a comparatively low level. In the Netherlands, this level is even below the "not pregnant" level.

Let us now look at the effects of pregnancy on the dissolution process of consensual unions, which are different in Germany and the Netherlands. In Germany, the dissolution risk is high as long as women are not pregnant. The rate drops strongly for some months when a woman is pregnant and then has her child (see also Blossfeld 2022). But 6 months after childbirth, the dissolution rate rises again. The strong rise 6 months after childbirth can also be observed in the Netherlands. In Germany and the Netherlands, women living in consensual unions with a non-marital child therefore not only have a comparatively low rate of entering into marriage but also a comparatively high dissolution risk. In other words, the rise in the number of consensual unions will certainly increase the number of female-headed single-parent families in both countries.

About a year after this comparative study, Blossfeld et al. (1999) wanted to examine whether these relationships could be reproduced with new data from Germany. They started to carry out some preliminary analyses and used only one dummy variable for the pregnancy process. However, the effect of this variable was nonsignificant. This was surprising, because results of the earlier comparative study were convincing in theoretical terms. However, after checking the input data and command files, the authors found that the programmer had accidentally switched the time-dependent dummy variable to the time of the birth of the child rather than the time when it was clear that conception had occurred (i.e., about 8 to 6 months before birth). Thus, a shift in the switch of a time-dependent covariate by 6 months made the effect of pregnancy disappear. This "finding" created a lot of confusion in the research group. What had happened to the pregnancy effect? After much discussion, one explanation seemed to unite theory and the seemingly contradictory results of the estimated models: The effect of pregnancy on entry into marriage must be strongly time-dependent. It must start to rise at some time shortly after conception, increase to a maximum during pregnancy, and then decrease again. Thus, when the time-dependent covariate was switched to the time of conception, the effect was strongly positive, because it compared the situation before the discovery of conception (which accumulates a period with a low marriage rate) with the situation after the discovery of conception (which accumulates a high marriage rate for some months). But when the time-dependent covariate was switched to the time of childbirth, a period with a low marriage rate up to the time of discovery of conception was mixed with a period with a high marriage rate during pregnancy (see Table 6.6.4). Thus the average tendency to marry before the birth of the child roughly corresponds to the average

Table 6.6.4 Piecewise constant exponential model for transitions from consensual unions to marriage and dissolution for West Germany (FRG) and the Netherlands (NL).

Variable	Entry into marriage FRG[3]	Entry into marriage NL[4]	Dissolution FRG[3]	Dissolution NL[4]
Constant	-2.79**	-4.01**	-10.60**	-4.92**
Duration				
Less than 2 years	0.08	-0.01	-0.49**	-0.18**
More than 2 years	-0.08	0.01	0.49**	0.18**
Birth cohort				
1950–53[1]	-0.09	0.07	0.37	-0.19
1954–58	0.01	0.16*	0.22	-0.13
1959–63	0.11	0.00	-0.68*	-0.15
1964–69[2]	-0.03	-0.23**	-0.09	0.46**
School enrolment				
At school	-0.16*	-0.36**	-0.40	0.11
Not at school	0.16*	0.36**	0.40	-0.11
Educational level				
Low	-0.17*	0.14*	0.03	-0.08
Medium	-0.09	-0.07	0.29	0.10
High	0.26	-0.07	-0.32	-0.03
Fertility				
Not pregnant	-1.19**	-0.43**	5.48**	-0.09
Pregnant	1.13**	1.19**	-5.45	0.17
First child birth	1.21**	0.21	-4.75	-0.69
Six months after birth	-1.15**	-0.98**	4.72*	0.61*
Sex				
Male	-0.08		0.09	
Female	0.08		-0.09	
Married before				
No	0.07		-0.01	
Yes	-0.07		0.01	

* Statistically significant at 0.1 level. ** Statistically significant at 0.05 level.
1) For West Germany, the birth cohort of 1949 was also included.
2) For West Germany, the birth cohort of 1969 was also included.
3) Men and women. 4) Only women.

Table 6.6.5 Exponential model for transitions from consensual unions to marriage with a series of dummy variables defined as lags/leads with regard to the month of childbirth.

Variable	Coefficient	*p* value
Intercept	-4.4852	0.0000
Dummy: -9 months	1.4094	0.0161
Dummy: -8 months	1.9515	0.0000
Dummy: -7 months	1.7443	0.0006
Dummy: -6 months	1.4567	0.0129
Dummy: -5 months	1.0039	0.1507
Dummy: -4 months	1.3071	0.0257
Dummy: -3 months	0.2657	0.7015
Dummy: -2 months	1.8605	0.0000
Dummy: -1 months	-8.4526	0.0145
Dummy: 0 months	0.2657	0.7015
Dummy: 1 months	-8.4526	0.9114
Dummy: 2 months	0.8743	0.2208
Dummy: 3 months	-0.3593	0.0881

tendency to marry after the child is born, and the estimated coefficient for the time-dependent covariate "childbirth" does not deviate significantly from zero. There is, of course, a simple way to test this hypothesis. Blossfeld et al. (1999) used a series of time-dependent dummy variables, each indicating one month since the occurrence of conception. And in fact, as shown in Table 6.6.5, the effects of the time-dependent dummy variables at first increase, reach a maximum at about 8 months (i.e., a month before the birth of a child), and then decrease. Thus, starting with conception, there is an increasing normative pressure to avoid an out-of-wedlock birth, which increases the marriage rate, particularly for people who are already "ready to marry." However, as the number of marriages increases, the composition of the group of still unmarried couples shifts toward couples being "less ready for marriage" or being "not ready for marriage," which, of course, decreases the pregnancy effect again.

This is not only an important result in substantive terms. The methodological lesson is also very revealing. One should be very careful in modeling qualitative parallel processes with only one dummy variable, particularly in situations where it is theoretically quite open at which point in time the value of the dummy variable must be switched (see also Blossfeld and Mills 2001; Mills and Trovato 2001).

Example 4: An Analysis of Diffusion of Cohabitation Among Young Women in Europe with Individual-Level Event History Models

Nazio and Blossfeld (2003) used methods of event history analysis with episode splitting to study the diffusion of cohabitation across successive generations of young women in Italy, capitalist West Germany, and socialist East Germany. In particular, they were interested in the normative shifts regarding consensual unions which have evolved over the last 30 years from a rare and deviant form of partnership to a common and socially accepted union. In their analysis, diffusion was considered as an individual-level process by which the practice of cohabitation is communicated through certain channels and adopted or rejected over time among the members of a society (Rogers 1983). The focus of the analysis was on cohabitation before first marriage (if at all) in order to better understand the dynamic changes during the family formation phase in the historical period. In the 1960s, cohabitation was clearly a social innovation. It was perceived as relatively new by most people in Europe, at least in the phase of family formation. This form of union then gradually became integrated into the process of family formation in varying degrees in most Northern European countries. Nazio and Blossfeld wanted to answer the question of what is driving the diffusion of consensual unions among young women in Italy, West Germany, and East Germany; and if there is convergence or divergence in this process between countries, they wanted to know what forces are responsible.

Using longitudinal data from the Fertility and Family Surveys (see Klijzing and Corijn 2001), Nazio and Blossfeld adopted an individual-level diffusion model (see Strang and Tuma 1993). In these models, the individual woman's rate of adoption of cohabitation can be estimated, among other factors, as a function of prior adoptions by other individuals in the social system. In methodological terms, these models are particularly attractive because they allow (1) to estimate a flexible individual-level analogue of the relatively limited standard population-level models of diffusion; (2) to incorporate the influence of time-constant and time-varying individual heterogeneity affecting the intrinsic propensity of women to adopt the cohabitation practice in different stages of their life course; (3) to take into account ideas about structures of communication, knowledge awareness, and structural equivalence; and (4) to estimate more complex models of temporal variation in the process of diffusion.

The diffusion of cohabitation among young women is a highly complex time-related process (see Figure 6.6.6). A characteristic of cohabitation before entry (if at all) into marriage is that the time span of potential adoption for each generation is highly concentrated on the period of transition from youth to adulthood. There is then a continuous succession of birth cohorts over time moving through this life-course window (see Figure 6.6.6). Earlier research has shown that in modern societies, the readiness of young women to form marital or nonmarital unions over the life course is governed to a large extent by organizational rules and institutional structures in the education and employment systems (Blossfeld 1995). At specific ages, women typically move from one institutional domain to another (e.g., from secondary school to vocational training, or from school to the labor market), and these transitions often serve as markers for the beginning of a life stage in which women form partnerships (Blossfeld and Nuthmann 1990; Huinink 1995; Klijzing and Corijn 2001; Blossfeld et al. 2005, Blossfeld 2022). It is well-known that finishing women's education counts as one of the most important transitions in the process of getting ready for entry into marriage (see Example 1; Blossfeld and Huinink 1991; Blossfeld 1995).

Figure 6.6.6 presents a stylized picture of the complex dynamics involved in the diffusion of premarital cohabitation among young women. There is a continuous inflow of birth cohorts entering into the life stage "ready for partnership formation" and thus becoming members of the risk set of potential adopters. At the same time, there is a continuous outflow of women from this risk set, not only because some young women adopt cohabitation but also because others marry and therefore also leave the risk set. In Figure 6.6.6, dotted lines describe the inflow to and outflow from the risk set given the variable ages at which these entries and exits take place. In other words, in the case of premarital cohabitation, potential adoption is typically limited to a specific window in the life course, and the population of potential adopters is highly dynamic over time. These idiosyncrasies of the adoption process have significant implications for the mechanisms driving the diffusion process of premartial cohabitation among young women over time.

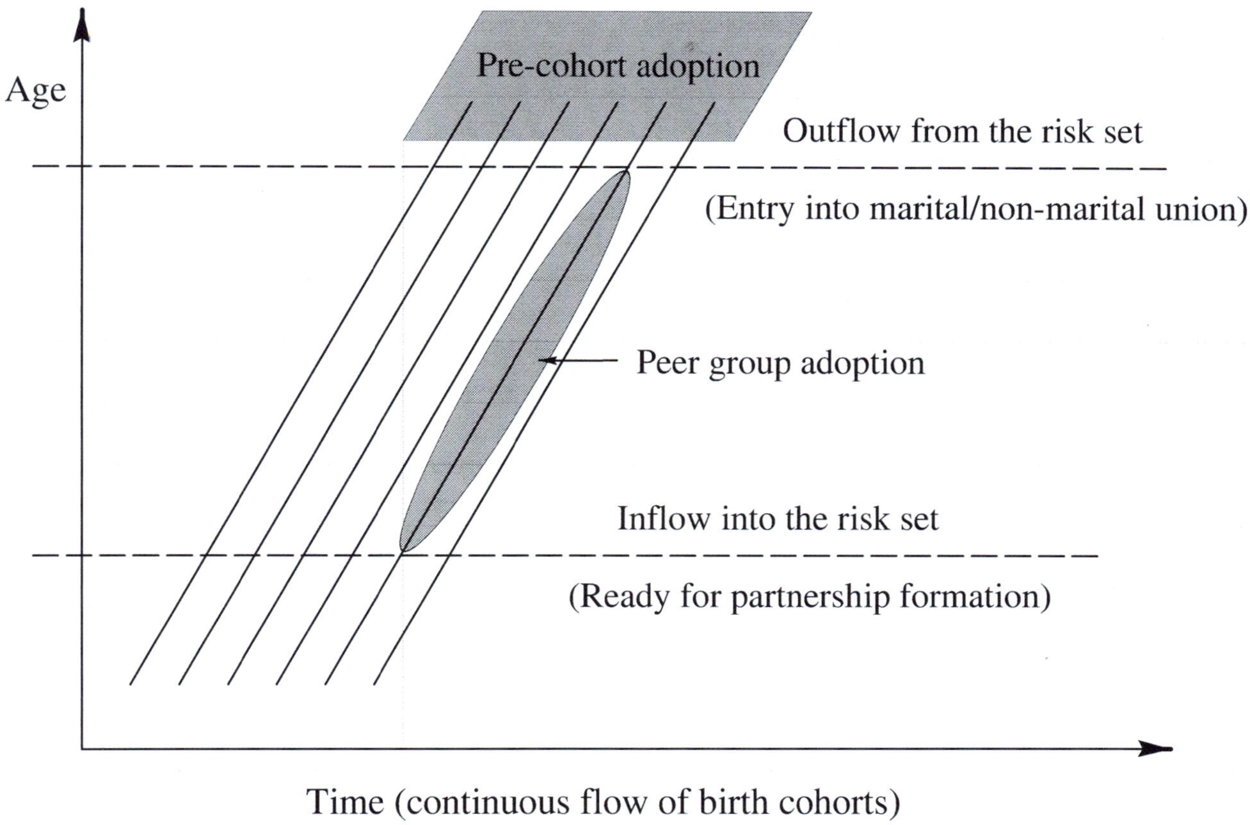

Figure 6.6.6 Time-related dimensions of the diffusion process of pre-marital cohabitation.

Each new cohort of women that enters into the phase of being ready for partnership formation will encounter an ever-increasing proportion of prior adopters from previous cohorts. Each new cohort of women will therefore gradually experience premarital cohabitation as a less deviant (or stigmatized) and more socially acceptable living arrangement right from the beginning. Newspapers, radio, television, and social media will increasingly spread knowledge awareness on the growing popularity of cohabitation among older birth cohorts and enhance social acceptability of nonmarital cohabitation. Nazio and Blossfeld (2003) therefore expected that the cumulative proportion of cohabitation adopters from previous cohorts would have a positive effect on the transmission of cohabitation to subsequent generations of women. This mechanism is illustrated in the trapezoidal area (Precohort adoption) in Figure 6.6.6.

However, young women often need to confirm their beliefs about cohabitation through more direct experience. They have to be persuaded by further evaluative information about the actual benefits and possible disadvantages of cohabitation through more concrete examples. These examples are most convincing when they come from other individuals like themselves who have previously adopted the innovation, and whose experiences can constitute a sort of vicarious trial for the newcomers (Strang 1991; Kohler 2001). It is therefore not only a question of conversation and personal contact with near peers, but also the perception of the practice proper for an individual of their position within the social structure (structural equivalence; see Burt 1987). Cohabitation of peers is therefore likely to constitute a particularly valuable trial example of the new living arrangement. This suggests that at the heart of the diffusion process, there is direct social modeling by peers who have already adopted the new way of life. This mechanism is represented in the oval area (Peer group adoption) in Figure 6.6.6.

Blossfeld and Nazio's description of the diffusion process showed that in East and West Germany, each successive birth cohort experienced not only an impressive rise in the proportions of cumulative precohort adoption (see Figure 6.6.7; a colored version of this figure can be downloaded from www.routledge.com/9781032657783) but also a steep increase in the cumulative proportions of peer group adoption (see Figure 6.6.8; a colored version of this figure can be downloaded from www.routledge.com/9781032657783) at each age. This suggests that the social acceptance of cohabitation has increased for each younger birth cohort in Germany to the extent that cohabitation has become a normal form of partnership in the process of starting a family. Among the youngest birth cohorts, about 50 % of women in West Germany and about 40 % in East Germany have adopted cohabitation before they eventually start a first marriage. In contrast, in Italy, even among the youngest birth cohorts, no more than about 10 % of women have adopted cohabitation before finally entering into a first marriage (see Figures 6.6.7 and 6.6.8).

Figure 6.6.7 Cumulative pre-cohort adoption in West Germany, East Germany and Italy

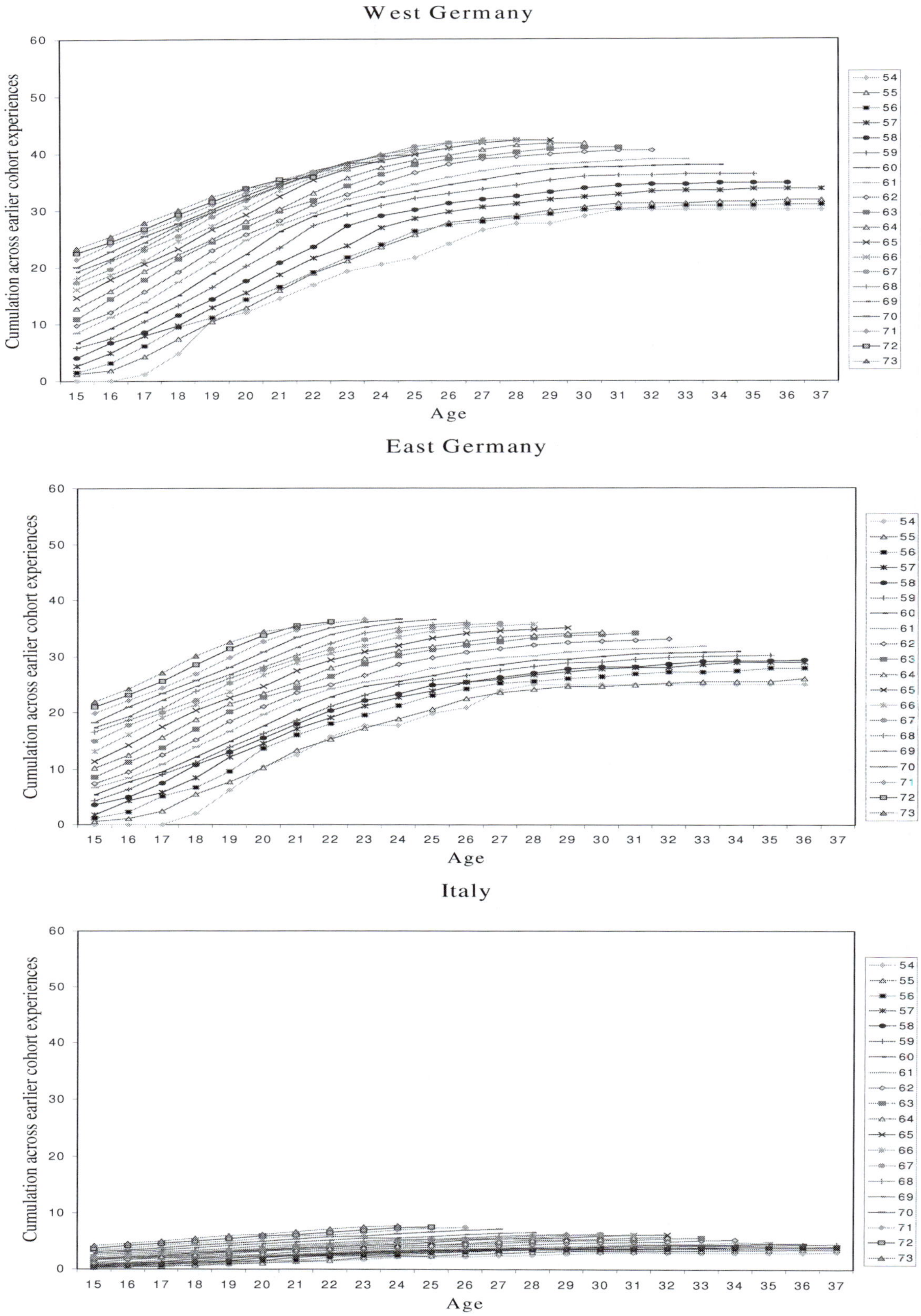

Figure 6.6.8 Cumulative peer group adoption in West Germany, East Germany and Italy

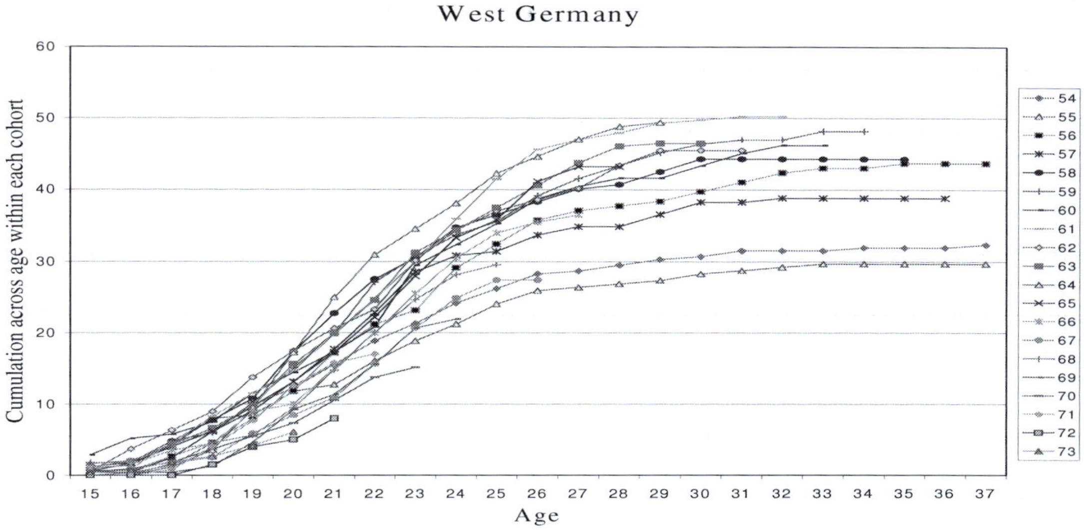

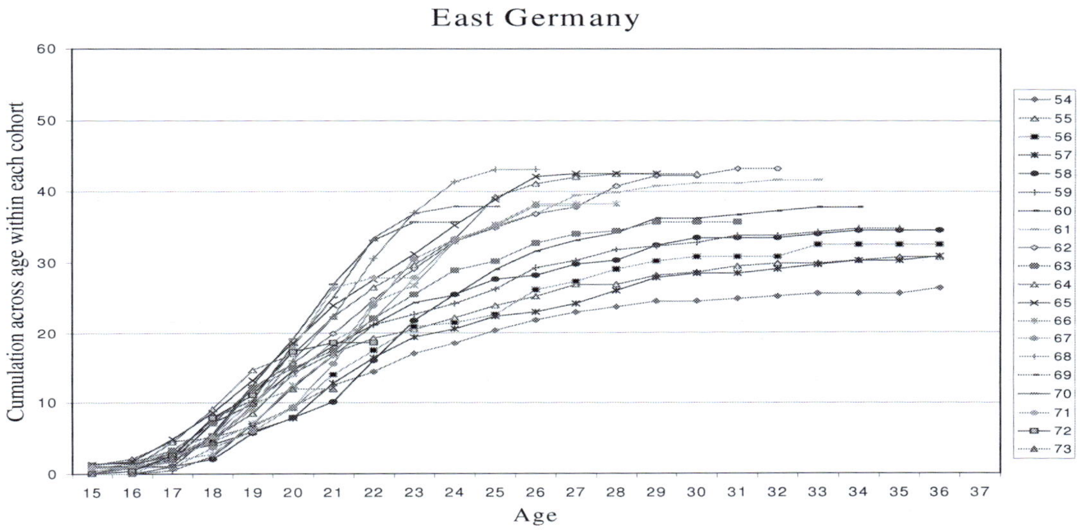

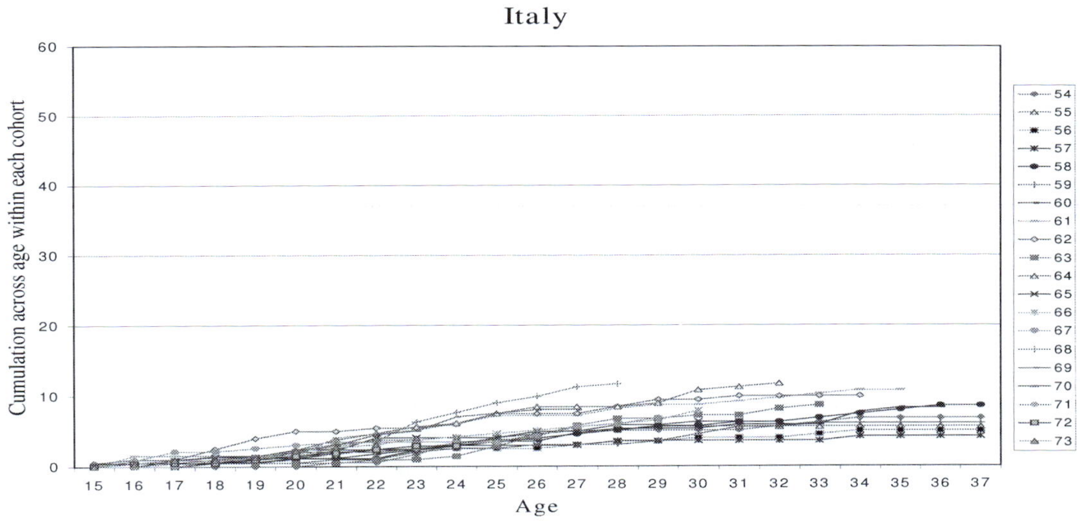

In the past, diffusion processes have generally been formulated in terms of population-level epidemic models (Diekmann 1989, 1992). This type of analysis assumes that all members of the population have the same chance of influencing and being influenced by each other (Strang and Tuma 1993). However, the assumption of homogeneous mixing often does not hold in empirical applications (Strang 1991). Therefore, Nazio and Blossfeld (2003) did not estimate a population-level model but turned to an individual-level model of diffusion as suggested by Strang and Tuma (1993). In these event history models, the individual's rate of adoption of cohabitation can be estimated as a function of prior adoptions by other actors in the social system. In methodological terms, these models are particularly attractive for studying the diffusion of cohabitation, because they allow the researcher to incorporate the effects of time-constant and time-varying individual heterogeneity on the intrinsic propensity of women to adopt cohabitation. They can also be used to test ideas about structures of communication and structural equivalence (see Strang and Tuma 1993) or, more precisely, the effects of knowledge awareness and direct social modeling. A simple individual-level based diffusion model might be formulated in the following way (see Strang and Tuma 1993):

$$r_n(t) = \exp\left(\alpha + \Sigma_{s \in G(t)} \beta\right)$$

where $r_n(t)$ is the propensity that an individual moves from nonadoption to adoption at time t, α represents the effect of individual characteristics, $G(t)$ consists of the theoretically relevant set of previous adopters, and β is the effect of the intrapopulation diffusion process on the rate of individual adoption. Thus, this model combines both individual heterogeneity and the contagious influences of previous adopters on nonadopters, and allows us to model diffusion within an event history framework.

Since the readiness to enter a union is highly time-dependent and governed to a large extent by women's age and by organizational rules and institutional structures of the education and employment systems, Nazio and Blossfeld controlled for the intrinsic propensity of women with a series of time-constant and time-dependent covariates. In particular, they took into consideration women's changing age, their time-dependent enrollment in the education system, the associated progressive upgrading of their educational attainment levels, and their changing employment participation. The α-term defined in the formula above is therefore replaced by $\alpha' x(t)$ in the estimated rate equation, incorporating time-constant and time-dependent individual's heterogeneity on women's likelihood to adopt cohabitation.

Since Nazio and Blossfeld assumed that knowledge awareness and direct social modeling are the two driving mechanisms of diffusion, knowledge awareness (P_c) was measured at each point in time by the cumulative proportion of prior adopters from previous birth cohorts at each age, and direct social modeling (P_g) was measured as the cumulative proportion of prior adopters belonging to the women's own birth cohort at each age (see Figures 6.6.7 and 6.6.8):

$$P_g = \frac{\Sigma_{i=c} \Sigma_{j<t} m_{ij}}{N_c} \cdot 100 \qquad P_c = \frac{\Sigma_{i<c} \Sigma_{j<t} n_{ij}}{N_p(t)} \cdot 100$$

where c indicates the birth cohort, t is woman's age, m_{ij} is the number of prior adopters within the woman's own birth cohort at age t, N_c is the total number of women in the woman's own birth cohort, n_{ij} is the number of prior adopters among older birth cohorts at age t, and $N_p(t)$ is the number of women belonging to older birth cohorts at age t. In order to allow for the nonlinearity in the relationships, Nazio and Blossfeld used a third-degree polynomial and replaced the A-term of the diffusion formula above by the following combination of these two factors:

$$\begin{aligned} r_n(t) = {} & \exp(\alpha' x(t)) \cdot \\ & \exp(\beta_1 P_g + \beta_2 P_g^2 + \beta_3 P_g^3) \exp(\gamma_1 P_c + \gamma_2 P_c^2 + \gamma_3 P_c^3). \end{aligned}$$

This individual-level diffusion model does not require the assumption that diffusion occurs only through interpersonal contacts. Thus, it is not only direct conversation and immediate personal contacts to near peers that count but also the perception of the practice proper for people at the same age (structural equivalence; see Burt 1987). There is accordingly no assumption of a complete mixing of social members that would imply that there is a complete, pairwise interaction between prior and potential adopters. Rather, in defining the diffusion covariates in the model, a distinction is made between individuals on the basis of their age and birth cohorts. Nazio and Blossfeld's model also does not assume that there is a constant and permanent ceiling on the number of potential adopters in the social system. The number of potential adopters is indeed changing over time, as there is a continuous inflow of new birth cohorts of women entering into the risk set. In addition, there is no implicit assumption that the innovative behavior does not change its meaning over the diffusion process. Instead, Nazio and Blossfeld assumed that the meaning and character of cohabitation is subject to change, and that knowledge awareness and direct social modeling might change their effects over time.

The application of event history methods to the data of the Fertility and Family Surveys in Italy, West Germany, and East Germany yielded interesting differences with regard to the diffusion of cohabitation. In many countries, the growing tendency of young adults to opt for informal cohabitation instead of marriage seems to be a rational response to the new challenges of a deteriorating labor market for young adults. Cohabitation offers the opportunity to postpone long-term commitments without incurring the penalties of heterosexual isolation or promiscuity. It also offers the pooling of resources and the economies of scale that living together provides.

Nazio and Blossfeld's empirical analysis shows that young people can choose between various alternatives: They might stay longer in the parental home (mainly in Italy), live in a single household (often the case in West Germany), get married, or cohabit. The degree to which cohabitation is perceived as more advantageous than its alternatives depends heavily on the cultural, structural, and political setting.

Nazio and Blossfeld's analysis shows that cohabitation in Italy is indeed not an attractive choice. First, young Italians have great difficulties in securing public housing. The rental market, which is squeezed by specific rental laws and distorted by a black housing market, is rather expensive in this country. Thus, the best choice for young people is often to live with their parents or to buy a house (see Kurz and Blossfeld 2004). However, buying a house not only requires a huge financial investment, but is also a long-term binding decision. For this reason, marital decisions would implicitly have very similar consequences for the life course of young Italians. Second, the Mediterranean welfare state provides only weak protection against the increasing labor market risks of the young generation, and the cohesive Mediterranean family is the relevant locus of social assistance: Parents are responsible for their children and vice versa. In other words, increasing youth unemployment and uncertainties of employment relationships together with the idiosyncrasies of the Italian housing market and Catholic familialism make it more attractive for young people to stay in the parental home for longer than cohabitation or living as a single person. As Nazio and Blossfeld's longitudinal analysis shows, cohabitation in Italy is therefore limited to small, highly selective groups of women who have a good reason to break with traditional gender roles and family models (see the results in Table 6.6.6). As a rule, these women are not religious, have left the educational system and are working, have grown up in an urban environment, and, in the case of Italy, live mainly in the North.

In East Germany, the relatively low average age at marriage was the result of a comparatively high level of individual life-course predictability in the socialist society and a consequence of a specific housing allocation policy. In the historical period between the mid-1970s and 1989, there was a strong incentive for young women to adopt cohabitation, even when they were still in school. After the breakdown of the socialist society in 1989, a historical period that could only be partially covered by the data, the institutional framework of West Germany was introduced in East Germany, and economic uncertainty and rising unemployment have increased dramatically. It is well known that these changes have resulted in rapidly declining nuptiality and fertility rates and increased the rate of cohabitation and out-of-wedlock births in East Germany.

In West Germany, the housing market has been accessible for young people for many decades until 2020. It was easy to rent a flat; and prices were, with the exception of a few expensive cities, generally affordable. In cross-national comparative terms, the proportion of homeowners in West Germany is also relatively low at 45 % (see Kurz and Blossfeld 2004). If young people do not work, they are normally supported by the conservative welfare state or by their parents. In this historical period, therefore, in the face of rising unemployment and increasing job insecurity, living together or living as a single person were attractive options in comparison to living with one's parents in West Germany.

An important result of Nazio and Blossfeld's study is that no autonomous cohort trend remains on women's rate of entry into cohabitation after controlling for cohort differences in educational participation and attainment levels as well as women's labor force participation. In addition, the diffusion analysis shows that the dissemination of abstract knowledge based on the cohabitation experiences of earlier generations does not seem to have any substantial influence on the diffusion of cohabitation in the population in the countries studied (see Table 6.6.6). Thus the diffusion of cohabitation does not imply a mechanism that links the experiences across generations. Rather, cohabitation seems to be driven mainly by direct social modeling of peers (see the significant estimates in Table 6.6.6). Concrete experiences of the same age group clearly constitute a sort of useful vicarious trial for potential adopters. In East and West Germany, the relationship between peer group adoption and the individual adoption rate is S-shaped and monotonically increasing (see Figure 6.6.9). Initially, each new cohort appears to begin within a few specific groups of early innovators who have a particular interest in adopting cohabitation. Then, with the gradual accumulation of peer-group experience, the diffusion spills over from these specific groups to broader sections of the entire birth cohort population. At the level of about 30 % (in West Germany) to 40 % (in East Germany) of cumulative peer group adoption, the effect of direct social modeling then increases even more rapidly. After these points, direct social modeling clearly becomes one of the most important driving forces for the individual adoption rate of cohabitation. Although East and West Germany show the same overall S-shaped diffusion pattern, the diffusion process is slower in East Germany and the acceleration phase is triggered at a

Table 6.6.6 Estimation of covariate effects on the diffusion of cohabitation for women (exponential model with time-constant and time-dependent covariates).

	West Germany	East Germany	Italy
Constant	-29.59**	-23.18**	-11.39**
Log(Age-15)	0.52**	0.68**	-0.25*
Log(39-Age)	3.79**	2.58**	0.45
Out of education	0.19*	-0.10	1.29**
(Ref. education: compulsory)	0	0	0
MOB	-0.20	-0.39**	–
HMB	0.50**	0.34*	–
MMB	0.68**	0.47**	–
ABI	0.44**	-0.29	–
FHS	0.22	0.82**	–
UNI	0.62**	0.14**	–
Education: secondary	–	–	-0.44**
Education: tertiary	–	–	0.08
Being employed	0.09	-0.05	0.32**
Religiosity	-0.20**	-0.23**	-1.10**
(Ref. residence: rural/small city)	0	0	0
Residence: middle-sized city	0.13*	0.09	-0.23
Residence: big city	0.01	0.00	0.29*
(Ref. region: North)	–	–	0
Region: Center	–	–	-0.38**
Region: South and Islands	–	–	-0.48**
Peer group adoption	0.15**	0.16**	1.22**
(Peer group adoption)^2	-0.00**	-0.01**	-0.18**
(Peer group adoption)^3	0.00**	0.00**	0.01**
Precohort adoption	-0.02	-0.03	0.69
(Precohort adoption)^2	0.00	0.00	-0.18
(Precohort adoption)^3	0.00	-0.00	0.01
Events	895	868	214
Log-likelihood	(-10321.4)	(-12461.6)	(-14626.0)
	-10014.8	-12100.8	-14493.6

* Statistically significant at 0.1 level. ** Statistically significant at 0.05 level.

comparatively later stage (see Figure 6.6.9). In socialist East Germany, there was obviously more resistance to the spread of cohabitation to the entire birth cohort population than in West Germany. In Italy, the diffusion of cohabitation to broader groups of the population appears to be blocked (see Figure 6.6.9). The social pioneers practicing cohabitation in Italy are very specific groups of people whose experiences obviously cannot serve as suitable models for their peers in other groups. Hence, the strengths-of-weak-ties mechanism does not work in the Italian diffusion process.

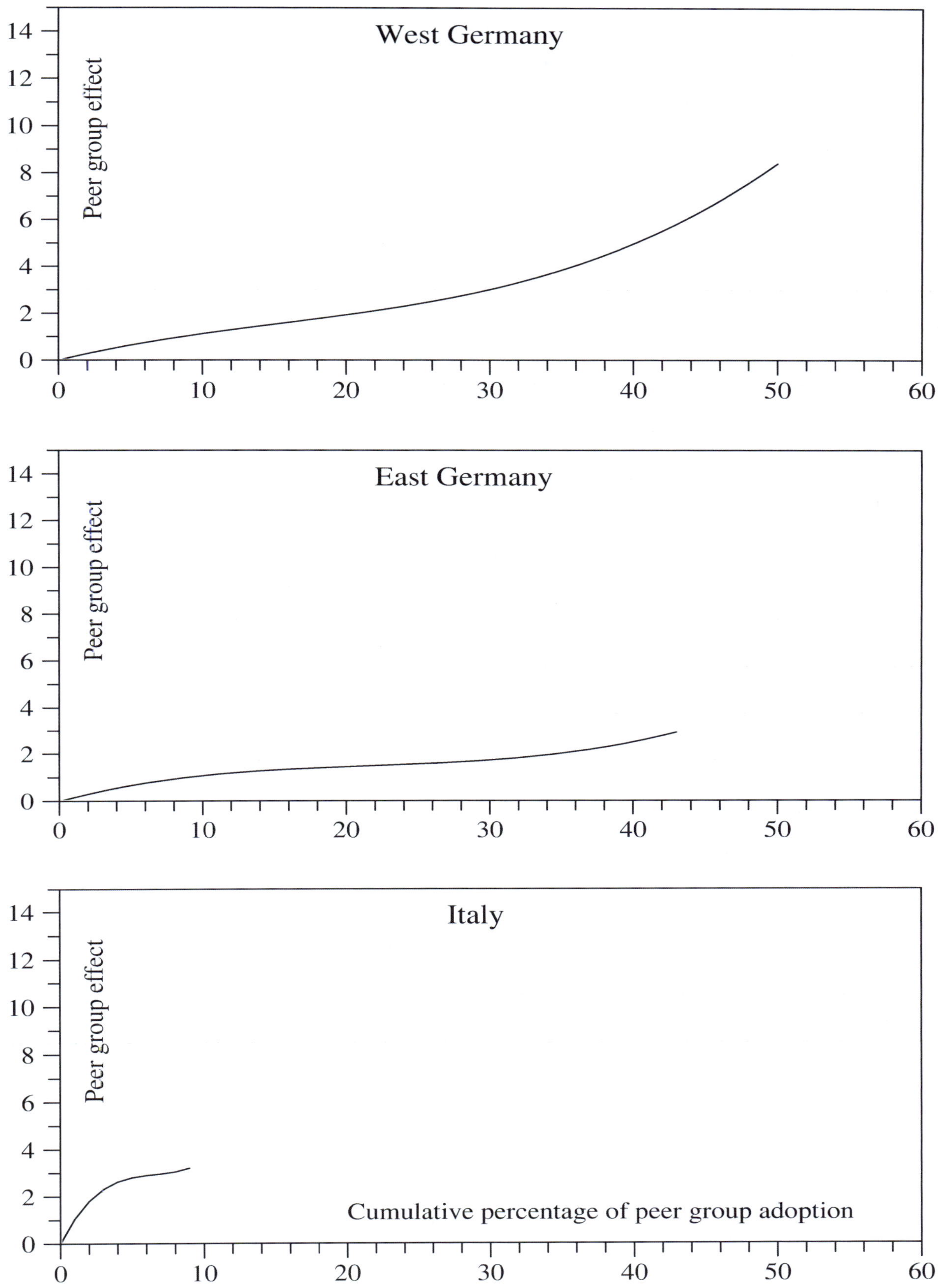

Figure 6.6.9 Effect of peer group adoption in West Germany, East Germany, and Italy.

Notes

[1]See, for example, Petersen 1986a, 1986b; Blossfeld, Hamerle, and Mayer 1989. The method of episode splitting can also be applied in parametric models of time dependence (see chapter 7) or semiparametric (Cox) models (see chapter 9). Since the logic of including time-dependent covariates is the same for all parametric models, we demonstrate this method only for exponential models.

[2]In this book, we focus only on continuous-time, discrete-state dependent processes. If you are interested in the analysis of discrete-time, discrete-state dependent processes see Blossfeld 2025.

[3]A complete history of state occupancies and times of changes is referred to as a *sample path* (Tuma and Hannan 1984).

[4]In other words, the value of the time-dependent covariate contains information about the state of the dependent process.

[5]As noted by Yamaguchi (1991), the effects of ancillary time-dependent covariates can lead to selection bias. For example, if regional unemployment rates or crime rates reflect the composition of the population in each region, a transition rate model will lead to biased estimates and erroneous conclusions if it does not account for (or control for) these differences.

[6]There are also other suggestions for analyzing dynamic systems based on this approach (e.g., Klijzing 1993).

[7]The basic model for the development of Y_t is a Markov model. It makes two assumptions: First, it assumes that the episodes defined with respect to Y_t are independent of previous history. Thus, if the history of the process makes the episodes dependent, it is crucial to include these dependencies as covariates in transition rate models at the beginning of each new episode of the Y_t process (Courgeau and Lelièvre 1992). Second, the model assumes that transitions to a destination state of the system are not allowed to depend on the episode's duration, but only on the type of states. However, this is not a necessary assumption. The model for the system could also be formulated as a semi-Markov model that takes into account the duration dependence in the various origin states.

[8]The number of distinct values L of the system variable Y_t is given by the product of the distinct values for each of the J variables. When the system is formed by J dichotomous variables, then the number of distinct values is $L = 2^J$. Of course, some of these combinations may not be possible and must then be excluded.

[9]If the modeling approach is based on a continuous time axis, this could be formally justified by the fact that the probability of simultaneous state changes is zero (see Coleman 1964).

[10]In this example, a problem could arise if the analysis is based only on observed behavior. For example, it could be that a couple first decides to marry, then, following this decision, the woman becomes pregnant, and finally the couple marries. In this case, we would observe pregnancy occurring before marriage and assume that pregnancy increases the probability of marriage. However, the time order between the processes is reversed: The couple decides to marry and then the woman gets pregnant. Since the time lag between decisions and behavior is probably not random and differs for various couples, an analysis using only behavioral observations may lead to incorrect conclusions. Courgeau and Lelièvre (1992) introduced the notion of "fuzzy time" for the time span between decisions and behavior. Note, however, that this problem does not alter the key temporal issues embedded within the causal logic (see section 1.2). There is clearly a time order with regard to decisions and behavior. However, as this example demonstrates, using only the temporal order of behavioral events without taking into account the timing of decisions could lead to serious misspecification.

[11]Courgeau and Lelièvre (1992) called this specific case "unilateral dependence" or "local dependence."

[12]Tuma and Hannan (1984) called this type of coupling between processes "cross-state dependence."

[13]In this respect, our approach to the analysis of interdependent systems differs significantly from the approach by Lillard (1993), Lillard and Waite (1993), Lillard, Brien, and Waite (1995); and Brien, Lillard, and Waite (1999) who estimate the hazard rate of a dependent process as a function of the *actual current state* of an independent process as well as on a (unobserved) *hazard rate*.

[14]This terminology is adapted from Gardner and Griffin (1986) and Pötter (1993).

[15]The mathematical steps that lead to this factorization are very simple in principle, but unfortunately require a complex terminology. The mathematical apparatus is therefore not given here. The mathematics can be found in Gardner and Griffin (1986), Pötter (1993), and Rohwer (1995). An important implication is that because not only the states, but also functions of time (e.g., duration) can be included conditionally, the distinction between state and rate dependence proposed by Tuma and Hannan (1984) loses its meaning (see Pötter 1993).

[16]Technically, thus, there was nothing wrong with the traditional approach that simply ignored the "feedback" effects and analyzed the impact of processes on the basis of time-dependent covariates as if they were external. However, from a theoretical perspective it was not necessary to "justify" (on theoretical grounds) or to "conclude" (from some preliminary empirical analyses) that a dependent process has only a small effect on the independent one(s) and that it can be ignored.

[17]$j_l = j$ for $l = 1, \ldots, L$; $k_l = j$ for $l = 1, \ldots, L-1$; and $k_L = k$.

[18]These formulations assume that there is only a single transition from a given origin state to one possible destination state. It is easy, however, to generalize the result for a situation with many possible destination states. The survivor function is then

$$\tilde{G}_k(t) = \prod_{l=1}^{L} \tilde{G}_k(t_l \mid s_l)$$

with conditional pseudosurvivor functions defined by

$$\tilde{G}_k(t_l \mid s_l) = \frac{\tilde{G}_k(t_l)}{\tilde{G}_k(s_l)} = \exp\left\{ -\int_{s_l}^{t_l} r_k(\tau)\, d\tau \right\}.$$

[19]Actually, this formula is also used by Stata.

[20]Consequently, using a data set of episode splits leads to identical estimation results if the same set of covariates is included in the model.

[21]The original variable `des` has received a system missing, which is equivalent to right censoring.

[22]Episode splitting in Stata is very flexible. If we assume that job changes are more likely around the date of marriage, we could immediately split at 3 months before and 6 months after marriage by putting -3 and 6 in the at-option instead of 0, that is: `stsplit ma_e, after(mdate) at(-3 6)`. The variable ma_e would contain three values: -4 if no marriage is in sight, -3 if the observed person in a current job is 3 months before and up to 6 months after marriage, and 6 if a person has been married for at least 6 months. In general, the new variable defined after `stplit` can contain all values specified in the `at`-option and, for all (sub)episodes prior to first splitting point, the new variable contains the lowest value in the `at`-option minus 1.

[23]$(\exp(-0.3447) - 1) \cdot 100\,\% \approx -29.2\,\%$.

[24]$(\exp(0.2735) - 1) \cdot 100\,\% \approx 31.5\,\%$.

[25]Event history analyses show that younger women change their employment behavior decreasingly at the time of marriage, but increasingly at the time of the birth of the first child. Thus, the important marker for employment has shifted across cohorts in the first family phase from marriage to the birth of the first child (see Blossfeld and Hakim 1997).

[26]$(\exp(0.2735 - 1.0231) - 1) \cdot 100\,\% \approx -52.7\,\%$.

[27]In these cases, the values of the causal variables are really known continuously; that is, at every moment in some interval (Tuma and Hannan 1984).

[28]"Continuous" measurement of quantitative processes usually means that variables are measured intermittently with a very small interval between measurements (see Tuma and Hannan 1984).

[29]Actually, as is demonstrated in section 7.2, we have only approximated a Gompertz model.

[30]These were (1) level of productivity, (2) national income per capita (deflated), (3) national income per economically active person (deflated), (4) private consumption (deflated), (5) proportion of expenditure on services in private consumption, (6) proportion of gainfully employed in public sector, (7) proportion of 13-year-old students attending German *Gymnasium* (upper secondary school), (8) proportion gainfully employed in service sector, (9) proportion of students in resident population, (10) proportion of civil servants in economically active population, (11) proportion of white-collar employees in economically active population, (12) unemployment rate, (13) proportion of gross national product invested in plant and equipment, and (14) proportion of registered vacancies of all jobs excluding self-employment.

[31]In the following formula, $\max(K(P_h, t))$ is the highest value that a woman can achieve in job h. The formula $K(P_h, t)$ then states that her resources correspond to a value that increases with time until the maximum level, $\max(K(Pt, t))$, is reached.

[32]Rate function coefficients and their standard errors are helpful in determining how women's educational and career investments influence first motherhood and first birth, in what direction, and at what level of significance. However, the magnitude of the effects and their substantive significance are easier to assess by examining the survivor functions for typical educational and occupational careers, which indicate the probability that a woman will remain unmarried or childless until age t.

[33]Figure 6.6.4 was obtained from the estimation of an exponential model that included an intercept and the time-dependent covariates log(current age - 15), log(45 - current age), "in training" and the "level of education". For details see Blossfeld and Huinink (1991).

[34]Figure 6.6.5 is the result of the estimation of an exponential model that included an intercept and the time-dependent covariates log(current age - 15), log(45 - current age), "in training", "level of education", and the "level of career resources." For details see Blossfeld and Huinink (1991).

[35]About 85 % of the entries into consensual unions between 1984 and 1989 were observed for these cohorts.

[36]Using effect coding for dummy variables, the differences between the levels of a qualitative variable are expressed as differences from a "virtual mean." The effect of the category chosen as the reference in the estimation can be computed as the negative sum of the effects of the dummy variables included in the model.

Chapter 7

Parametric Models of Time Dependence

In some sociological applications, substantive theory or previous empirical research suggests a specific shape of time dependence for the transition rate. If this time path can be modeled in terms of a tractable waiting time distribution, the likelihood can be calculated by using its density $f(t)$ and survivor function $G(t)$—as we have shown for the exponential model in section 4.1. An obvious estimation strategy is then to use any computer program or software package (e.g., BMDP, GAUSS, LIMDEP, SAS, TDA, R, Phyton, Stata) that offers function optimization routines and to maximize the likelihood function with respect to the unknown parameters.[1] This strategy is very flexible, as researchers can create their own parametric models, but it requires mathematical and programming skills. However, our experience shows that applied social scientists usually do not want to be bothered with writing programs and maximizing likelihood functions. Stata therefore offers the opportunity to estimate widely used parametric models with a single command. In Stata, `streg` performs maximum likelihood estimation for parametric event history models with

```
streg varlist, distribution(distname)
```

where `distname` is one of the possible distribution types shown in Box 7.1. The command can be used with single- or multiple-record or single- or multiple-failure data. You must specify the command `stset` before using `streg`. In the examples in this chapter, we assume that the transition rate depends only on the time spent in an episode and that these durations are independently and identically distributed. In other words, we study so-called *semi-Markov models* or *Markov renewal models* (see Tuma and Hannan 1984). In the case of discrete-time event history models, the estimation of parametric event history models is described in Blossfeld (2025).

In this chapter, we begin by discussing the substantive meaning of time dependence from a causal-analytical point of view. Then we give an overview of models with a monotonically decreasing or increasing transition rate, for which the Gompertz (–Makeham) and the Weibull distribution are usually applied. Finally, we demonstrate the log-logistic and the log-normal model that are widely used when the transition rate at first increases and then decreases after reaching a maximum. In our examples, these parametric models are estimated as single time-dependent rates[2] as well as models with covariates linked to different parameters of the distributions (see also Blossfeld, Hamerle, and Mayer 1989).

7.1 Interpretation of Time Dependence

From a causal-analytical point of view, it is important to realize that although the models in this chapter are called time-dependent, time itself is not a causal factor. Therefore, time cannot serve as a variable to explain the observed pattern of time dependence. Thus, what does it mean when we say that a transition rate is time-dependent?

The interpretation of time dependence in parametric transition rate models can be approached from three different angles: First, the observed time dependence can be considered as *spurious*, that is, it may simply be the result of *unobserved heterogeneity*. We discuss this perspective in detail in chapter 10. Second, time dependence could be seen as the result of a *diffusion process* that reflects the changing relationship over time between a set of interdependent individual units in a dynamic system (Diekmann 1989). And third, time dependence can be interpreted as an expression of a causal process taking place in time between

Box 7.1 Stata models with a time-dependent transition rate

```
Exponential model:                 streg varlist, distribution(exponential)
Weibull model:                     streg varlist, distribution(weibull)
Gompertz model:                    streg varlist, distribution(gompertz)
Log-Normal model:                  streg varlist, distribution(lognormal)
                                   streg varlist, distribution(lnormal)
Log-Logistic model:                streg varlist, distribution(loglogistic)
                                   streg varlist, distribution(llogistic)
Generalized gamma survival model: streg varlist, distribution(ggamma)
    *Stata version 13 or earlier* streg varlist, distribution(gamma)
```

a continuously changing quantitative factor and the qualitative outcome variable under study (Tuma and Hannan 1984; Blossfeld, Hamerle, and Mayer 1989).

We first provide an overview of the perspective of *diffusion models* on time dependence. These models are usually based on the idea that some sort of contagion, infection, imitation, or simply social pressure drives the process under study (Rogers 1983; Mahajan and Peterson 1985; Strang 1991; Strang and Tuma 1993; Greve, Strang and Tuma 1995). Hernes (1972), for example, suggested a diffusion model to explain the bell-shaped transition rate of entry into marriage. In this model, he assumes two competing structural processes explaining the time dependence. On the one hand, with rising age t, the proportion $F(t)$ of a cohort that has already entered into a first marriage increases. This, in turn, enhances the pressure to marry on those who are still unmarried. On the other hand, there is some sort of decreasing social attractiveness and, more importantly, a declining chance $s(t)$ of contact between unmarried peers with increasing time t. In particular, Diekmann (1989) has shown that the differential equation

$$\frac{\mathrm{d}F(t)}{\mathrm{d}\,t} = s(t)\,F(t)\,(1 - F(t))$$

with $s(t) = m\exp(-c(t-1))$ corresponds to a bell-shaped transition rate, $r(t) = s(t)\,F(t)$, of entry into marriage (see also Wu 1990).

Although interpretations of time dependence based on the idea of a diffusion process are illuminating in theoretical terms, they are still rarely used in the social sciences (see Example 4 in section 6.6). The standard modeling strategy is to use measures of time as proxies for time-varying causal factors that are difficult to observe directly (see Tuma and Hannan 1984). As discussed earlier, measures of time may serve as proxy variables, for example, for "the amount of marriage-specific investments" in divorce studies, "the stock of job-specific labor force experience" in mobility analyses, "the intensity of information flow" in organizational studies, or "the intensity of ties between a person and a locale" in migration studies. Given the difficulty of measuring such theoretically important concepts over time, time-dependent parametric models become a useful tool for modeling the changes in the transition rates over time.

There is a natural link between the exponential model and some time-dependent parametric models (Tuma and Hannan 1984). It can be shown, for example, that a metric causal variable increasing linearly over duration leads to a Gompertz model when it is included in an exponential model on the basis of the episode-splitting method. Take the application in section 6.5 in which we argued that the unobserved variable "job-specific labor force experience," $x_{\mathrm{LFX}}(t)$, starts at 0 at the beginning of each new job and increases linearly over the duration t a person spends in a job. The rate equation with this latent variable $x_{\mathrm{lfx}}(t)$, controlling for a vector of other covariates, x, would therefore be as follows

$$r(t) = \exp\left(x\beta + x_{\mathrm{lfx}}(t)\,\beta_{\mathrm{lfx}}\right).$$

An observable proxy variable for this latent factor is $t \simeq x_{\mathrm{lfx}}$. Applying the method of episode splitting by measuring the proxy variable t intermittently at the beginning of arbitrarily defined subepisodes within each job leads to the following model:

$$r(t) = \exp\left(x\beta + t\,\beta_{\mathrm{lfx}}\right)$$

that can be rewritten

$$r(t) = \exp\left(x\beta\right)\,\exp\left(t\,\beta_{\mathrm{lfx}}\right).$$

With $b = \exp(x\beta)$ and $c = \beta_{\mathrm{lfx}}$, this is actually a Gompertz model

$$r(t) = b\,\exp\left(c\,t\right)$$

which will be explained in more detail in section 7.2. If we therefore hypothesize a causal mechanism that a linearly increasing stock of "job-specific labor force experience" leads to a monotonic decline in the job-exit rate and include the time path of the duration via episode splitting in an exponential model, this actually corresponds to an *approximation* of a Gompertz model.

A Weibull model, conversely, would be approximated if the stock of "job-specific labor force experience" is assumed to change as a *logarithmic function of duration* ($x_{\mathrm{lfx}}(t) = \log(t)$):

$$\begin{aligned} r(t) &= \exp\left(x\beta + \log(x_{\mathrm{lfx}})\beta_{\mathrm{lfx}(t)}\right), \\ &\simeq \exp\left(x\beta + \log(t)\beta_{\mathrm{lfx}}\right), \\ &= \exp(x\beta)\,t^{\beta_{\mathrm{lfx}}}. \end{aligned}$$

With $a = \exp(x\beta)$ and $b = \beta_{\mathrm{lfx}}$, one obtains a special Weibull model, as can be seen in section 7.3.[3]

To say that the stock of "job-specific labor force experience" affects the job-exit rate *logarithmically* instead of linearly with increasing job duration means that the *same* amount of job duration, t, leads to a *smaller* stock of job-specific experience, $\log(t)$. Of course, this raises the question of which of these two causal models (Gompertz or Weibull distribution) is the correct one. Since available theory in the social sciences typically provides little or no guidance for choosing one parameterization over another, the problem usually boils down to the question of which model provides a better fit to the data. Chapter 8 discusses some widely used methods for assessing the fit of parametric models.

However, given the problem of unobserved heterogeneity (chapter 10) and the difficulties in evaluating which of the models provides the relatively best fit, we recommend that parametric models of time dependence should be applied only with extreme caution. You should always be aware of the fact that you are dealing with *assumed* causal factors that are (usually) latent or unobserved and are only measured on the basis of assumed proxies. Thus, any substantive conclusion based on a time-dependent parametric model necessarily rests on a series of untested and/or untestable assumptions: (1) that there is indeed the supposed causal factor, (2) that this factor does in fact lead to a specific shape of the transition rate (*via* an assumed causal mechanism), and (3) that the measure of time is a reasonably good proxy for the path of the unobserved factor. In summary, a strong theory is needed for these empirical applications.

In the following examples, we do not pretend to have such a strong theory. Our goal is merely didactic. We demonstrate from a substantive point of view how different time-dependent parametric models can be applied with Stata in empirical research and how to interpret the results of the estimation.

7.2 Gompertz Models

We begin the discussion of parametric models of time dependence with the Gompertz distribution. The "Gompertz law," which states that the transition rate decreases monotonically with time, has been successfully applied, for example, in studying the lifetime of organizations (see, e.g., Carroll and Delacroix 1982; Freeman, Carroll, and Hannan 1983; Hannan and Freeman 1989; Lomi 1995; Carroll and Hannan 2000) or job tenure (see, e.g., Sørensen and Tuma 1981; Blossfeld and Mayer 1988). The transition rate is given by the formula

$$r(t) = b\,\exp(c\,t) \qquad b \geq 0. \tag{7.1}$$

The corresponding survivor and density functions are

$$\begin{aligned} f(t) &= \exp\left\{-\frac{b}{c}\left(\exp(c\,t)-1\right)\right\}\left(b\,\exp(c\,t)\right), \\ G(t) &= \exp\left\{-\frac{b}{c}\left(\exp(c\,t)-1\right)\right\}. \end{aligned}$$

In the case of $c = 0$, it is assumed that these expressions reduce to the density and survivor function for a simple exponential model. Figure 7.2.1 shows graphs of the transition rate function for $b = 1$ and some values of c.

The Gompertz distribution has two parameters to include covariates. Stata uses an exponential link function for the b parameter and a linear link function for the c parameter. The model formulation for the transition rate from a given origin state to destination state k is

$$\begin{aligned} r_k(t) &= b_k\,\exp(c_k\,t), \\ b_k &= \exp(B^{(k)}\beta^{(k)}), \\ c_k &= C^{(k)}\gamma^{(k)}. \end{aligned}$$

It is assumed that the first component of each of the covariate (row) vectors $B^{(k)}$ and $C^{(k)}$ is a constant equal to one. The associated coefficient vectors, $\beta^{(k)}$ and $\gamma^{(k)}$, are the model parameters to be estimated. In Stata, the Gompertz model can be requested with the command

```
streg varlist, d(gomp)
```

Linking of the ancillary parameter c can be specified with the option
`ancillary(varlist)`.

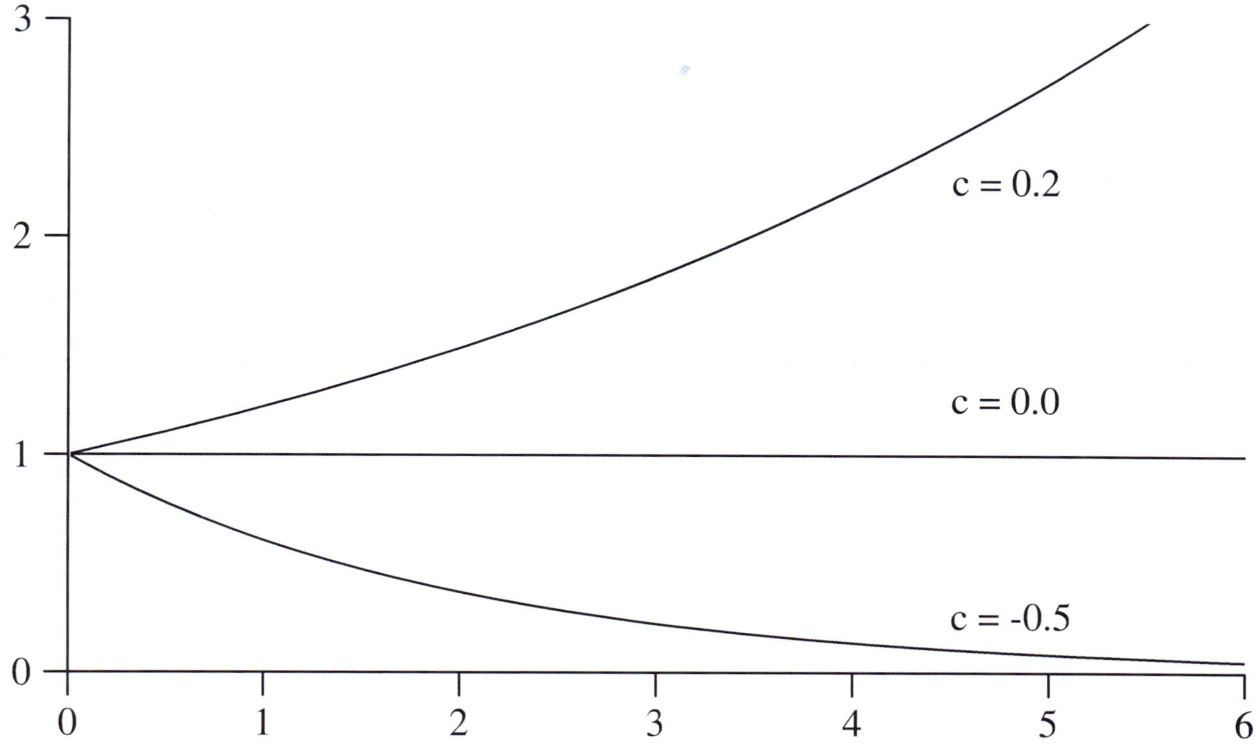

Figure 7.2.1 Gompertz transition rate ($b = 1$).

Models Without Covariates (Single Time-Dependent Rate)

In order to demonstrate the estimation and interpretation of the Gompertz model, we first specify a model without covariates:

$$r(t) = b\exp(ct)\,,\ b = \exp(\beta_0)\,,\ c = \gamma_0.$$

We again analyze the job-exit rate from "being in a job" (origin state $= 0$) to "having left the job" (destination state $= 1$) and regard duration in a job as a proxy variable for the stock of job-specific skills acquired in each new job. This implies that job-specific experience—as the duration itself—starts with each new job at $t = 0$ and then rises linearly with the time spent in a job. Since on-the-job training is an investment that is more or less lost when a job is changed, we hypothesize a causal mechanism that with increasing job-specific labor force experience, the transition rate will decline monotonically. Given the Gompertz model, this means that the estimated parameter for c should be significant and negative (see Figure 7.2.1).

The do-file to estimate this Gompertz model, shown in Box 7.2.1 differs from the do-file for an exponential model in Box 4.1.1 only in that the model selection command has been changed to

```
streg, dist(gomp) nohr
```

The estimation results are shown in Box 7.2.2. The notation `_cons` refers to parameter `b`, and `gamma` corresponds to parameter `c` in 7.1. A comparison of this model with the exponential model in Box 4.1.2, based on a likelihood ratio test, shows that the Gompertz model provides a significant improvement:

$$\mathrm{LR} = 2\left((-898.45) - (-937.97)\right) = 79.04.$$

The estimated parameters are $\hat{b} = \exp(\hat{\beta}_0) = \exp(-4.0729) = 0.017$ and $\hat{c} = \hat{\gamma}_0 = -0.0067$. Thus, as hypothesized, the estimated c is negative and significant. We therefore conclude that increasing job-specific labor force experience leads to a declining rate of moving out of the job. The estimated rate for this model can be plotted with the command `stcurve, hazard` (see Box 7.2.1). The resulting plot is shown in Figure 7.2.2. Note that it is also possible to plot the survival function and the cumulative hazard function. To do this, you only need to specify the option `survival` for the survival function and `cumhaz` for the cumulative hazard function.

We give examples of how to interpret the results: If you compare, for example, an employee who has just started a new job ($\hat{r}(0) = 0.017$) with an employee who has already been working for 10 years (120 months) in the same job ($\hat{r}(120) = 0.017\exp(-0.0067 \cdot 120) = 0.0076$), then the tendency of the second employee to change her or his job has been reduced by about 55 % due to the accumulation of job-specific skills.[4]

Box 7.2.1 Do-file `ehg1.do` (Gompertz model without covariates)

```
version 19
set scheme sj
capture log close
set more off
log using ehg1.log, replace

use rrdat1, clear

gen des = tfin ~= ti                /*destination state*/
gen tf  = tfin - tstart + 1         /*ending time*/

stset tf, failure(des)              /*define single episode data*/

streg, dist(gomp) nohr              /*fit parametric survival model*/

stcurve, hazard ytick(0(0.005)0.02) ylabel(0(0.01)0.02)
/*gompertz transition rate*/

log close
```

Box 7.2.2 Stata's output using `ehg1.do` (Box 7.2.1)

```
       failure _d:  des
 analysis time _t:  tf

Iteration 0:   log likelihood = -39205.948
Iteration 1:   log likelihood = -957.17646
Iteration 2:   log likelihood = -904.45409
Iteration 3:   log likelihood =  -898.4633
Iteration 4:   log likelihood = -898.45391
Iteration 5:   log likelihood = -898.45391

Gompertz regression -- log relative-hazard form

No. of subjects =          600     Number of obs    =        600
No. of failures =          458
Time at risk    =        40782
                                   Wald chi2(0)     =          .
Log likelihood  =   -898.45391                Prob > chi2      =          .

------------------------------------------------------------------------
    _t |      Coef.   Std. Err.      z    P>|z|     [95% Conf. Interval]
------+-----------------------------------------------------------------
_cons |  -4.072898    .0633252   -64.32   0.000    -4.197013   -3.948783
------+-----------------------------------------------------------------
gamma |  -.0066923    .0008669    -7.72   0.000    -.0083913   -.0049933
------------------------------------------------------------------------
```

Based on the survivor function of the Gompertz model

$$G(t) = \exp\left\{-\frac{b}{c}\left(\exp(c\,t) - 1\right)\right\},$$

it is possible to estimate the median M of the job duration, defined by $G(M) = 0.5$, as

$$G(\hat{M}) = \exp\left\{-\frac{0.017}{-0.0067}\left(\exp(-0.0067\,\hat{M}) - 1\right)\right\} = 0.5.$$

The estimated median is about 47.6 months. Thus the Gompertz model estimates a smaller median than the exponential model ($\hat{M} = 61.9$) in section 4.1.2.[5]

Based on the survivor function, you can also compute the probability that an employee is still working in the same job after a period of, say, 10 years:

$$G(120) = \exp\left\{-\frac{0.017}{-0.0067}\left(\exp(-0.0067 \cdot 120) - 1\right)\right\} \approx 0.246,$$

or about 25 %. This estimate therefore almost corresponds to that for the exponential model in section 4.1.2, where the estimate for this example was 26 %.

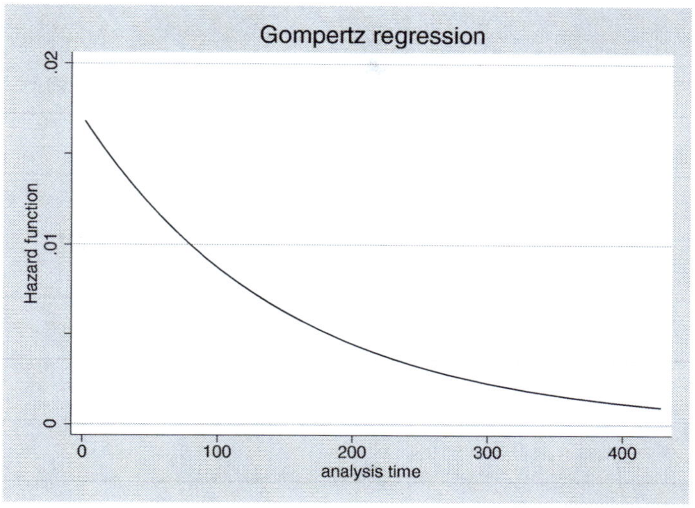

Figure 7.2.2 Gompertz transition rate estimated with do-file `ehg1.do` (Box 7.2.1). The plot was generated with do-file `ehg1.do`.

However, using this simple model without covariates assumes that there is no important heterogeneity across individuals and job episodes. This assumption is, in fact, not valid, as we have already shown in the discussion of the exponential and piecewise constant exponential models in chapters 4 and 5. Moreover, ignoring heterogeneity in the present example is particularly problematic, as a declining duration dependency may be the result of heterogeneity that is not explicitly taken into account.[6] Thus, the conclusion that the transition rate decreases due to the job-specific accumulation of skills could be wrong. It could be a consequence of unobserved heterogeneity. Therefore, in the next step, we estimate a Gompertz model controlling for a set of time-constant covariates.

Models with Covariates Linked to the b Parameter

There are several options for linking covariates to the parameters of the Gompertz model. We demonstrate all of them. First, time-constant covariates can be linked to the b parameter, so that we can estimate the following model:

$$r(t) = b \exp(ct), \; b = \exp(B\beta), \; c = \gamma_0.$$

The do-file for this model, `ehg2.do`, is shown in Box 7.2.3. This do-file is almost identical to the do-file in Box 7.2.1. There is only one difference: A list of time-constant covariates has been added to the instruction.

Results of the model estimation are shown in Box 7.2.4. The value of the log likelihood for this model is -861.9230. Thus, if this Gompertz model is compared with a Gompertz model without covariates (Box 7.2.2), the likelihood ratio test leads to a χ^2–value of

$$\text{LR} = 2\left((-861.9230) - (-898.45391)\right) = 73.06$$

with six degrees of freedom. The model with covariates obviously provides a better fit.

In addition to this overall goodness-of-fit check, you can use the information provided in the standard errors of the estimation procedure. Assuming a sufficiently large sample, the estimated coefficients divided by their standard errors (provided in column `Std. Err.`) have an approximately standard normal distribution. Using this test statistic, one can particularly assess whether the coefficients differ significantly from zero. Box 7.2.4 provides basically the same picture as the estimation results for the simple exponential model (Box 4.1.4). With the exception of `pnoj`, all variables appear to have a statistically significant impact on the transition rate for leaving a job (given a significance level of $\alpha = 0.05$).

The estimate of c, $\hat{c} = \hat{\gamma}_0 = -0.0059$ (the `gamma` coefficient in Box 7.2.4) is still significant and negative. Hence, the time-constant factors linked to the b parameter cannot fully explain the declining transition rate. The c coefficient is only slightly smaller than that in Box 7.2.2.

It is interesting to compare the estimates of this Gompertz model with the estimates in Box 6.5.4, where "job-specific labor force experience" in each of the jobs is approximated with the help of the episode-splitting method. The estimates for the models with the maximum subepisode length of 24 and 12 months are almost identical with the Gompertz model in Box 7.2.4. Thus, the method of episode splitting provides a quite good approximation of the Gompertz model.

Box 7.2.3 Do-file `ehg2.do` (Gompertz model with covariates)

```
version 19
capture log close
set more off
log using ehg2.log, replace

use rrdat1, clear

gen des = tfin ~= ti                /*destination state*/
gen tf  = tfin - tstart + 1         /*ending time*/

stset tf, failure(des)              /*define single episode data*/

gen coho2 = tb>=468 & tb<= 504      /*cohort 2*/
gen coho3 = tb>=588 & tb<= 624      /*cohort 3*/
gen lfx   = tstart - te             /*labor force experience*/
gen pnoj  = noj - 1                 /*previous number of jobs*/

streg edu coho2 coho3 lfx pnoj pres, dist(gomp) nohr
/*fit parametric survival model*/

log close
```

Box 7.2.4 Stata's standard output using `ehg2.do` (Box 7.2.3)

```
       failure _d:  des
 analysis time _t:  tf

Fitting constant-only model:

Iteration 0:   log likelihood =  -937.9681
Iteration 1:   log likelihood = -900.98488
Iteration 2:   log likelihood = -898.46151
Iteration 3:   log likelihood = -898.45391
Iteration 4:   log likelihood = -898.45391

Fitting full model:

Iteration 0:   log likelihood = -898.45391
Iteration 1:   log likelihood = -864.11298
Iteration 2:   log likelihood = -861.92627
Iteration 3:   log likelihood = -861.92295
Iteration 4:   log likelihood = -861.92295

Gompertz regression -- log relative-hazard form

No. of subjects =          600       Number of obs    =        600
No. of failures =          458
Time at risk    =        40782
                                     LR chi2(6)       =      73.06
Log likelihood  =   -861.92295       Prob > chi2      =     0.0000

----------------------------------------------------------------------
    _t |      Coef.   Std. Err.      z    P>|z|     [95% Conf. Interval]
-------+--------------------------------------------------------------
   edu |   .0633836    .024793     2.56   0.011     .0147902    .1119771
 coho2 |   .4129613   .1151944     3.58   0.000     .1871843    .6387383
 coho3 |   .3118589   .1222395     2.55   0.011     .0722738    .5514439
   lfx |  -.0038652   .0009288    -4.16   0.000    -.0056857   -.0020447
  pnoj |   .0627439   .0440582     1.42   0.154    -.0236086    .1490964
  pres |  -.0250146   .0054506    -4.59   0.000    -.0356976   -.0143317
 _cons |  -3.910537   .2813632   -13.90   0.000    -4.461999   -3.359076
-------+--------------------------------------------------------------
 gamma |   -.005876   .0008735    -6.73   0.000    -.0075879    -.004164
```

Models with Covariates Linked to the b and c Parameters

When examining the exponential model with period-specific effects in section 5.4, we have already seen that the effect of time-constant covariates varies with duration. The hypothesis is that the effects of some of

Box 7.2.5 Do-file `ehg3.do` (Gompertz model with covariates)

```
version 19
capture log close
set more off
log using ehg3.log, replace

use rrdat1, clear

gen des = tfin ~= ti                /*destination state*/
gen tf  = tfin - tstart + 1         /*ending time*/

stset tf, failure(des)              /*define single episode data*/

gen coho2 = tb>=468 & tb<= 504      /*cohort 2*/
gen coho3 = tb>=588 & tb<= 624      /*cohort 3*/
gen lfx   = tstart - te             /*labor force experience*/
gen pnoj  = noj - 1                 /*previous number of jobs*/

streg edu coho2 coho3 lfx pnoj pres, dist(gomp) nohr ///
ancillary(edu coho2 coho3 lfx pnoj pres) /*fit parametric survival model*/

log close
```

the time-constant covariates serving as signals will decline with job duration (Arrow 1973; Spence 1973, 1974). This hypothesis is supported empirically in Box 5.4.2. A similar test, in whcih the parameter c of the Gompertz model is made dependent on time-constant covariates, can now be conducted with the Gompertz model. The model to be used is

$$r(t) = b\,\exp(ct)\,,\; b = \exp\left(B\beta\right)\;,\; c = C\gamma.$$

As shown in Box 7.2.5, this model can be estimated by simply adding the option `ancillary` to link a set of covariates with the c parameter of the model. The estimation results are shown in Box 7.2.6.

Looking at the log-likelihood value in Box 7.2.6, which is -857.33, it becomes clear that this model provides only a slight improvement over the model in Box 7.2.4. Thus, if this model is compared with the Gompertz model with covariates linked only to the b parameter (Box 7.2.4), the likelihood ratio test statistic is not significant with

$$\mathrm{LR} = 2\left((-857.33) - (-861.92)\right) = 9.18$$

and six degrees of freedom. However, we briefly want to discuss the results for the c parameter. An inspection of the single estimated coefficients shows that only the effect of `pnoj` ("number of previously held jobs") is significant. Its effect has a declining relevance over job duration. In contrast to the signal or filter theory, however, the effects of prestige (`pres`), educational attainment (`edu`), and general labor force experience at the beginning of each job (`lfx`) have no effect on the "shape parameter" c (the `gamma` coefficients in Box 7.2.6). Thus, the result of this model contradicts the result of the exponential model with period-specific effects in section 5.4. This suggests that caution is required when dealing with highly sophisticated parametric models linking many covariates to various parameters.

Alternatively, a comparison with the results of the exponential model with period-specific effects in Box 5.4.2 shows that the parameters for the first period from 0 to 24 months are indeed very similar to the coefficients for the b parameter in Box 7.2.6. Thus, the b parameter effects of the covariates in the Gompertz model seem to describe well the effects of the covariates at a starting phase of each new job.

7.3 Weibull Models

This section describes the Weibull model in the accelerated failure time metric. All formulas would be slightly different in the case of proportional hazard models with a Weibull distribution for the episode durations. In the single transition case, the density function, survivor function, and the transition rate are each given by

$$\begin{aligned} f(t) &= b\,a^b\,t^{b-1}\exp\left\{-(at)^b\right\} \qquad a,b>0,\\ G(t) &= \exp\left\{-(at)^b\right\},\\ r(t) &= b\,a^b\,t^{b-1}. \end{aligned}$$

Box 7.2.6 Stata's output using `ehg3.do` (Box 7.2.5)

```
         failure _d:  des
   analysis time _t:  tf

 Fitting constant-only model:

 Iteration 0:   log likelihood =  -937.9681
 Iteration 1:   log likelihood = -889.13039
 Iteration 2:   log likelihood = -883.48448
 Iteration 3:   log likelihood = -883.46275
 Iteration 4:   log likelihood = -883.46274

 Fitting full model:

 Iteration 0:   log likelihood = -883.46274
 Iteration 1:   log likelihood = -858.61766
 Iteration 2:   log likelihood = -857.33758
 Iteration 3:   log likelihood = -857.33467
 Iteration 4:   log likelihood = -857.33467


 Gompertz regression -- log relative-hazard form

 No. of subjects =          600          Number of obs    =        600
 No. of failures =          458
 Time at risk    =        40782
                                          LR chi2(6)       =      52.26
 Log likelihood  =   -857.33467          Prob > chi2      =     0.0000

 ------------------------------------------------------------------------------
         |      Coef.   Std. Err.       z    P>|z|     [95% Conf. Interval]
 -------+----------------------------------------------------------------------
 _t      |
     edu |   .091787   .0318233     2.88   0.004     .0294145    .1541596
   coho2 |  .5073593   .1547354     3.28   0.001     .2040835     .810635
   coho3 |  .2069269   .1686586     1.23   0.220     -.123638    .5374918
     lfx | -.0051039   .0013032    -3.92   0.000     -.007658   -.0025497
    pnoj |   .151307   .0603776     2.51   0.012      .032969     .269645
    pres | -.0322516   .0071172    -4.53   0.000    -.0462011   -.0183022
   _cons | -4.016663   .3663095   -10.97   0.000    -4.734616   -3.298709
 -------+----------------------------------------------------------------------
 gamma   |
     edu | -.0005215   .0004568    -1.14   0.254    -.0014168    .0003738
   coho2 |  -.001869   .0021518    -0.87   0.385    -.0060864    .0023485
   coho3 |  .0029796   .0028039     1.06   0.288    -.0025161    .0084752
     lfx |  .0000258   .0000188     1.37   0.170     -.000011    .0000625
    pnoj | -.0019492   .0009624    -2.03   0.043    -.0038354    -.000063
    pres |  .0001677   .0001044     1.61   0.108     -.000037    .0003724
   _cons | -.0053782   .0056633    -0.95   0.342     -.016478    .0057217
 ------------------------------------------------------------------------------
```

Figure 7.3.1 shows graphs of the transition rate for $a = 1$ and different values of b.

The Weibull model is also flexible and appropriate for a wide variety of situations (e.g., Carroll and Hannan 2000). Like the Gompertz model, the Weibull model can also be used to model a monotonically falling ($0 < b < 1$) or monotonically increasing rate ($b > 1$); see Figure 7.3.1. For the special case of $b = 1$, you obtain the exponential model. It is therefore possible to test the hypothesis of a constant risk against the alternative of $b \neq 1$.

The Weibull model has two parameters, so you have two possibilities when it comes to including covariates. Stata uses the following parameterization:[7]

$$a = \exp(-A\alpha), \quad b = \exp(B\beta).$$

It is assumed that the first component of each of the covariate (row) vectors A and B is a constant equal to one. The associated coefficient vectors, α and β, are the model parameters to be estimated.
The Stata command to request estimation of a Weibull model is

```
streg, dist(weibull)
```

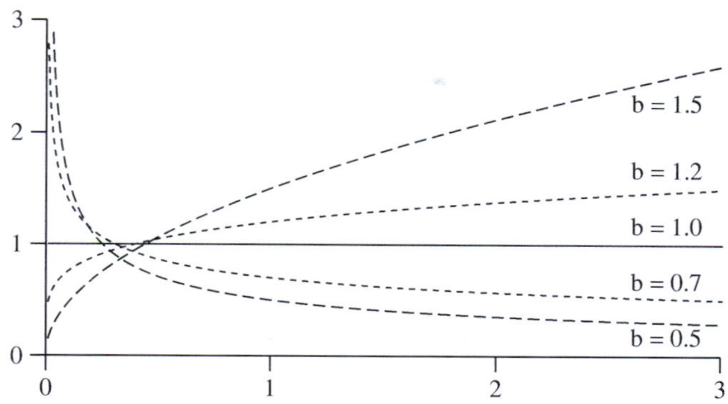

Figure 7.3.1 Weibull transition rates ($a = 1$).

Note that, by default, the model is parameterized as a proportional hazard model. To specify the accelerated failure time version, which we assumed earlier, you have to add the option `time`.

Models Without Covariates (Single Time-Dependent Rate)

In order to demonstrate the estimation and interpretation of the Weibull model, we first specify a model without covariates:

$$r(t) = b\,a\,t^{b-1}, \quad a = \exp\left(-\alpha_0\right), \quad b = \exp\left(\beta_0\right).$$

Again, we use the job-exit example of movement from "being in a job" (origin state = 0) to "having left the job" (destination state = 1), but now assume that the *logarithm* of the duration in a job is a proxy variable for the change in the stock of job-specific skills acquired in each new job. This means that job-specific experience—like the duration itself—starts in each new job and then rises as a logarithmic function of the time spent in the job. Again, we hypothesize that with increasing job-specific labor force experience, the transition rate will decline monotonically. Given the Weibull model, this suggests that the estimated parameter b is significant, and its size is between 0 and 1 (see Figure 7.3.1).

The Stata do-file (`ehg5.do`) to estimate this Weibull model, shown in Box 7.3.1, differs from the do-file for the Gompertz model in Box 7.2.1 only in that the distribution has now been changed to `weibull`, and we use the option `time` to select the accelerated failure time parameterization.

The estimation results are shown in Box 7.3.2. The α_0 parameter is called `_cons` in the Stata output, and the β_0 parameter is called `/ln_p`. A comparison with the exponential model in Box 4.1.2 leads to a highly significant likelihood ratio test statistic: 18.24 with one degree of freedom. However, there is little improvement compared to the Gompertz model (see Box 7.2.2). Hence, the Gompertz model appears to provide a better fit with the observed data. This highlights an advantage of a linear as opposed to a log-linear specification of the accumulation of job-specific labor force experience over duration. For the a term, the estimated parameter is

$$\hat{a} = \exp(-\hat{\alpha}_0) = \exp(-4.4616) = 0.0115.$$

Box 7.3.1 Do-file `ehg5.do` (Weibull model without covariates)

```
version 19
set scheme sj
capture log close
set more off
log using ehg5.log, replace

use rrdat1, clear

gen des = tfin ~= ti               /*destination state*/
gen tf  = tfin - tstart + 1        /*ending time*/

stset tf, failure(des)             /*define single episode data*/

streg, dist(w) time                /*fit parametric survival model*/

stcurve, hazard ytick(0(0.005)0.02) ylabel(0(0.01)0.02)

log close
```

Box 7.3.2 Stata's output using `ehg5.do` (Box 7.3.1)

```
  failure _d:  des
  analysis time _t:   tf

 Fitting constant-only model:

 Iteration 0:   log likelihood =  -937.9681
 Iteration 1:   log likelihood = -928.85373
 Iteration 2:   log likelihood = -928.84734
 Iteration 3:   log likelihood = -928.84734

 Fitting full model:
 Iteration 0:   log likelihood = -928.84734

 Weibull regression -- accelerated failure-time form

 No. of subjects =          600        Number of obs    =        600
 No. of failures =          458
 Time at risk    =        40782        LR chi2(0)       =       0.00
 Log likelihood  =   -928.84734        Prob > chi2      =          .

 ----------------------------------------------------------------------
      _t |      Coef.   Std. Err.      z    P>|z|     [95% Conf. Interval]
 -------+--------------------------------------------------------------
   _cons |   4.461634    .054394    82.02   0.000     4.355023    4.568244
 -------+--------------------------------------------------------------
   /ln_p |  -.1477012    .0360025   -4.10   0.000    -.2182648   -.0771377
 -------+--------------------------------------------------------------
       p |   .8626888    .0310589                      .8039126    .9257624
     1/p |   1.159167    .0417329                      1.080191    1.243916
```

For the b term, the estimated parameter is

$$\hat{b} = \exp(\hat{\beta}_0) = \exp(-0.1477) = 0.8627.$$

As expected, the estimate of b is significant, positive, and smaller than 1. Thus, the Weibull model also predicts a decreasing rate for moving out of the job with increasing job-specific labor force experience.

The estimated rate for this model is plotted in Figure 7.3.2. Compared to the Gompertz rate in Figure 7.2.2, the Weibull rate is flat and more similar to the exponential model. This could be the reason why the likelihood ratio test statistic for the Weibull model is so small compared to the Gompertz model likelihood ratio test statistic. Again, let's compare an employee who has just started a new job:

$$\hat{r}(0) = 0.8627 \cdot 0.0115^{0.8627} = 0.018$$

with an employee who has already been working in the same job for 10 years (or 120 months):

$$\hat{r}(120) = 0.8627 \cdot 0.0115^{0.8627} \cdot 120^{0.8627-1} = 0.0095,$$

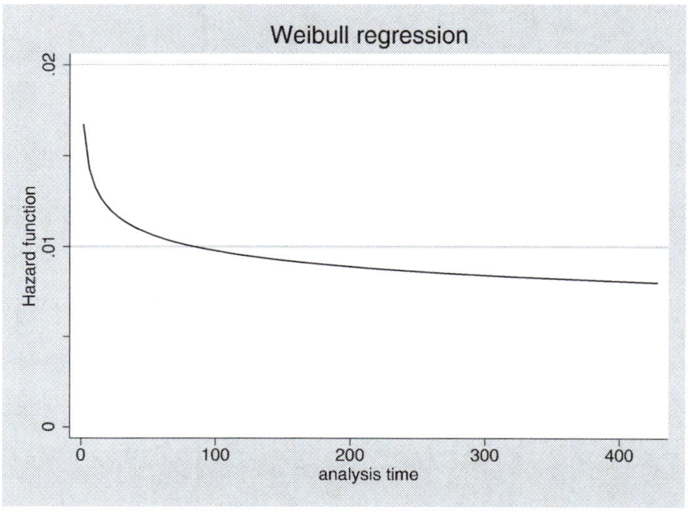

Figure 7.3.2 Weibull transition rate estimated with do-file `ehg5.do` (Box 7.3.1). The plot was generated with do-file `ehg5.do`.

Box 7.3.3 Do-file `ehg6.do` (Weibull model with covariates)

```
version 19
capture log close
set more off
log using ehg6.log, replace

use rrdat1, clear

gen des = tfin ~= ti               /*destination state*/
gen tf  = tfin - tstart + 1        /*ending time*/

stset tf, failure(des)             /*define single episode data*/

gen coho2 = tb>=468 & tb<= 504     /*cohort 2*/
gen coho3 = tb>=588 & tb<= 624     /*cohort 3*/
gen lfx   = tstart - te            /*labor force experience*/
gen pnoj  = noj - 1                /*previous number of jobs*/

streg edu coho2 coho3 lfx pnoj pres, dist(w) time
/*fit parametric survival model*/

log close
```

then the tendency of the second employee to change jobs has been reduced by about 48 % through the accumulation of job-specific skills . Based on the survivor function of the Weibull model, that is

$$G(t) = \exp\left(-(at)^b\right),$$

it is again possible to estimate the median, $\hat{M}$, of the job duration by

$$G(\hat{M}) = \exp\left(-(0.0115\, M)^{0.8627}\right) = 0.5,$$

resulting in $\hat{M} = 56.9$ months. Thus the median estimated with the Weibull model is greater than that estimated with the Gompertz model ($\hat{M} = 47.7$; see section 7.2) and smaller than the estimate resulting from an exponential model ($\hat{M} = 61.9$; see section 4.1.2).

In the next step, we estimate the Weibull model and include our known set of time-constant covariates.

Models with Covariates Linked to the a Parameter

The Weibull distribution also offers two parameters for including covariates. In a first step, we link the time-constant covariates only to the a parameter of the Weibull distribution and estimate the following model:

$$r(t) = b\, a^b\, t^{b-1}, \ a = \exp\left(-A\alpha\right), \ b = \exp\left(\beta_0\right).$$

The do-file for this model, `ehg6.do`, is shown in Box 7.3.3. This do-file differs from the do-file for the Gompertz model in Box 7.2.3 only in the model selection, which is now `dist(w)`; and the option `time`, which is used to select the accelerated failure time parameterization.

The estimation results are presented in Box 7.3.4. As can be seen in the box, the value of the log-likelihood function for this model is −886.71. Thus, if we compare this model with the Weibull model without covariates, we obtain a significant likelihood ratio statistic of 84.28 with six degrees of freedom. Hence, the covariates seem to have an important impact on job duration. This is confirmed by looking at the estimated standard errors. In particular, the estimate of b, $\hat{b} = \exp(-0.0903) = 0.9137$ (see `p` in Box 7.3.4), is still significant, positive, and less than 1. Again, the time-constant variables included here cannot explain the declining transition rate. But note that all coefficients linked to the a parameter (see `_t` in Box 7.3.4) have to be multiplied by -1 in order to compare these estimates with the estimates of the exponential and Gompertz models.

Models with Covariates Linked to the a and b Parameters

Finally, we also link covariates to the b parameter of the Weibull distribution. The model specification now becomes

$$r(t) = b\, a^b\, t^{b-1}, \ a = \exp\left(-A\alpha\right), \ b = \exp\left(B\beta\right).$$

Box 7.3.4 Stata's output using `ehg6.do` (Box 7.3.3)

```
       failure _d:  des
 analysis time _t:  tf

Fitting constant-only model:
Iteration 0:   log likelihood =  -937.9681
Iteration 1:   log likelihood = -928.85373
Iteration 2:   log likelihood = -928.84734
Iteration 3:   log likelihood = -928.84734

Fitting full model:
Iteration 0:   log likelihood = -928.84734
Iteration 1:   log likelihood = -889.15193
Iteration 2:   log likelihood = -886.71221
Iteration 3:   log likelihood = -886.70676
Iteration 4:   log likelihood = -886.70676

Weibull regression -- accelerated failure-time form

No. of subjects =          600         Number of obs   =        600
No. of failures =          458
Time at risk    =        40782
                                       LR chi2(6)      =      84.28
Log likelihood  =   -886.70676         Prob > chi2     =     0.0000

------------------------------------------------------------------------------
      _t |      Coef.   Std. Err.      z    P>|z|     [95% Conf. Interval]
-------+----------------------------------------------------------------------
     edu |   -.077931    .027014    -2.88   0.004    -.1308776   -.0249845
   coho2 |  -.6062033   .1241684    -4.88   0.000    -.8495689   -.3628377
   coho3 |  -.5776778   .1303482    -4.43   0.000    -.8331555      -.3222
     lfx |   .0036419   .0010393     3.50   0.000     .0016048    .0056789
    pnoj |  -.0637472   .0483515    -1.32   0.187    -.1585144      .03102
    pres |   .0291634    .006026     4.84   0.000     .0173527    .0409741
   _cons |   4.406308   .3060021    14.40   0.000     3.806555    5.006062
-------+----------------------------------------------------------------------
   /ln_p |  -.0902791   .0363595    -2.48   0.013    -.1615424   -.0190157
-------+----------------------------------------------------------------------
       p |   .9136762   .0332208                      .8508304     .981164
     1/p |    1.09448   .0397948                      1.019198    1.175322
```

Box 7.3.5 Do-file `ehg7.do` (Weibull model with covariates)

```
version 19
capture log close
set more off
log using ehg7.log, replace

use rrdat1, clear

gen des = tfin ~= ti                /*destination state*/
gen tf  = tfin - tstart + 1         /*ending time*/

stset tf, failure(des)              /*define single episode data*/

gen coho2 = tb>=468 & tb<= 504      /*cohort 2*/
gen coho3 = tb>=588 & tb<= 624      /*cohort 3*/
gen lfx   = tstart - te             /*labor force experience*/
gen pnoj  = noj - 1                 /*previous number of jobs*/

streg edu coho2 coho3 lfx pnoj pres, dist(w) time ///
ancillary(edu coho2 coho3 lfx pnoj pres) /*fit parametric survival model*/

log close
```

As shown in Box 7.3.5, this model can be estimated by simply adding the `ancillary` option to link covariates to the parameter b. Using the modified do-file (`ehg7.do`), the estimation results are shown in Box 7.3.6.

The log-likelihood value is -883.54 and is only slightly better than that for the model with a single constant in the b parameter (Box 7.3.4). The likelihood ratio test statistic to compare the two models is

Box 7.3.6 Stata's output using `ehg7.do` (Box 7.3.5)

```
      failure _d:  des
analysis time _t:  tf

Fitting constant-only model:
Iteration 0:   log likelihood = -928.84734
Iteration 1:   log likelihood = -924.09756
Iteration 2:   log likelihood = -924.04382
Iteration 3:   log likelihood = -924.04305
Iteration 4:   log likelihood = -924.04304

Fitting full model:
Iteration 0:   log likelihood = -924.04304
Iteration 1:   log likelihood = -889.85953
Iteration 2:   log likelihood =  -883.8695
Iteration 3:   log likelihood = -883.54403
Iteration 4:   log likelihood = -883.54308
Iteration 5:   log likelihood = -883.54307

Weibull regression -- accelerated failure-time form

No. of subjects =          600          Number of obs    =       600
No. of failures =          458
Time at risk    =        40782
                                        LR chi2(6)       =     81.00
Log likelihood  =   -883.54307          Prob > chi2      =    0.0000

------------------------------------------------------------------------
      |      Coef.   Std. Err.       z    P>|z|     [95% Conf. Interval]
------+-----------------------------------------------------------------
_t    |
  edu |  -.0857577   .0273797    -3.13   0.002    -.1394208   -.0320946
coho2 |  -.5984699   .1284651    -4.66   0.000    -.8502568    -.346683
coho3 |  -.5581806   .1297498    -4.30   0.000    -.8124856   -.3038756
  lfx |   .0031906   .0010914     2.92   0.003     .0010516    .0053297
 pnoj |  -.0570866   .0496505    -1.15   0.250    -.1543999    .0402266
 pres |   .0289681   .0061361     4.72   0.000     .0169416    .0409947
_cons |   4.495128   .3075437    14.62   0.000     3.892354    5.097903
------------------------------------------------------------------------
ln_p
  edu |  -.0027124   .0184046    -0.15   0.883    -.0387848      .03336
coho2 |   .0835133   .0854602     0.98   0.328    -.0839856    .2510122
coho3 |   .2016667   .0914015     2.21   0.027      .022523    .3808103
  lfx |   .0010158   .0007358     1.38   0.167    -.0004264     .002458
 pnoj |  -.0385058   .0358827    -1.07   0.283    -.1088346     .031823
 pres |    .000506   .0040108     0.13   0.900    -.0073549    .0083669
_cons |  -.1613031   .2129186    -0.76   0.449    -.5786159    .2560097
------------------------------------------------------------------------
```

5.72 with six degrees of freedom and does not show a significant improvement in model fit. If we look at the estimated coefficients of the b parameter, then only the effect of the cohort dummy variable `coho3` is significant. Its effect has an increasing relevance over the job duration. Again, the result that so few parameters are not significant could also be a consequence of the small number of events in the example data set.

7.4 Log-Logistic Models

This section describes the standard log-logistic model.[8] In the single transition case, the standard log-logistic model is based on the assumption that the duration of episodes follows a log-logistic distribution. The density, survivor, and rate functions for this distribution are

$$
\begin{aligned}
f(t) &= \frac{b\,a^b\,t^{b-1}}{[1+(at)^b]^2} \qquad a, b > 0, \\
G(t) &= \frac{1}{1+(at)^b}, \\
r(t) &= \frac{b\,a^b\,t^{b-1}}{1+(at)^b}.
\end{aligned}
$$

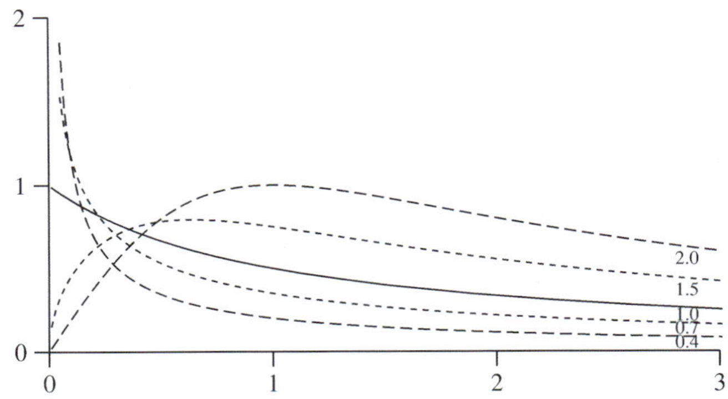

Figure 7.4.1 Log-logistic transition rates ($\lambda = 1$).

Figure 7.4.1 shows graphs of the rate function for $a = \lambda = 1$ and different values of b. The time t_{max} when the rate reaches its maximum, $r_{\max}$, is given by

$$t_{\max} = \frac{1}{a}(b-1)^{\frac{1}{b}} \qquad r_{\max} = a\,(b-1)^{1-\frac{1}{b}}.$$

Obviously, the log-logistic distribution is even more flexible than the Gompertz and Weibull distributions. As Figure 7.4.1 shows, for $b \leq 1$, we obtain a monotonically declining transition rate; and for $b > 1$, the transition rate first rises monotonically up to a maximum and then falls monotonically. Thus, this model can be used to test a monotonically declining time dependence against a nonmonotonic pattern. In the literature, the log-logistic model along with the log-normal and the sickle distributions are the most commonly recommended models if the transition rate is somehow bell-shaped.[9] This is often the case in divorce, marriage, or childbirth studies (see Diekmann 1989; Blossfeld 1995) and research on corporate demography (Brüderl 1991b; Brüderl and Diekmann 1995; Carroll and Hannan 2000).

Stata uses the following parameterization of the distribution parameters:

$$a = \exp(-A\,\alpha) \quad \text{and} \quad b = \exp(-B\beta).$$

It is assumed that the first component of the covariate row vectors A and B are constants equal to one and that both α and β are the model parameters to be estimated. To estimate a log-logistic model with Stata, enter:

```
streg, dist(loglogistic)
```

Models Without Covariates (Single Time-Dependent Rate)

To illustrate the application of the log-logistic distribution, we begin with a model without covariates:

$$r(t) = \frac{b\,a^b\,t^{b-1}}{1 + (at)^b}, \quad a = \exp(-\alpha_0), \quad b = \exp(-\beta_0).$$

Using the job-exit example again, we can examine whether the rate from "being in a job" (origin state = 0) to "having left the job" (destination state = 1) declines monotonically ($b \leq 1$) or is bell-shaped ($b > 1$) over the duration in a job. As discussed in section 5.2, a bell-shaped transition rate is quite plausible in theoretical terms. It could be the result of an interplay of two contradictory causal forces (increases in job-specific investments and decreases in the need to resolve mismatches) that cannot be measured easily, so that the duration in a job has to serve as a proxy variable for them. Empirical evidence for this thesis has already been provided in Figure 5.2.1.

A Stata do-file to estimate this standard log-logistic model is given in Box 7.4.1. The model selection command is now `streg, dist(logl)`. Estimation results are shown in Box 7.4.2. In Stata's standard output, the parameter α_0 is called `_cons`, and β_0 is called `/ln_gam`.

The estimated parameters are $\hat{a} = \exp(-3.8434) = 0.0214$ and $\hat{b} = \exp(0.2918) = 1.34$. As expected on the basis of the results of the piecewise constant exponential model in section 5.2, our estimate of b is greater than 1. Thus, we conclude that increasing job-specific labor force experience leads to an increasing and then decreasing rate of job exits. This result demonstrates that the Gompertz and the Weibull model are unable to reflect the increasing rate at the beginning of each job. They can only capture a monotonic decline or a monotonic increase over time. Since the job exit rate at the beginning of each job only increases

Box 7.4.1 Do-file `ehg8.do` (Log-logistic model without covariates)

```
version 19
set scheme sj
capture log close
set more off
log using ehg8.log, replace

use rrdat1, clear

gen des = tfin ~= ti             /*destination state*/
gen tf  = tfin - tstart + 1      /*ending time*/

stset tf, failure(des)           /*define single episode data*/

streg, dist(logl)                /*fit parametric survival model*/

stcurve, hazard  ytick(0(0.005)0.02) ylabel(0(0.01)0.02)

log close
```

Box 7.4.2 Stata's output using `ehg8.do` (Box 7.4.1)

```
       failure _d:  des
  analysis time _t:  tf

Iteration 0:   log likelihood = -2241.4588  (not concave)
Iteration 1:   log likelihood = -943.12654
Iteration 2:   log likelihood = -884.57982
Iteration 3:   log likelihood =  -884.4348
Iteration 4:   log likelihood = -884.43476

Log-logistic regression -- accelerated failure-time form

No. of subjects =          600       Number of obs    =        600
No. of failures =          458
Time at risk    =        40782
                                     Wald chi2(0)     =          .
Log likelihood  =   -884.43476       Prob > chi2      =          .

----------------------------------------------------------------------
      _t |     Coef.   Std. Err.      z    P>|z|     [95% Conf. Interval]
---------+------------------------------------------------------------
   _cons |  3.84342   .0548579    70.06   0.000     3.735901    3.95094
---------+------------------------------------------------------------
 /ln_gam |-.2917997   .0385254    -7.57   0.000    -.3673081   -.2162913
---------+------------------------------------------------------------
   gamma | .7469181   .0287753                       .6925962    .8055007
----------------------------------------------------------------------
```

for a few months and then decreases sharply, both models fit a model with a decreasing rate to the data. Thus, after "the first job phase," the log-logistic model arrives at the same substantive interpretation as the Gompertz and the Weibull models. However, the log-logistic model offers a more appropriate modeling in the first phase, which is demonstrated in the comparatively high LR test statistic.

The maximum of the log-logistic distribution is reached after a duration of about 21 months:

$$(1/0.0214)(1.34 - 1)^{1/1.34} = 20.89.$$

Thus, there appears to be an adjustment process leading to a rising rate up to this job duration. However, after this point in time, further increases in job-specific investments increasingly outweigh the force of resolving mismatches, so that the job-exit rate declines with increasing duration. In do-file `ehg8.do` of Box 7.4.1, `stcurve` is used after `streg` to plot the hazard function. The resulting plot is shown in Figure 7.4.2. With the exception of the first part, this plot is very similar to the Gompertz model plot in Figure 7.2.2.

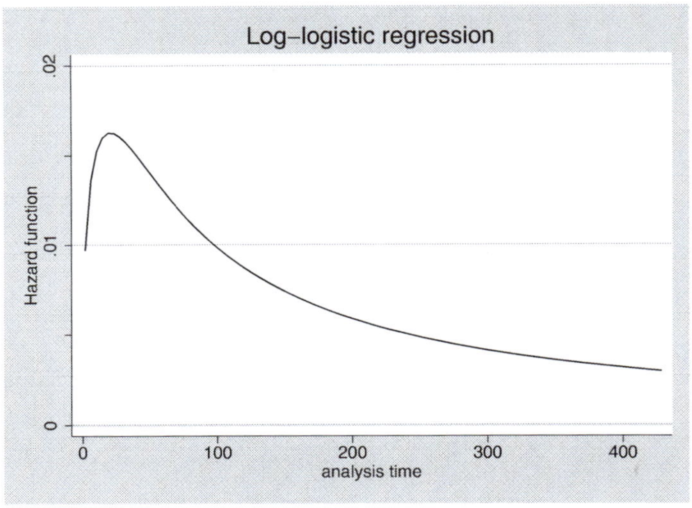

Figure 7.4.2 Log-logistic transition rate estimated with do-file `ehg8.do`, Box 7.4.1. The plot was generated with do-file `ehg8.do`.

Models with Covariates Linked to the a Parameter

The log-logistic distribution has two parameters with which covariates can be linked. We begin by linking covariates to the a parameter only. Thus we estimate the following model:

$$r(t) = \frac{b\, a^b\, t^{b-1}}{1 + (at)^b}, \quad a = \exp\left(-A\alpha\right), \quad b = \exp\left(-\beta_0\right).$$

A do-file to estimate this model is presented in Box 7.4.3. It largely corresponds to the previously used do-files, with the exception of the distribution, which was changed to `loglogistic`.

The estimation results are shown in Box 7.4.4. The value of the likelihood ratio statistic for comparison with a model without covariates is 83.16 with six degrees of freedom. Hence, the model fit improved considerably by including the covariates. Again, all single coefficients are significant and are quite similar to the parameter estimates for the exponential model.[10] There is only one difference: The effect of `pnoj`, the number of previously held jobs, is now also significant.

Also, the estimated shape parameter, $\hat{b} = \exp(0.3615) = 1.44$, is still significant and greater than 1. However, the time-constant covariates included here have increased the b value compared to the model without covariates, and the nonmonotonic pattern is therefore steeper and more skewed to the left. Note that a model with covariates that are also linked to the b parameter would provide basically the same results: Estimates of the α coefficients are almost identical, and the β coefficients are not significant. Therefore, we do not present this model here.

Box 7.4.3 Do-file `ehg9.do` (Log-logistic model with covariates)

```
version 19
capture log close
set more off
log using ehg9.log, replace

use rrdat1, clear

gen des = tfin ~= ti              /*destination state*/
gen tf  = tfin - tstart + 1       /*ending time*/

stset tf, failure(des)            /*define single episode data*/

gen coho2 = tb>=468 & tb<= 504    /*cohort 2*/
gen coho3 = tb>=588 & tb<= 624    /*cohort 3*/
gen lfx   = tstart - te           /*labor force experience*/
gen pnoj  = noj - 1               /*previous number of jobs*/

streg edu coho2 coho3 lfx pnoj pres, dist(logl)
/*fit parametric survival model*/

log close
```

Box 7.4.4 Stata's output using **ehg9.do** (Box 7.4.3)

```
Iteration 4:    log likelihood =  -842.8523

Log-logistic regression -- accelerated failure-time form

No. of subjects =          600          Number of obs    =        600
No. of failures =          458
Time at risk    =        40782
                                        LR chi2(6)       =      83.16
Log likelihood  =    -842.8523          Prob > chi2      =     0.0000

------------------------------------------------------------------------
      _t |      Coef.   Std. Err.      z    P>|z|     [95% Conf. Interval]
---------+--------------------------------------------------------------
     edu | -.0818697    .026856   -3.05   0.002    -.1345066   -.0292328
   coho2 | -.5338245   .1249407   -4.27   0.000    -.7787037   -.2889453
   coho3 | -.4004827   .1318179   -3.04   0.002    -.6588411   -.1421244
     lfx |  .0042518   .0009131    4.66   0.000     .0024622    .0060413
    pnoj | -.0970352   .0476403   -2.04   0.042    -.1904085   -.0036618
    pres |  .0296742   .0056095    5.29   0.000     .0186799    .0406686
   _cons |   3.77846   .2925781   12.91   0.000     3.205017    4.351902
---------+--------------------------------------------------------------
 /ln_gam | -.3614768   .0385635   -9.37   0.000    -.4370598   -.2858938
---------+--------------------------------------------------------------
   gamma |  .6966468   .0268651                      .6459328    .7513424
```

7.5 Log-Normal Models

Like the log-logistic model, the log-normal model is also a widely used model of time dependence that implies a nonmonotonic relationship between the transition rate and the duration: The transition rate initially increases to a maximum and then decreases (see Figure 7.5.1).

In this section, two versions of the log-normal model are described: a standard log-normal model and a model with an additional shift parameter. The models correspond to the two- and three-parameter log-normal distributions as described by, for instance, Aitchison and Brown (1973). Descriptions of log-normal rate models are given by Lawless (1982, p. 313) and Lancaster (1990, p. 47).

In the single transition case, the standard (two-parameter) log-normal model is derived by assuming that the logarithm of the episode durations follows a normal distribution or, equivalently, that the durations follow a log-normal distribution with the density function

$$f(t) = \frac{1}{b\,t}\,\phi\left(\frac{\log(t) - a}{b}\right) \qquad b > 0.$$

ϕ and Φ are used, respectively, to denote the standard normal density and distribution functions:

$$\phi(t) = \frac{1}{\sqrt{2\pi}}\exp\left(-\frac{t^2}{2}\right) \quad \text{and} \quad \Phi(t) = \int_0^t \phi(\tau)\,\mathrm{d}\tau.$$

The survivor function is

$$G(t) = 1 - \Phi\left(\frac{\log(t) - a}{b}\right),$$

and the transition rate can be written as

$$r(t) = \frac{1}{b\,t}\,\frac{\phi(z_t)}{1 - \Phi(z_t)} \quad \text{with} \quad z_t = \frac{\log(t) - a}{b}.$$

Figure 7.5.1 shows the graphs of the rate function for $a = 0$ and some different values of b. As can be seen, the graphs are very similar for the log-normal and the log-logistic models, provided, in the latter case, that $b > 1$.

The standard log-normal distribution has two parameters, so there are two options for including covariates. Stata uses a linear link function for a, but also provides the possibility of linking covariates to the dispersion parameter, b, via an exponential link function. Hence, we get the following parameterization of the model:

$$a = A\alpha \quad \text{and} \quad b = \exp(B\beta).$$

It is assumed that the first component of each of the covariate (row) vectors A and B is a constant equal to one. The associated coefficient vectors α and β are the model parameters to be estimated.[11]

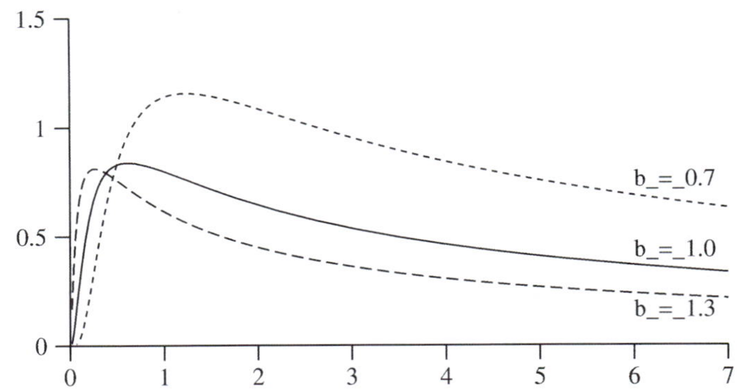

Figure 7.5.1 Log-normal transition rates ($a = 0$).

Models Without Covariates (Single Time-Dependent Rate)

To demonstrate the application of the log-normal distribution, we first estimate a model without covariates:

$$r(t) = \frac{1}{b\,t} \frac{\phi(z_t)}{1 - \Phi(z_t)}, \; z_t = \frac{\log(t) - a}{b}, \; a = \alpha_0, \; b = \exp(\beta_0).$$

Using the job-exit example, we can examine whether the rate from "being in a job" (origin state = 0) to "having left the job" (destination state = 1) is bell-shaped over the duration in a job. The substantive interpretation is the same as for the log-logistic model in the previous section.

The Stata do-file to estimate this standard log-normal model is shown in Box 7.5.1. Part of the estimation results is shown in Box 7.5.2. The estimated parameters are $\hat{\alpha_0} = 3.8852$ (`_cons` in the Stata output) and $\hat{\beta_0} = 0.24355$ (`/ln_sig` in the Stata output). Therefore, $\hat{b} = \exp(0.2436) = 1.28$ (`sigma` in the Stata output). The estimated rate for this model is shown in Figure 7.5.2. The shape of the rate is very similar to the shape of the log-logistic rate in section 7.4 (see Figure 7.4.2). Thus, the interpretation of this model is basically identical to the interpretation of the log-logistic model. In the next step, we again include our set of time-constant covariates.

Models with Covariates Linked to the a Parameter

We now link covariates to the a parameter and estimate the following model: $a = A\alpha$ and $b = \exp(\beta_0)$. The Stata do-file for this model is presented in Box 7.5.3. The estimation results are shown in Box 7.5.4. The likelihood ratio test statistic for the comparison with a log-normal model without covariates (Box 7.5.2) is 83.16 with six degrees of freedom, again showing a significant improvement of the model fit. In terms of the statistical significance and direction of influence of the covariates, the result for the log-normal model is basically the same as the result for an exponential model. With regard to the log-logistic model, there is, however, one difference: The effect of the number of previously held jobs (`pnoj`) is not significant.

Box 7.5.1 Do-file `ehg10.do` (Log-normal model without covariates)

```
version 19
set scheme sj
capture log close
set more off
log using ehg10.log, replace

use rrdat1, clear

gen des = tfin ~= ti             /*destination state*/
gen tf  = tfin - tstart + 1      /*ending time*/

stset tf, failure(des)           /*define single episode data*/

streg, dist(ln) time             /*fit parametric survival model*/

stcurve, hazard ytick(0(0.005)0.02) ylabel(0(0.01)0.02)
/*log-normal transition rate*/

log close
```

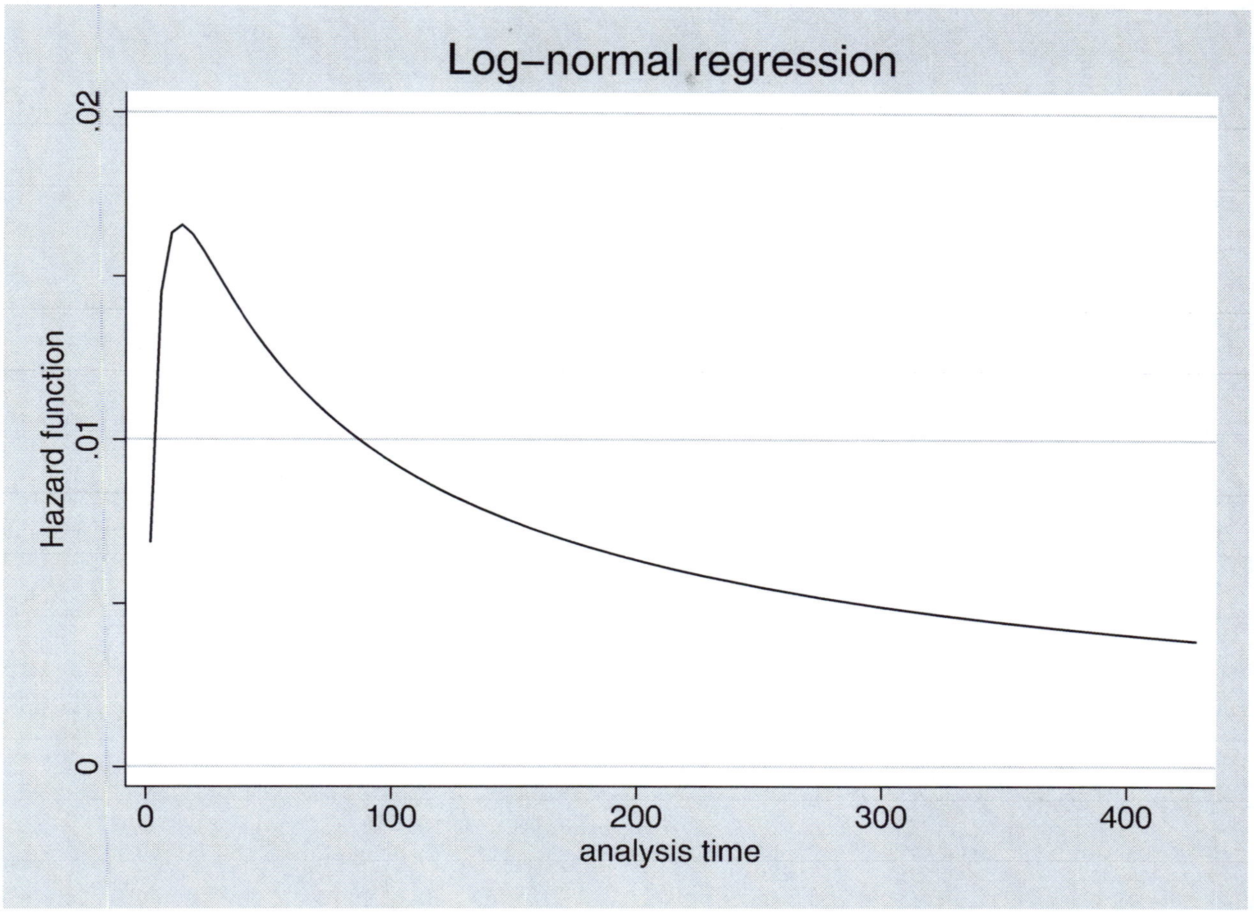

Figure 7.5.2 Log-normal transition rate estimated with do-file `ehg10.do`, Box 7.5.1. The plot was generated with do-file `ehg10.do`.

Box 7.5.2 Stata's output using `ehg10.do` (Box 7.5.1)

```
      failure _d:  des
analysis time _t:  tf

Iteration 0:   log likelihood = -5130.4981  (not concave)
Iteration 1:   log likelihood = -1186.5402
Iteration 2:   log likelihood = -1094.3427
Iteration 3:   log likelihood = -882.42882
Iteration 4:   log likelihood =  -880.1065
Iteration 5:   log likelihood = -880.08561
Iteration 6:   log likelihood = -880.08561

Log-normal regression -- accelerated failure-time form

No. of subjects =          600             Number of obs    =        600
No. of failures =          458
Time at risk    =        40782
                                           Wald chi2(0)     =          .
Log likelihood  =   -880.08561             Prob > chi2      =          .

------------------------------------------------------------------------------
      _t |      Coef.   Std. Err.      z    P>|z|     [95% Conf. Interval]
---------+--------------------------------------------------------------------
   _cons |   3.885249   .0548329    70.86   0.000     3.777778    3.992719
---------+--------------------------------------------------------------------
 /ln_sig |   .2435508   .0341814     7.13   0.000     .1765564    .3105451
---------+--------------------------------------------------------------------
   sigma |   1.275771   .0436076                      1.193102    1.364168
```

The estimated coefficient $\hat{b} = \exp(0.1794) = 1.20$ is still significant and greater than 1. Due to the time-constant variables included here, the non-monotonic pattern is even steeper and more skewed to the left

Box 7.5.3 Do-file `ehg11.do` (Log-normal model)

```
version 19
capture log close
set more off
log using ehg11.log, replace

use rrdat1, clear

gen des = tfin ~= ti              /*destination state*/
gen tf  = tfin - tstart + 1       /*ending time*/

stset tf, failure(des)            /*define single episode data*/

gen coho2 = tb>=468 & tb<= 504    /*cohort 2*/
gen coho3 = tb>=588 & tb<= 624    /*cohort 3*/
gen lfx   = tstart - te           /*labor force experience*/
gen pnoj  = noj - 1               /*previous number of jobs*/

streg edu coho2 coho3 lfx pnoj pres, dist(ln)
/*fit parametric survival model*/

log close
```

Box 7.5.4 Stata's output using `ehg11.do` (Box 7.5.3)

```
      failure _d:  des
analysis time _t:  tf

Fitting constant-only model:

Iteration 0:   log likelihood = -917.42548
Iteration 1:   log likelihood = -880.25604
Iteration 2:   log likelihood = -880.08576
Iteration 3:   log likelihood = -880.08561
Iteration 4:   log likelihood = -880.08561

Fitting full model:

Iteration 0:   log likelihood = -880.08561
Iteration 1:   log likelihood = -854.64507
Iteration 2:   log likelihood = -839.93906
Iteration 3:   log likelihood = -839.82987
Iteration 4:   log likelihood = -839.82982

Log-normal regression -- accelerated failure-time form

No. of subjects =          600     Number of obs   =        600
No. of failures =          458
Time at risk    =        40782
                                   LR chi2(6)      =      80.51
Log likelihood  =   -839.82982     Prob > chi2     =     0.0000

------------------------------------------------------------------------
      _t |     Coef.   Std. Err.      z    P>|z|     [95% Conf. Interval]
---------+--------------------------------------------------------------
     edu | -.0812716   .0269766   -3.01   0.003    -.1341448   -.0283983
   coho2 | -.5164192   .1246737   -4.14   0.000    -.7607751   -.2720633
   coho3 | -.4514127   .1294633   -3.49   0.000     -.705156   -.1976693
     lfx |  .0038863    .000875    4.44   0.000     .0021713    .0056013
    pnoj | -.0848716   .0463878   -1.83   0.067      -.17579    .0060468
    pres |  .0287017   .0055916    5.13   0.000     .0177423    .0396611
   _cons |  3.856658   .2915528   13.23   0.000     3.285225    4.428091
---------+--------------------------------------------------------------
 /ln_sig |  .1793801   .0342088    5.24   0.000      .112332    .2464282
---------+--------------------------------------------------------------
   sigma |  1.196475     .04093                     1.118884    1.279447
```

than was the case for the log-logistic model.

Again, it would also be possible to link covariates to the b parameter, but this model does not provide different results. We therefore do not present them here.

Conclusion. As shown in this chapter, Stata can be used to estimate a variety of transition rate models with a time-dependent transition rate. It is easy to select different model types and specify different ways of including covariates. However, we have already noted that using parametric transition rate models with only limited ability to fit a given data set can lead to misleading results—for instance, using a Weibull model whose transition rate is, in fact, bell-shaped. One precaution against this danger is to estimate a variety of different models and see if the estimation results—for the most interesting covariates—are robust. We will return to this topic at the end of chapter 10.

Notes

[1]This strategy was demonstrated in Blossfeld, Hamerle, and Mayer (1989) using the function optimization subprogram P3R of BMDP and the P3RFUN subroutine written by Petersen (1986a, 1986b, 1988b).

[2]That is, a transition rate model without covariates.

[3]See Blossfeld, Hamerle, and Mayer (1989) for the various forms of the Weibull model.

[4]Stata calculates and shows you the results with the `display`:

```
display exp(_b[_cons])                    display exp(_b[_cons])*exp(e(gamma)*120)
```

[5]In Stata, you can calculate the median with `predict median, median time`.

[6]This is discussed in chapter 10.

[7]This is the so-called accelerated failure time formulation. Note the minus sign in the link function for the a parameter.

[8]A sometimes useful extension of the standard log-logistic model was proposed by Brüderl (1991a; see also Brüderl and Diekmann 1994, 1995).

[9]As demonstrated in the Application Example 2 of section 6.6, another flexible strategy to model bell-shaped transition rates is to use a combination of two time-dependent variables.

[10]Due of the parameterization of this model in Stata, the signs of the coefficients have to be reversed.

[11]You can use the option `ancillary(varlist)` to link covariates to the b parameter.

Chapter 8
Methods for Testing Parametric Assumptions

As discussed in the previous chapter, the standard strategy in using parametric models of time-dependence is to consider measures of time as proxies for time-varying causal factors that are difficult to observe directly. However, the available theory in the social sciences usually provides little or no guidance on choosing one parametric model over another. For example, as shown in the previous chapter, there is little theoretical basis for deciding whether job-specific labor force experience changes linearly (Gompertz model) or log-linearly (Weibull model) over time. It is therefore important to empirically test the appropriateness of models on which inferences are based. One way to do this was demonstrated in the previous chapter by using likelihood ratio tests as a tool to compare the improvement in goodness of fit of alternative models. This method is limited, however, to nested models.

In this chapter, we deal with two different approaches for testing the suitability of parametric models. First, we present an informal method for assessing the fit of parametric models by comparing *transformations of nonparametric estimates of survivor functions* with the predictions from parametric models. Second, we demonstrate how *pseudoresiduals*, also often referred to as *generalized residuals*, can be calculated and used to evaluate distributional assumptions. Although both approaches can give some indications of the appropriateness of model specification, as we demonstrate in this chapter, they have only the character of *heuristic tools*.

8.1 Simple Graphical Methods

Several authors have proposed simple graphical methods for identifying systematic deviations of parametric models from the data.[1] The basic idea is to produce plots that should be approximately linear if the assumed family of models is appropriate, since deviations from linearity can be easily detected by eye. Most of these approaches begin with a nonparametric estimation of a survivor function using the life table method or, preferably, the product limit (Kaplan-Meier) estimator. Then, given a parametric assumption about the distribution of waiting times, we try to find a suitable transformation of the survivor function so that the result becomes a linear function ($y = a + bx$) that can be plotted for visual inspection. In addition, this approach often provides first estimates for the parameters of the assumed distribution, which can then be used as starting values for fitting the model using maximum likelihood.[2] We will briefly describe this method for four parametric distributions.

Exponential Model. This model was discussed in chapter 4. The survivor function for the basic exponential distribution is

$$G(t) = \exp(-rt).$$

Taking the (always natural) logarithm, we get a straight line

$$\log(G(t)) = -r\,t.$$

If the exponential model holds, a plot of $\log(\hat{G}(t))$ against t, using the estimated survivor function $\hat{G}(t)$, should result in an approximately linear graph that runs through the origin. The negative slope of this line is an estimate for the transition rate r.

Weibull Model. This model was discussed in section 7.3. The survivor function of the Weibull model is

$$G(t) = \exp\left(-(at)^b\right).$$

Taking the logarithm, we get

$$\log(G(t)) = -(at)^b,$$

and taking the logarithm again results in the linear function

$$\log\left(\log(-G(t))\right) = b\log(a) + b\log(t).$$

If the Weibull model holds, a plot of $\log(\log(-\hat{G}(t)))$ against $\log(t)$ should therefore be approximately linear. The slope of this line is an estimate of b.

Log-Logistic Model. This model was discussed in section 7.4. The survivor function for the basic (type I) version of the log-logistic model is

$$G(t) = \frac{1}{1 + (at)^b}.$$

This can be transformed in the following way. First, we get

$$1 - G(t) = \frac{(at)^b}{1 + (at)^b}$$

and, divided by $G(t)$, we get

$$\frac{1 - G(t)}{G(t)} = (at)^b.$$

Then, taking logarithms results in the linear function

$$\log\left(\frac{1 - G(t)}{G(t)}\right) = b\log(a) + b\log(t).$$

Therefore, if the log-logistic model holds, a plot of $\log((1-\hat{G}(t))/\hat{G}(t))$ against $\log(t)$ should be approximately linear. The slope of this line is an estimate of b.

Log-Normal Model. This model was discussed in section 7.5. The survivor function for the basic log-normal model is

$$G(t) = 1 - \Phi\left(\frac{\log(t) - a}{b}\right)$$

with Φ used to denote the standard normal distribution function. Using Φ^{-1} to denote the inverse of this function, we obtain the linear function

$$\Phi^{-1}(1 - G(t)) = -\frac{a}{b} + \frac{1}{b}\log(t).$$

Therefore, if the log-normal model holds, a plot of $\Phi^{-1}(1 - \hat{G}(t))$ against $\log(t)$ should be approximately linear.

Application examples of simple graphical models are not presented here. They can be found in Blossfeld and Rohwer (2002), Chapter 8.

8.2 Pseudoresiduals

Additional information for model selection can be obtained by using *pseudoresiduals*, also called *generalized residuals*. In OLS regression analyses, the traditional and perhaps best method to evaluate possible violations of the underlying model assumptions is to examine the residuals directly. Residuals are deviations of the *observed* values of the dependent variable from the values *estimated* under the assumptions of a specific model. In transition rate models, the "dependent variable" is the transition rate, which is, however, *not observable*. Thus, it is not possible to compute residuals by comparing the observed versus the predicted transition rates for each unit or episode.

Nonetheless, there is a similar approach that can be used for transition rate models. It is based on the *pseudoresiduals* (generalized residuals) suggested by Cox and Snell (1968). To apply this method to transition rate models, see, for instance, Blossfeld, Hamerle, and Mayer (1989, p. 82), and Lancaster (1985).

The definition is as follows: Let $\hat{r}(t; x)$ denote the estimated rate, depending on time t and on a vector of covariates, x. The estimation is based on a random sample of individuals $i = 1, \ldots, N$, with duration t_i and covariate vectors x_i. Pseudoresiduals are then defined as cumulative transition rates, evaluated for the given sample observations, that is,

$$\hat{e}_i = \int_0^{t_i} \hat{r}(\tau; x_i)\, d\tau \qquad i = 1, \ldots, N.$$

The reasoning behind this definition is that if the model is appropriate and if there are no censored observations, the residuals should approximately behave like a sample from a standard exponential distribution. If some of the observations are censored, the residuals can be regarded as a censored sample from a standard exponential distribution. In either case, one can calculate a product-limit estimate of the survivor function of the residuals, say $G_{\hat{e}}(e)$; and a plot of $-\log(G_{\hat{e}}(\hat{e}_i))$ against $\hat{e}_i$ can be used to test whether the residuals actually follow a standard exponential distribution.

Pseudoresiduals are calculated according to

$$\hat{e}_i = \int_{s_i}^{t_i} \hat{r}(\tau; x_i)\, d\tau \qquad i = 1, \ldots, N$$

with s_i and t_i denoting the starting and ending times of the episode. In addition, the survivor functions are dependent on starting times, which means that the program calculates and outputs the following

$$\hat{G}(t_i \mid s_i; x_i) = \exp\left\{ - \int_{s_i}^{t_i} \hat{r}(\tau; x_i)\, d\tau \right\} \qquad i = 1, \ldots, N.$$

Therefore, if the starting times are not zero, especially if the method of episode splitting is used, the output file does not yet appropriate information about the residuals.

To demonstrate the application of these graphical tests with Stata, we use our standard example data (`rrdat1`). There are three steps:

Step 1. In the first step, we need to `stset` our data. The first part of the do-file `ehh2.do` (8.2.1) is similar to `ehd2.do` (Box 4.1.3): `stset tf, failure(des)`. We estimate an exponential model with `streg, dist(e)`. We can then specify a postestimation command `predict cs, csnell` to calculate the generalized Cox-Snell residuals. Stata will generate a new variable, `cs`, containing the Cox-Snell residuals.

We could also include two additional commands, `predict hazard, haz` and `predict survival, sur`, to calculate the predicted hazard and each observation's predicted survivor probability. A list of the first 10 records of our data is shown in Box 8.2.2. In this file, the estimated rate (Column 5) and the pseudoresiduals (Column 7) are presented for each episode. For example, for Job Episode 1, the covariates might have the following values (see Box 2.2.2): `lfx` = 0, `pnoj` = 0, `pres` = 34, `coho2` = 0, `coho3` = 0, and `edu` = 17. Thus, the estimated rate for this job episode, based on the estimates for the exponential model in Box 4.1.4, is

$$\hat{r}(t \mid x) = \exp\big(-4.4894 + 0.0773 \cdot 17 - 0.0280 \cdot 34\big) = 0.0161.$$

This estimate is shown in the column `hazard` in Box 8.2.2. The corresponding pseudoresidual, for the first episode with a duration of 428 months is calculated as follows:

$$\hat{e} = -\log\left(\hat{G}(t \mid x)\right) = -\log\left(\exp(-0.0161 \cdot 428)\right) = 6.9004.$$

Step 2. If the model fits the data, these residuals should have a standard exponential distribution. One way to check this assumption is to calculate an empirical estimate of the cumulative hazard function. To do this, we use the Kaplan-Meier survival estimates. To this end, we first `stset` the data, specifying the Cox-Snell residuals as time variable and keep our censoring variable `des` as before. Then we generate two new variables. First, we specify the `sts generate` command to create a variable `km` with the Kaplan-Meier survival estimates. Second, we generate a new variable `cumhaz` containing the cumulative hazard.

Step 3. Finally, we perform a graphical test of the distribution of the pseudoresiduals. We plot the cumulative hazard against the Cox-Snell residuals. If the model fits the data, the plot should be approximately a straight line with a slope of 1 passing through the origin.

The commands are shown in Box 8.2.1. The same procedure, following the three steps, can be used for all other transition rate models. Figure 8.2.1 shows the plots for four different models which are described in chapters 4 and 7. As can be seen in this figure, the plot for the exponential model has the poorest fit with respect to a straight line with a slope of 1. The plot for the Weibull distribution also deviates strongly from linearity. The log-logistic and the log-normal distributions, on the other hand, provide better fits. But which distribution fits best? The log-logistic and log-normal models are very similar. The estimated shapes of the transition rates are also almost identical. These models estimate a nonmonotonic hazard rate. Thus, in substantive terms, it would not matter which of them we choose. However, it is important to point out that the graphical inspection of pseudoresiduals should not be understood as a goodness-of-fit *test*. They merely provide some clues for model selection, and the final selection should also be based on theoretical considerations.

Box 8.2.1 Do-file ehh2.do

```
version 19
set scheme sj
capture log close
set more off
log using ehh2.log, replace

use rrdat1, clear

gen des = tfin~=ti                  /*destination state*/
gen tf  = tfin - tstart + 1         /*ending time*/

gen coho2 = tb>=468 & tb<= 504      /*cohort 2*/
gen coho3 = tb>=588 & tb<= 624      /*cohort 3*/
gen lfx   = tstart - te             /*labor force experience*/
gen pnoj  = noj - 1                 /*previous number of jobs*/

stset tf, failure(des)              /*define single episode data*/

streg edu coho2 coho3 lfx pnoj pres, dist(e)   /*Exponential model*/
predict residual, csnell            /*Cox-Snell residuals*/
stset residual, failure(des)        /*define single episode data*/
sts gen km = s                      /*generate Kaplan-Meier survivor function*/
generate cumhaz = -ln(km)           /*generate cumulative hazard*/
line cumhaz residual residual, sort name("exponential, replace") ///
legend(off) t1(Exponential) xtitle("")
drop cumhaz residual km

streg edu coho2 coho3 lfx pnoj pres, dist(weibull) /*Weibull model*/
predict residual, csnell

stset residual, failure(des)
sts gen km = s
generate cumhaz = -ln(km)
line cumhaz residual residual, sort name("weibull, replace") ///
legend(off) t1(Weibull) xtitle("")
drop cumhaz residual km

streg edu coho2 coho3 lfx pnoj pres, dist(llogistic) /*Log-logistic model*/
predict residual, csnell
stset residual, failure(des)
sts gen km = s
generate cumhaz = -ln(km)
line cumhaz residual residual, sort name("llogistic, replace") ///
legend(off) t1(Log-logistic) xtitle("")
drop cumhaz residual km

streg edu coho2 coho3 lfx pnoj pres, dist(lognormal) /*Log-normal model*/
predict residual, csnell

stset residual, failure(des)
sts gen km = s
generate cumhaz = -ln(km)
line cumhaz residual residual, sort name("lognormal, replace") ///
legend(off) t1(Log-Normal) xtitle("")

drop cumhaz residual km

graph combine exponential weibull llogistic lognormal, ///
l1(-ln(Kaplan-Meier)) b1(Cox-Snell residual) saving("Figure 8_1_1")

log close
```

Box 8.2.2 Pseudoresiduals for exponential model

	id	des	tstart	tf	hazard	survival	residual
1.	1	0	555	428	.0161225	.0010074	6.900412
2.	2	1	593	46	.0131337	.5465387	.6041502
3.	2	1	639	34	.0061498	.8113198	.209093
4.	2	1	673	220	.0058589	.2755576	1.288959
5.	3	1	688	12	.0153083	.8321857	.1836997
6.	3	1	700	30	.0156408	.6254874	.4692241
7.	3	1	730	12	.0138755	.8466178	.1665059
8.	3	1	742	75	.0141769	.3453265	1.063265
9.	3	1	817	12	.0118557	.8673888	.1422679
10.	4	1	872	55	.0121056	.5138569	.6658105

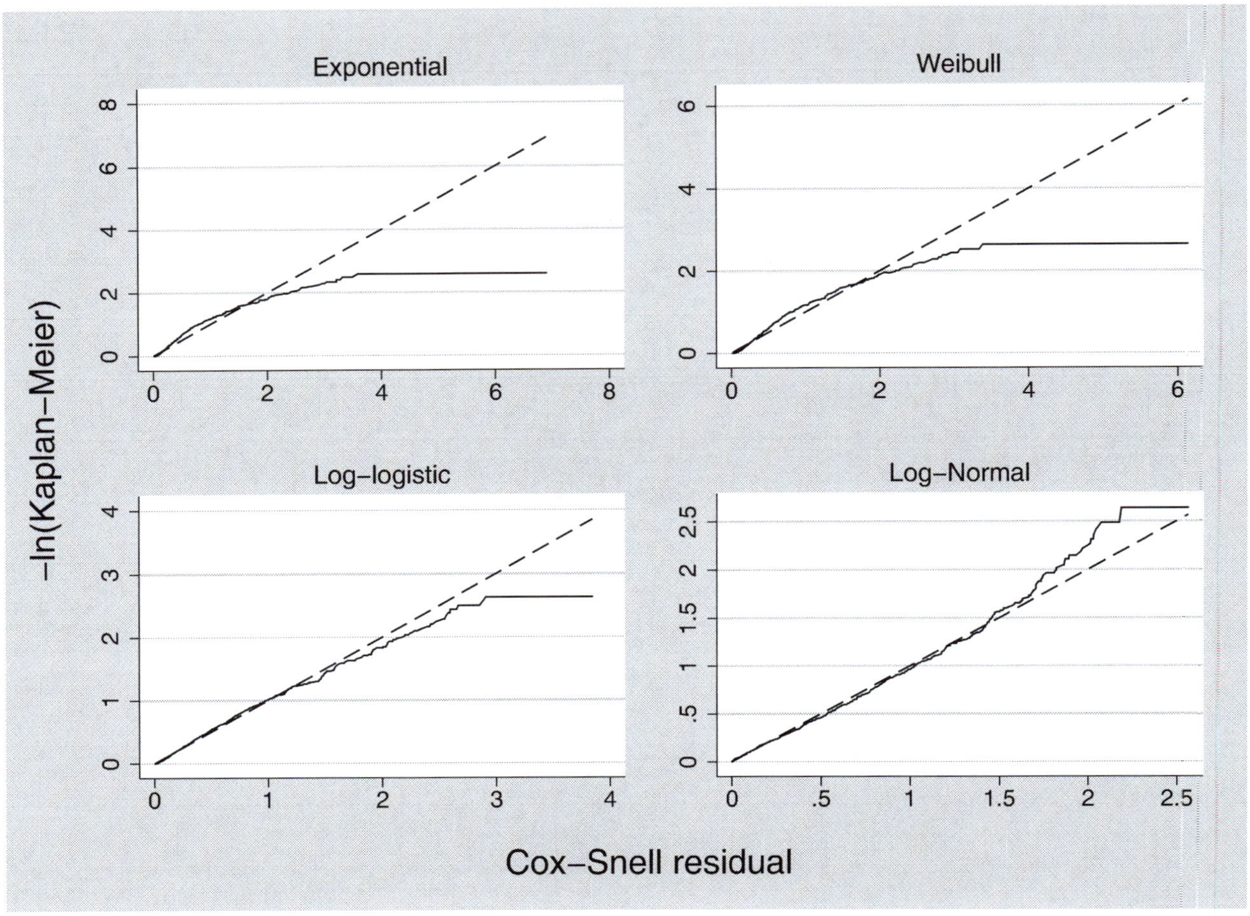

Figure 8.2.1 Graphic check of pseudoresiduals. Plots of logarithm of survivor functions (of residuals) versus residuals.

Notes

[1]See, among others, Lawless (1982), Wu (1989, 1990), and Wu and Tuma (1990).

[2]Blossfeld, Hamerle, and Mayer (1989) suggested complementing the visual inspection of transformed survivor curves by fitting an ordinary least squares regression line (OLS). However, using OLS can also be considered only as a heuristic approach in this case as the data are heteroscedastic and the residuals are highly correlated.

Chapter 9

Semiparametric Transition Rate Models

Models with a time-dependent transition rate, as described in chapter 7, are based on specific parametric assumptions about the distribution of durations and allow for straightforward maximum likelihood estimations. As demonstrated in detail, time in these models usually serves as a proxy variable for a latent causal factor that is difficult to measure directly. The problem with substantive applications is that theories in the social sciences, at least at their current stage of development, rarely offer convincing arguments in favor of a specific parametric model. We therefore advised in chapter 7 that these models should be used with extreme caution. Estimating a variety of model specifications and comparing their results seems to be an appropriate strategy. However, this leads to another problem: The adequacy of alternative parametric models of time dependence can only be assessed with heuristic tools, as shown in chapter 8. Although these goodness-of-fit tests can provide some rough indications of which classes of models are preferable, they cannot serve as rigorous tests to support a particular parametric model. Therefore, an interesting alternative strategy is to specify only a functional form for the influence of covariates, but leave the shape of the transition rate as unspecified as possible. Such models are known as semiparametric models.

The most widely used semiparametric model is the proportional hazards model proposed by Cox (1972), or, as stated in the literature, the *Cox model*. It may be written as

$$r(t) = h(t) \exp\big(A(t)\alpha\big). \tag{9.1}$$

The transition rate, $r(t)$, is the product of an unspecified *baseline rate*, $h(t)$, and a second term specifying the possible influences of a covariate vector $A(t)$ on the transition rate.[1]

Model (9.1) is a special case of a so-called *proportional* transition rate model, since the effects of the covariates can only induce proportional shifts in the transition rate, but cannot change its shape. Therefore, the Cox model should only be used if this proportionality assumption is justified.

The Cox model is widely used, although the proportionality assumption restricts its range of possible empirical applications. It is particularly attractive when the researcher (1) has no clear idea about the shape of timedependence; or (2) has only a weak theory supporting a specific parametric model; or (3) knows the time path of the process, but is unable to adequately model its fluctuations with a tractable waiting time distribution; or (4) is interested only in the magnitude and direction of the effects of observed covariates, controlling for time dependence.

The Cox model (9.1) can be extended not only for multiple origin and destination states but also multi-episode data. This chapter describes how to estimate such models with Stata. First, we explain the partial likelihood estimation of the model and present an example. Second, we discuss the use of time-dependent covariates based on partial likelihood estimation; and, in particular, the application of the method of episode splitting with Cox models. Third, we describe some methods for testing the proportionality assumption required in Cox models. In particular, we show how to estimate stratified Cox models to satisfy the proportionality requirement. Fourth, we describe how to estimate rate and survivor functions for the Cox model. Finally, we give some application examples.

9.1 Partial Likelihood Estimation

The implementation of the Cox model in Stata is based on the following model formulation

$$r_k(t) = h_k(t) \exp\left\{A^{(k)}(t)\,\alpha^{(k)}\right\}$$

where $r_k(t)$ is the transition rate at time t for the transition from the given origin state to destination state k; $h_k(t)$ is the unspecified baseline rate for the same transition;[2] $A^{(k)}(t)$ is a (row) vector of covariates, specified for the transition to destination state k; and $\alpha^{(k)}$ is a vector of associated coefficients. The covariates can have time-dependent values; examples are given in section 9.2.

It is possible to explain each transition by a specific set of covariates. Note, however, that model estimation requires *all* covariates to be defined for all members of a sample of episodes. This is necessary to calculate the risks for all transitions regardless of the transition that actually occurs at the end of an episode. It should also be noted that at least one covariate must be included for each transition, as Cox models do not include intercept terms; all constant effects are included in the baseline rates.

The model estimation is based on the method of partial likelihood (Cox 1972, 1975). To explain this method, we first consider the case of a single transition to destination state k. To simplify notation, the indices relating to the transitions are omitted. Assuming that there are no ties (i.e., all ending times in the sample are different), the partial likelihood, L^p, may be written as

$$L^p = \prod_{i \in \mathcal{E}} \frac{\exp(A_i(t_i)\,\alpha)}{\sum_{l \in \mathcal{R}(t_i)} \exp(A_l(t_i)\,\alpha)} \tag{9.2}$$

where $\mathcal{E}$ denotes the set of all uncensored episodes with destination state k and $\mathcal{R}(t_i)$ is the risk set at the ending time, say t_i, of the ith episode contained in $\mathcal{E}$. The definition of the risk set is exactly the same as that for the product-limit estimator: the set of all episodes, exiting with an event or being censored, with starting time less than t_i and ending time equal to or greater than t_i. $A_i(t_i)$ is the (row) vector of covariates for the ith episode (evaluated at t_i), and α is the vector of associated coefficients to be estimated.

Given this notation, the calculation of the partial likelihood is easy to understand: Consider the ending times of episodes that end with an event; then create the risk set for each of these time points; and finally, evaluate the expression on the right-hand side of (9.2). This expression can be interpreted as the probability that it is just the ith episode to end at this point in time, given the risk set containing all episodes that could end with an event at this point in time. Note that when calculating these probabilities, time-dependent covariates can easily be taken into account: At each successive point in time, the actual covariate values can be used.

As Cox and other authors have shown, the partial likelihood defined in (9.2) can be treated as if it were a standard likelihood. Estimates of model coefficients attained by maximizing the partial likelihood have similar properties to the standard maximum likelihood estimates.

Difficulties arise only if there are tied ending times. Then, calculating the partial likelihood exactly becomes more complicated. Several simplifying approximations have been proposed. Like BMDP and SAS, the Stata algorithm uses an approximation proposed by Breslow (1974) for the partial likelihood calculation.[3] With the `breslow` (Breslow's method), `efron` (Efron's method), `exactm` (exact marginal-likelihood), and `exactp` options (exact partial-likelihood method), you specify the method for handling tied ending times. By default, Stata uses the Breslow method.

Assume that there are d_i episodes whose indices are contained in a set D_i, which all have the same ending time, t_i. The partial likelihood factor for this point in time is then approximated by

$$\frac{\exp(G_i(t_i)\,\alpha)}{\left[\sum_{l \in \mathcal{R}(t_i)} \exp(A_l(t_i)\,\alpha)\right]^{d_i}} = \prod_{j \in D_i} \frac{\exp(A_j(t_i)\,\alpha)}{\sum_{l \in \mathcal{R}(t_i)} \exp(A_l(t_i)\,\alpha)}$$

with $G_i(t_i)$ defined as the sum of the covariate vectors for the d_i episodes with events at t_i (evaluated at t_i). As is easy to see, the use of this approximation is already accounted for in the formulation of the partial likelihood in (9.2), because in this formulation, the product includes all episodes, whether tied or not, contained in $\mathcal{E}$. As is generally assumed, this approximation is sufficient if there are only "relatively few" ties in a sample of episode data. If there are "too many" ties, it seems better to use a discrete time model.

A detailed explanation of how the partial likelihood is calculated with Stata can be found in the Stata manual. An example of a partial likelihood estimation of the Cox model with Stata is shown in Box 9.1.1. The do-file, `ehi1.do`, is almost identical to the do-file for the exponential model in Box 4.1.4. Only the model selection command has been changed to

```
stcox edu coho2 coho3 lfx pnoj pres, nohr
```

The results of the estimation are shown in Box 9.1.2. The estimated coefficients surprisingly resemble the results obtained for the exponential model in Box 4.1.4. Thus, roughly the same substantive conclusions must be drawn (at least in terms of direction of influence and statistical significance of effects) with regard to the covariates included. It should be noted that there is no longer any estimate for the constant in Box 9.1.2. It cannot be identified in the Cox model, because all members of a risk set have the same value for each event time, which removes it from the numerator and denominator of each term in the partial likelihood. Thus, the constant becomes part of the baseline hazard rate in the Cox model.

After estimating our Cox model, we now visually test how well the model fits the data. Here, too, we use the Nelson–Aalen estimator (Nelson 1972; Aalen 1978) and the minus log of the Kaplan–Meier estimator (Kaplan and Meier 1958). The closer the plotted line is to the 45° line, the better the fit (Cox and Snell 1968). When comparing the estimated lines in Figures 9.1.1 and 9.1.2 with the 45° line, we see that our Cox model fits the data well, much better than the exponential model 4, in particular. The estimated line still deviates from the reference line towards the end of the process. However, this is not unusual since there are fewer observations for longer job durations.

Box 9.1.1 Do-file `ehi1.do` (Cox model)

```
version 19
capture log close
set more off
log using ehi1.log, replace

use rrdat1, clear

gen des    = tfin ~= ti             /*destination state*/
gen tf     = tfin - tstart + 1      /*ending time*/

gen coho2 = tb>=468 & tb<= 504      /*cohort 2*/
gen coho3 = tb>=588 & tb<= 624      /*cohort 3*/
gen lfx    = tstart - te            /*labor force experience*/
gen pnoj   = noj - 1                /*previous number of jobs*/

stset tf, failure(des)              /*define single episode data*/

stcox edu coho2 coho3 lfx pnoj pres, nohr
/*fit Cox proportional hazards model*/

estat gofplot   /* Nelson-Aalen goodness-of-fit visual test */
estat gofplot, km    /* Kaplan-Meier goodness-of-fit visual test */

log close
```

Box 9.1.2 Estimation results using do-file `ehi1.do` (Box 9.1.1)

```
       failure _d:  des == 1
 analysis time _t:  tf

Iteration 0:   log likelihood = -2584.5701
Iteration 1:   log likelihood = -2548.0198
Iteration 2:   log likelihood = -2546.7823
Iteration 3:   log likelihood = -2546.7756
Iteration 4:   log likelihood = -2546.7756
Refining estimates:
Iteration 0:   log likelihood = -2546.7756

Cox regression -- Breslow method for ties

No. of subjects =          600         Number of obs    =       600
No. of failures =          458
Time at risk    =        40782
                                       LR chi2(6)       =     75.59
Log likelihood  =   -2546.7756         Prob > chi2      =    0.0000

---------------------------------------------------------------------
   _t |      Coef.   Std. Err.      z    P>|z|     [95% Conf. Interval]
------+--------------------------------------------------------------
  edu |   .0668593   .0249149     2.68   0.007      .018027    .1156916
coho2 |   .4113074    .115347     3.57   0.000     .1852314    .6373834
coho3 |   .3052514   .1219834     2.50   0.012     .0661684    .5443344
  lfx |   -.003989   .0009324    -4.28   0.000    -.0058166   -.0021615
 pnoj |   .0686267   .0441925     1.55   0.120    -.0179891    .1552426
 pres |  -.0261678   .0055004    -4.76   0.000    -.0369484   -.0153871
```

Multiple Origin and Destination States

In order to extend the Cox model to the case of multiple origin and destination states (see, e.g., Blossfeld and Hamerle 1989b), we follow the general remarks in chapter 4. First, it should be restated that in the context of duration analysis, all estimations performed are conditional on given origin states. Therefore, given two or more origin states, the (partial) likelihood for all these states can be constructed separately; each single term can be maximized separately, or the product of all origin-specific terms can be taken and maximized simultaneously.

The case of two or more transitions from a given origin state is also simple (see Kalbfleisch and Prentice

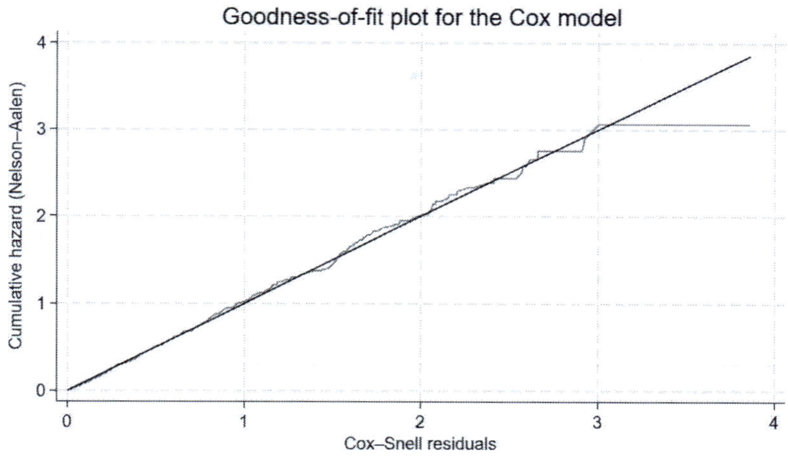

Figure 9.1.1 Nelson-Aalen plot of the goodness-of-fit (GOF) for the Cox model, generated with do-file `ehi1.do`.

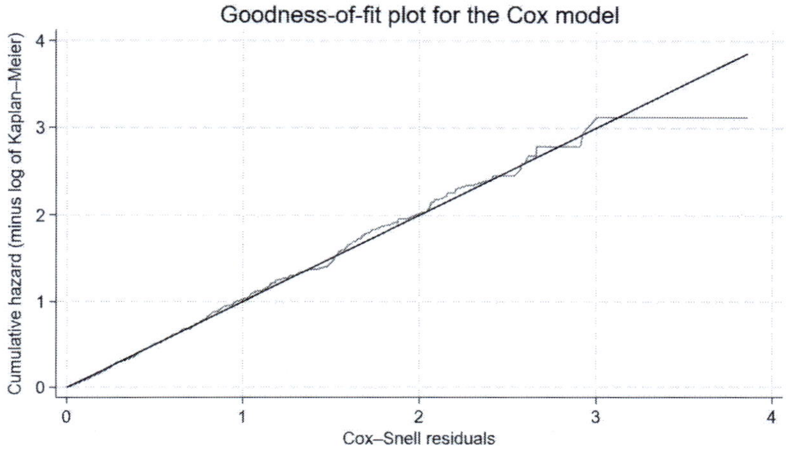

Figure 9.1.2 Kaplan-Meier plot of the goodness-of-fit (GOF) for the Cox model, generated with do-file `ehi1.do`.

1980, p. 169). You only need an appropriate definition of the risk set: $\mathcal{R}(t)$ is defined as the set of *all* episodes with the given origin state whose starting time is less than t and whose ending time is equal to, or greater than, t. The partial likelihood may then be written as

$$L^p = \prod_{k \in \mathcal{D}} \prod_{i \in \mathcal{E}_k} \frac{\exp\left(A^{(k)}(t_i)\,\alpha^{(k)}\right)}{\sum_{l \in \mathcal{R}(t_i)} \exp\left(A^{(k)}(t_l)\,\alpha^{(k)}\right)}.$$

In contrast to other software programs, such as *Transition Data Analysis* (TDA) (Rohwer and Pötter 2007), Stata only estimates the effects on the transition to one destination by default. These are the effects of the covariates on the rate to the destination state defined in `stset`. If there are multiple destination states, you can repeatedly "`stset`" your data with different destination states and estimate a model each time after `stset`. This is the same procedure that was used in chapter 4.2 to estimate competing risks models assuming an exponential distribution of the duration time till event. With some data manipulation and the use of the option strata (for stratification see chapter 9.3), Cox models with competing risks can also be estimated in one step (for details see Cleves et al. 2016, pp. 395–399). Since Stata version 11, semiparametric models for competing risks are implemented in the way proposed by Fine and Gray (1999). Note that the results differ from Cox models. For more details, type `help stcrreg` in the Stata command line or see Cleves et al. (2016, p. 399).

9.2 Time-Dependent Covariates

Cox models offer an easy way to include time-dependent covariates. The standard method is based on the fact that the partial likelihood calculation runs stepwise through all time points at which at least one of the uncensored episodes has an event. Consequently, it is possible to re-evaluate the values of the time-dependent covariates at these points in time. This is already provided for in the partial likelihood formula (9.2).

Box 9.2.1 Do-file `ehi2.do` (time-dependent covariates)

```
version 19
capture log close
set more off
log using ehi2.log, replace

use rrdat1, clear

gen des = tfin ~= ti                 /*destination state*/
gen ts  = 0                          /*starting time*/
gen tf  = tfin - tstart + 1          /*ending time*/

gen coho2 = tb>=468 & tb<=504        /*cohort 2*/
gen coho3 = tb>=588 & tb<=624        /*cohort 3*/
gen lfx   = tstart - te              /*labor force experience*/
gen pnoj  = noj - 1                  /*previous number of jobs*/

/*variable for episode splitting*/
gen     mdate = tmar - tstart if tmar >0
replace mdate = 9999          if tmar==0

gen newid = _n
stset tf, failure(des) id(newid)
stsplit ma_d , after(mdate) at(0)

generate marr = 1 if ma_d ==  0      /*time-dependent dummy variable*/
replace  marr = 0 if ma_d == -1      /*0 for single, 1 for married*/

stcox edu coho2 coho3 lfx pnoj pres marr        , nohr
/*fit Cox proportional hazards model - model 1*/

gen marrmen = sex==1 & marr==1       /*interaction variable*/
stcox edu coho2 coho3 lfx pnoj pres marr marrmen, nohr
/*fit Cox proportional hazards model - model 2*/

log close
```

If the time-dependent covariates only change their values at certain discrete points in time, the method of episode splitting can be used. The original episodes are split at every point in time at which one of the time-dependent covariates changes its value. Each of the original episodes is replaced by a contiguous set of subepisodes (splits) with respective values of the covariates. The last of these splits has the same exit status as the original episode; all other splits are regarded as right-censored. To use this method with partial likelihood maximization, you only need to define the risk set precisely. At every point in time, t, the risk set should contain all episodes, or splits, whose starting time is less than t and whose ending time is greater than or equal to t regardless of their exit status. Consequently, the risk set at every point in time contains only a single split associated with appropriate values of the time-dependent covariates for each of the original episodes. Clearly, it is necessary for the partial likelihood algorithm to take into account both the different ending and the different starting times of the episodes (splits).

To illustrate the inclusion of a time-dependent covariate in a Cox model, we again examine whether first marriage has an effect on the job-exit rate (see section 6.4). Therefore, we have to create a time-dependent dummy variable `marr` ("marriage"). Do-file `ehi2.do` (Box 9.2.1) shows how to do this. First, as described for the exponential model in Box 6.4.4, we create a variable `mdate` that indicates the marriage date relative to the beginning time of each of the jobs in an occupational career. The variable is negative if the marriage took place before the individual entered the respective job; it is zero if the marriage took place in the month in which the individual entered the respective job; and it is positive if the marriage took place after the individual entered the respective job. If an individual has not married by the time of the interview, we assign a high value that has to be greater than the longest job duration. Second, we `stset` our data, using the option id. Then we split episodes at the time of marriage using the following syntax: `stsplit ma_d, after(mdate) at(0)`. The new variable `ma_d` contains -1 for (sub-)episodes before, and 0 for (sub-)episodes after the marriage. This information is used to generate the time-dependent variable `marr`.[4]

The estimation results of the first model using do-file `ehi2.do` are presented in Box 9.2.2. The time-dependent covariate "marriage" does not appear to be significant, and entry into marriage does not seem to affect the job-exit rate in the Cox model. This conclusion is in sharp contrast to that drawn for the equivalent exponential model in Box 6.4.5, in which the variable `marr` ("marriage") was found to be significant. Thus controlling for an unspecified baseline hazard rate in the Cox model appears to have a strong effect on

Box 9.2.2 Estimation results of do-file `ehi2.do` (Box 9.2.1)

```
         failure _d:  des
   analysis time _t:  tf
                 id:  newid

Iteration 0:   log likelihood = -2584.5701
Iteration 1:   log likelihood = -2547.9248
Iteration 2:   log likelihood = -2546.6296
Iteration 3:   log likelihood = -2546.6222
Iteration 4:   log likelihood = -2546.6222
Refining estimates:
Iteration 0:   log likelihood = -2546.6222

Cox regression -- Breslow method for ties

No. of subjects =          600         Number of obs   =        761
No. of failures =          458
Time at risk    =        40782
                                       LR chi2(7)      =      75.90
Log likelihood  =   -2546.6222         Prob > chi2     =     0.0000

---------------------------------------------------------------------
   _t |      Coef.   Std. Err.      z    P>|z|     [95% Conf. Interval]
------+--------------------------------------------------------------
  edu |   .0664825   .0249497     2.66   0.008     .0175821     .115383
coho2 |   .4055774   .1157551     3.50   0.000     .1787015    .6324532
coho3 |   .2901494   .1249284     2.32   0.020     .0452943    .5350046
  lfx |  -.0041937   .0010035    -4.18   0.000    -.0061606   -.0022268
 pnoj |   .0673956    .044281     1.52   0.128    -.0193935    .1541847
 pres |  -.0262885   .0054992    -4.78   0.000    -.0370667   -.0155102
 marr |   .0660238   .1191552     0.55   0.580     -.167516    .2995636
```

Box 9.2.3 Estimation results with do-file `ehi2.do` (Box 9.2.1)

```
         failure _d:  des
   analysis time _t:  tf
                 id:  newid

Iteration 0:   log likelihood = -2584.5701
Iteration 1:   log likelihood = -2530.8076
Iteration 2:   log likelihood = -2528.4647
Iteration 3:   log likelihood = -2528.4461
Iteration 4:   log likelihood = -2528.4461
Refining estimates:
Iteration 0:   log likelihood = -2528.4461

Cox regression -- Breslow method for ties

No. of subjects =          600     Number of obs   =        761
No. of failures =          458
Time at risk    =        40782
                                   LR chi2(8)      =     112.25
Log likelihood  =   -2528.4461     Prob > chi2     =     0.0000

-----------------------------------------------------------------------
     _t |      Coef.   Std. Err.      z    P>|z|     [95% Conf. Interval]
--------+--------------------------------------------------------------
    edu |    .079178   .0240905     3.29   0.001     .0319615    .1263945
  coho2 |   .3767436   .1157905     3.25   0.001     .1497983    .6036888
  coho3 |   .2927654    .124286     2.36   0.018     .0491693    .5363616
    lfx |  -.0044539   .0010029    -4.44   0.000    -.0064196   -.0024882
   pnoj |   .1011551   .0449099     2.25   0.024     .0131332    .1891769
   pres |  -.0251975   .0053774    -4.69   0.000     -.035737    -.014658
   marr |   .5045789   .1302713     3.87   0.000     .2492518    .7599061
marrmen |  -.8608364   .1419966    -6.06   0.000    -1.139145   -.5825282
```

the coefficient of this time-dependent covariate.[5] However, we already know that the effect of marriage is different for men and women, so we should take this into account in an appropriate model specification.

We add the time-dependent interaction variable `marrmen` to capture possible interaction effects of marriage and sex (see lower part of do-file `ehi2.do` in Box 9.2.1). This is basically the same approach as already demonstrated in section 6.4 (see do-file `ehf3.do` in Box 6.4.4). The estimated parameters in Box 9.2.3 are also in line with our expectations and broadly consitent with the results of the exponential model in Box 6.4.6. The coefficient of marriage on the rate of moving out of the job is still positive for women. This effect is, however, larger than in the exponential model. In the Cox model, marriage *increases* the job-exit rate for women by about 66 %;[6] in the exponential model it is only 31 %. For men, the effect of marriage on the job change rate is negative. Marriage *decreases* the rate of moving out of the job by about 30 %.[7] In the exponential, it was 52 %. Marriage therefore actually makes men less mobile. To summarize, it can be said that the exponential, as compared to the Cox model, underestimates the impact of marriage on women and overestimates it for men.[8] This is because the Cox model controls for time dependence, which is negative after the initial phase.

9.3 The Proportionality Assumption

A basic feature of the Cox model is that the transition rates for different values of covariates are proportional. If, for instance, the transition rates, $r(t)$ and $r'(t)$, corresponding with the values of the ith covariate, A_i and A'_i, are compared, then

$$\frac{r(t)}{r'(t)} = \exp\{(A_i - A'_i)\,\alpha_i\}\,.$$

Models with this feature are called *proportional transition rate models*. This feature is not specific to the Cox model, but is implied in at least some of the many parametric models shown in chapters 4, 5, 6, and 7. This section discusses methods to test whether a sample of episode data satisfies the proportionality assumption.

Box 9.3.1 Do-file `ehi3.do` (graphical proportionality check)

```
version 19
set scheme sj
capture log close
set more off
log using ehi3.log, replace

use rrdat1, clear

gen des    = tfin ~= ti              /*destination state*/
gen tf     = tfin - tstart + 1       /*ending time*/

gen women = sex==2                   /*women*/

stset tf, failure(des)

stphplot, by(women) saving("Figure 9_3_1",replace)

stcoxkm,  by(women) saving("Figure 9_3_2",replace)

log close
```

Graphical Methods

If there are only a few categorical covariates in a model, it is possible to perform simple graphical tests to assess the proportionality assumption. Let us assume that this is the case, and that A and A' are two different vectors of covariate values. The sample can then be split into two groups, and nonparametric, Kaplan-Meier estimates of the survivor function, say $G(t)$ and $G'(t)$, can be calculated for these two groups. The proportionality assumption implies that

$$\frac{\log\left(G(t)\right)}{\log\left(G'(t)\right)} = \exp\{(A - A')\,\alpha\}\,.$$

Or, to take the (always natural) logarithms again, this becomes

$$\log\{-\log\left(G(t)\right)\} = \log\{-\log\left(G'(t)\right)\} + (A - A')\,\alpha.$$

Therefore, a plot of the transformed survivor functions can be used to check whether the proportionality assumption is (almost) correct.

In Stata, two commands, `stphplot` and `stcoxkm`, provide a graphical test of the proportionality assumption. The command `stphplot` generates plots of $-\log(-\log(\texttt{survival}))$ against $\log(\texttt{analysis time})$. Reasonably parallel lines indicate that the proportional hazards assumption has not been violated. Another graphical method uses the command `stcoxkm` to plot the observed Kaplan-Meier survival curves and compare them with the predicted curves from the Cox regression. Similarity between these two curves supports the proportional hazards assumption. We use our example data `rrdat1` and examine whether the proportional hazards assumption holds for men and women. To illustrate these diagnostic plots in Stata, we use `stphplot`, followed by `stcoxkm` in Box 9.3.1.

The plot generated with the `stphplot` command is shown in Figure 9.3.1 (Figure 9.3.2. not shown). The figure shows that the lines are not parallel, but cross each other. Nevertheless, the difference between observed and predicted values does not appear to be large. This raises the question of whether the difference is large enough to violate the proportionality assumption. A visual inspection of the plotted curves can hardly answer this question. It is more appropriate to carry out a statistical test.

Testing Time Dependence

This test is based on the idea that if the proportionality assumption is correct, there should be no interaction effect between the covariates and process time. Assuming that the ith component of the covariate vector A is to be tested, a model such as the following could be set up and estimated:

$$r(t) = h(t)\,\exp\{A\,\alpha + A_i\,(\log(\texttt{time}))\,\alpha_i'\}\,. \tag{9.3}$$

If the added interaction variable has a significant coefficient (i.e., if there is a significant interaction effect), the proportionality assumption for this covariate is probably not valid.

To illustrate this test, we re-examine whether men and women have proportional transition rates. The do-file used here is `ehi4.do`, shown in Box 9.3.2. The command includes two new options: `tvc(varlist)` and `texp(exp)`. The first is used to specify variables that vary continuously with respect to time; the

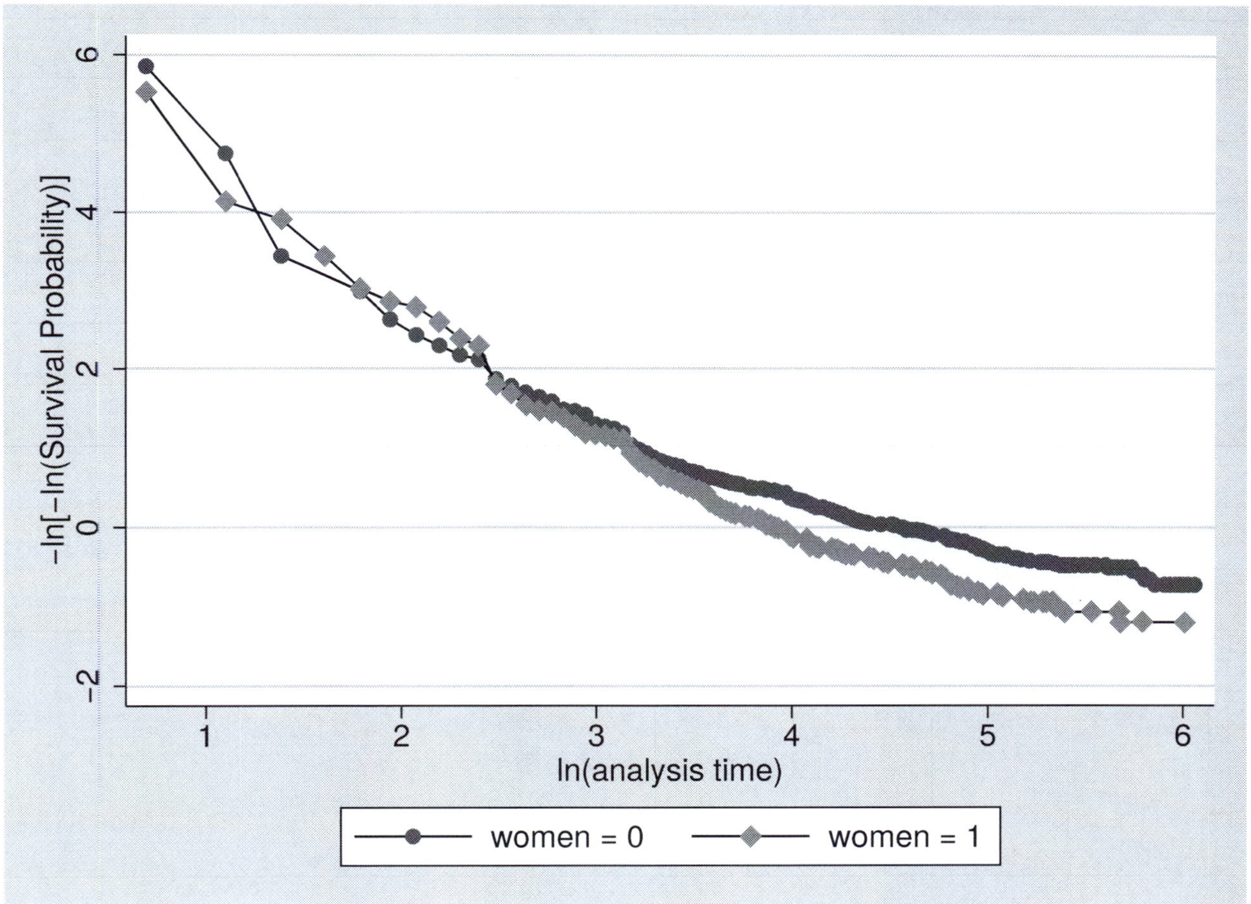

Figure 9.3.1 Graphic check for proportionality of men and women. Plot of $-\log(-\log(\hat{G}(t)))$ against $\log(t)$ (in months).

Box 9.3.2 Do-file `ehi4.do` (test of proportionality)

```
version 19
capture log close
set more off
log using ehi4.log, replace

use rrdat1, clear

gen des   = tfin ~= ti             /*destination state*/
gen tf    = tfin - tstart + 1      /*ending time*/

gen coho2 = tb>=468 & tb<= 504     /*cohort 2*/
gen coho3 = tb>=588 & tb<= 624     /*cohort 3*/
gen lfx   = tstart - te            /*labor force experience*/
gen pnoj  = noj - 1                /*previous number of jobs*/
gen women = sex==2                 /*women*/

stset tf, failure(des)

stcox edu coho2 coho3 lfx pnoj pres women, ///
tvc(women) texp(ln(tf)) nohr

log close
```

Box 9.3.3 Estimation results using do-file `ehi4.do` (Box 9.3.2)

```
         failure _d:  des
   analysis time _t:  tf

Iteration 0:   log likelihood = -2584.5701
Iteration 1:   log likelihood =  -2537.698
Iteration 2:   log likelihood = -2536.2794
Iteration 3:   log likelihood = -2536.2717
Iteration 4:   log likelihood = -2536.2717
Refining estimates:
Iteration 0:   log likelihood = -2536.2717

Cox regression -- Breslow method for ties

No. of subjects =          600        Number of obs   =       600
No. of failures =          458
Time at risk    =        40782
                                      LR chi2(8)      =     96.60
Log likelihood  =   -2536.2717        Prob > chi2     =    0.0000

----------------------------------------------------------------------
         _t |   Coef.    Std. Err.     z    P>|z|    [95% Conf. Interval]
-----------+----------------------------------------------------------
main       |
        edu | .0803298    .024667    3.26   0.001    .0319834    .1286762
      coho2 | .3936793   .1154586    3.41   0.001    .1673847    .6199739
      coho3 | .2715523   .1221511    2.22   0.026    .0321406     .510964
        lfx |-.0040585   .0009296   -4.37   0.000   -.0058806   -.0022365
       pnoj |  .088772   .0446729    1.99   0.047    .0012147    .1763293
       pres |-.0258837   .0054316   -4.77   0.000   -.0365295   -.0152379
      women | .7110886     .16072    4.42   0.000    .3960831    1.026094
-----------+----------------------------------------------------------
tvc        |
      women | .2587299   .0987092    2.62   0.009    .0652634    .4521965
----------------------------------------------------------------------

Note: Variables in tvc equation interacted with ln(tf).
```

second allows the specification of the function of analysis time that should be multiplied by the time-varying covariates. According to (9.3), we type:

```
stcox edu coho2 coho3 lfx pnoj pres women,
tvc(women) texp(ln(tf)) nohr
```

The estimation results of this model specification are shown in Box 9.3.3. The first part of the equation, main, reports the results for the time-constant covariates; the second part of the equation, tvc, reports the results for our time-varying covariate. The time-varying covariate women is obviously significant. In other words, the assumption that men and women have proportional job-exit risks is not justified.

What is the consequence of this result? There are at least two possible responses: First, the researcher can estimate a stratified Cox model with *group-specific baseline transition rates* for men and women (see the following section); or the researcher can simply use a *nonproportional* Cox model (i.e., the model just discussed) that includes the interaction effect and automatically corrects the violation of the proportionality assumption.

9.4 Stratification with Covariates and for Multiepisode Data

If the proportionality assumption is not acceptable for some covariates, it is sometimes useful to estimate a stratified model. This is possible if the covariates are categorical.[9] The entire sample can then be split into groups (strata), with one group for each of the possible combinations of categories. In the simplest case, when there is only a single dummy variable, there are only two groups. Let $\mathcal{G}$ denote the set of groups. The model should then be specified so that the baseline rate can be different for each group. There are two possibilities: The first is to define a different model for each group. Then, not only the baseline rates but also the covariate effects can vary across all groups.

Box 9.4.1 Do-file ehi5.do (stratification)

```
version 19
capture log close
set more off
log using ehi9.log, replace

use rrdat1, clear

gen des    = tfin ~= ti              /*destination state*/
gen tf     = tfin - tstart + 1       /*ending time*/

gen coho2 = tb>=468 & tb<= 504       /*cohort 2*/
gen coho3 = tb>=588 & tb<= 624       /*cohort 3*/
gen lfx    = tstart - te             /*labor force experience*/
gen pnoj   = noj - 1                 /*previous number of jobs*/

stset tf, failure(des)

stcox edu coho2 coho3 lfx pnoj pres, nohr strata(sex)

log close
```

This approach is, of course, limited to fairly large sets of episode data. If data sets are relatively small, or when there are no significant interaction effects, it may be more appropriate to assume that the covariate effects are the same in all groups.[10] Based on this assumption, a model for the single transition case can be written as

$$r_g(t) = h_g(t)\,\exp(A\,\alpha) \qquad g \in \mathcal{G}.$$

This model must be estimated for all groups simultaneously. Consequently, the estimation approach given in (9.2) requires a slight modification: The partial likelihood must be calculated as the product of the group-specific likelihoods. This can be written as

$$L^p = \prod_{k\in\mathcal{D}} \prod_{g\in\mathcal{G}} \prod_{i\in\mathcal{E}_{k,g}} \frac{\exp\left(A^{(k)}(t_i)\,\alpha^{(k)}\right)}{\sum_{l\in\mathcal{R}_g(t_i)} \exp\left(A^{(k)}(t_l)\,\alpha^{(k)}\right)} \tag{9.4}$$

with $\mathcal{G}$ denoting the set of groups of episodes with the given origin state, and $\mathcal{E}_{k,g}$ is the set of all uncensored episodes with destination state k *and* belonging to group g. Accordingly, $\mathcal{R}_g(t)$ is the risk set at time t which contains all episodes whose starting time is less than t and whose ending time is equal to or greater than t, and which belong to group g.

To illustrate the estimation of a stratified Cox model with Stata, we use our standard example. The do-file is ehi5.do (Box 9.4.1). The sample is stratified by sex (there are two groups: men and women). This

Box 9.4.2 Estimation results using do-file `ehi5.do` (Box 9.4.1)

```
         failure _d:  des == 1
   analysis time _t:  tf

Iteration 0:   log likelihood = -2259.5669
Iteration 1:   log likelihood =   -2225.69
Iteration 2:   log likelihood = -2224.6371
Iteration 3:   log likelihood = -2224.6321
Iteration 4:   log likelihood = -2224.6321
Refining estimates:
Iteration 0:   log likelihood = -2224.6321

Stratified Cox regr. -- Breslow method for ties

No. of subjects =          600          Number of obs   =        600
No. of failures =          458
Time at risk    =        40782
                                        LR chi2(6)      =      69.87
Log likelihood  =   -2224.6321          Prob > chi2     =     0.0000

------------------------------------------------------------------------
    _t |      Coef.   Std. Err.      z    P>|z|     [95% Conf. Interval]
-------+----------------------------------------------------------------
   edu |   .0793089   .0247736     3.20   0.001     .0307535    .1278642
 coho2 |   .3974667   .1155737     3.44   0.001     .1709465    .6239869
 coho3 |   .2801267   .1226056     2.28   0.022     .0398241    .5204292
   lfx |  -.0040179    .000931    -4.32   0.000    -.0058425   -.0021932
  pnoj |   .0888606   .0447533     1.99   0.047     .0011458    .1765754
  pres |  -.0257537   .0054493    -4.73   0.000    -.0364341   -.0150734
------------------------------------------------------------------------
                                                 Stratified by men women
```

is achieved by using the `strata` option. Estimation results are shown in Box 9.4.2. They are quite similar to the estimation results shown in Box 9.1.2, where the same model was estimated without stratification by sex.

Stratification can also be used to analyze multiepisode data. Then each individual can have its own intensity to make the transition from the origin to the destination state. In such a model, no person-specific, that is, overall time-constant characteristic can be included as a covariate, because all such factors are already controlled for in the fixed effect partial likelihood (Allison 1996). Coefficients can only be estimated for variables that change their values between episodes of a person or within episodes. The focus of this model specification is on the effects of variations within individuals on the transition rate and not on differences between individuals. Following the notions often used in multilevel analyses: at the first level, are job episodes; and at the second level, persons. In a fixed effect model, person-specific characteristics can be taken into account with so-called cross-level interaction terms.[11]

Declaring survival time date and episode splitting both work in the way described before. Instead of using sex as a group variable, we now use the variable `id` (not `newid!`). As covariates, we use prestige (`pres`), which can vary between episodes of one and the same person, and the marriage variable (`marr`) (see Box 9.4.3). The results are shown in the upper part of Box 9.4.4 and show that an increase in job prestige and marriage makes leaving jobs more difficult. This is in line with the previous results.

In a further model, we include a cross-level interaction effect for marriage and men (syntax in the middle of Box 9.4.3). The result is shown in the same box as the previous one. The results differ somewhat from previous results, such as those in Box 9.2.3. According to the current model, women do not differ in their job-exit rates before and after marriage. For men, we see similar effects as in previous models. After marriage, their jobs are much more stable.

The do-file `ehi6.do` also provides the syntax for the re-estimation of the last model without persons who have only one job episode (lower part of Box 9.4.3). We can identify these cases with the command `bysort id: gen max = _N` (see also chapter 2.2). If we drop these episodes, the number of subjects, observations, and failures is slightly lower. However, the estimated results (not shown in Box 9.4.4) are exactly the same as in the previous model. All persons with only one (job) episode do not contribute to the model estimation. This is easy to see if we look at the group-specific partial likelihood defined in 9.4.

Box 9.4.3 Do-file `ehi6.do` (stratification - multi episode fixed effects model)

```
version 19
capture log close
set more off
log using ehi6.log, replace

use rrdat1, clear

gen des = tfin ~= ti               /*destination state*/
gen ts  = 0                        /*starting time*/
gen tf  = tfin - tstart + 1        /*ending time*/

gen lfx   = tstart - te            /*labor force experience*/
gen pnoj  = noj - 1                /*previous number of jobs*/

/*variable for episode splitting*/
gen     mdate = tmar - tstart if tmar >0
replace mdate = 9999          if tmar==0

gen newid = _n
stset tf, failure(des) id(newid)
stsplit ma_d , after(mdate) at(0)

generate marr = 1 if ma_d ==  0  /*time-dependent dummy variable*/
replace  marr = 0 if ma_d == -1  /*0 for single, 1 for married*/
gen marrmen = sex==1 & marr==1   /*interaction variable*/

stcox pres marr                   , nohr strata(id)
stcox pres marr marrmen           , nohr strata(id)

bysort id: gen max = _N
stcox pres marr marrmen if max > 1, nohr strata(id)

log close
```

9.5 Baseline Rates and Survivor Functions

The partial likelihood method provides estimates of the parameters of a Cox model, but not a direct estimate of the underlying baseline rate. Of course, it would be useful to have such estimates, because they could give some information about the type of time dependence in a set of episode data, and this could give us some insight into whether a fully parametrical model would be appropriate.

There are different proposals for estimating the baseline rate. One approach is based on a proposal made by Breslow (1974), also discussed in Blossfeld, Hamerle, and Mayer (1989). To explain this method, we first assume a single transition, and that all episodes have the starting time zero. Then, in the usual notation, $\mathcal{E}$ denotes the set of, say q, not censored episodes, and the ordered ending times can be denoted by $\tau_1 < \tau_2 < \ldots < \tau_q$.

Let E_i be the number of events at τ_i and let $\mathcal{R}_i$ be the risk set at τ_i. Furthermore, let $\hat{\alpha}$ be the partial likelihood estimate of the model parameters. Then, defining $\tau_0 = 0$, we can consider

$$\hat{h}_i^b = \frac{E_i}{(\tau_i - \tau_{i-1}) \sum_{l \in \mathcal{R}_i} \exp(A_l \, \hat{\alpha})} \qquad i = 1, \ldots, q$$

to be an estimate of the baseline rate of the model. This is a constant during each interval $(\tau_{i-1}, \tau_i]$, resulting in a step function at the times when at least one event occurs. This step function can now be integrated to obtain an estimate of the cumulative baseline rate:

$$\hat{H}^b(t) = \sum_{l=1}^{i} (\tau_l - \tau_{l-1}) \, \hat{h}_l + (t - \tau_i) \, \hat{h}_{i+1} \qquad \tau_i < t \le \tau_{i+1}.$$

An estimate of the baseline survivor function can then be calculated by

$$\hat{G}^b(t) = \exp\left\{-\hat{H}^b(t)\right\}.$$

Finally, due to the assumption of proportional effects, the cumulative transition rate for an arbitrary covariate vector A can be estimated by

$$\hat{H}(t) = \hat{H}^b(t) \exp\left(A\hat{\alpha}\right), \tag{9.5}$$

Box 9.4.4 Estimation results using do-file **ehi6.do** (Box 9.4.3)

```
stcox pres marr                          , nohr strata(id)

failure _d:  des
analysis time _t:  tf
id:  newid
/*....*/
Stratified Cox regr. -- Breslow method for ties

No. of subjects =          600                 Number of obs    =       761
No. of failures =          458
Time at risk    =        40782
LR chi2(2)      =        30.11
Log likelihood  =   -369.67483                 Prob > chi2      =    0.0000

------------------------------------------------------------------------------
_t    |       Coef.   Std. Err.       z    P>|z|     [95% Conf. Interval]
-----+------------------------------------------------------------------------
pres |  -.0367026   .0093694    -3.92   0.000    -.0550662   -.0183391
marr |  -.5312283   .1610036    -3.30   0.001    -.8467896    -.215667
------------------------------------------------------------------------------
Stratified by id

. stcox pres marr marrmen                  , nohr strata(id)

failure _d:  des
analysis time _t:  tf
id:  newid
/*....*/
Stratified Cox regr. -- Breslow method for ties

No. of subjects =          600                 Number of obs    =       761
No. of failures =          458
Time at risk    =        40782
LR chi2(3)      =        38.81
Log likelihood  =   -365.32384                 Prob > chi2      =    0.0000

------------------------------------------------------------------------------
_t       |       Coef.   Std. Err.       z    P>|z|     [95% Conf. Interval]
--------+---------------------------------------------------------------------
pres     |  -.0334026   .0094729    -3.53   0.000    -.0519692   -.0148361
marr     |   .0127095    .244301     0.05   0.959    -.4661116    .4915305
marrmen  |  -.9770766   .3345288    -2.92   0.003    -1.632741   -.3214123
------------------------------------------------------------------------------
Stratified by id
```

and the corresponding survivor function estimate is

$$\hat{G}(t) = \exp\left\{-\hat{H}^b(t)\,\exp\left(A\hat{\alpha}\right)\right\} = \hat{G}^b(t)^{\exp(A\hat{\alpha})}. \tag{9.6}$$

Using these formulas, we can estimate the cumulative baseline transition rate and the corresponding baseline survivor function. A generalization to the case of multiple origin and destination states can be performed analogously: Each transition is treated separately. For instance, with respect to the transition (j, k), the procedure described above is applied to all episodes with the origin state j. All episodes that do not end in the destination state k are regarded as censored. The resulting estimates are then pseudosurvivor functions and transition-specific cumulative rates, respectively.

We now illustrate the calculation of baseline rates with Stata (using a slightly different approach). The do-file is `ehi7.do` (Box 9.5.1), and is essentially identical to the do-file `ehi1.do` shown in Box 9.1.1. We only add three options

```
stcox edu coho2 coho3 lfx pnoj pres, nohr
basesurv(s) basehc(h) basec(c)
```

to the `stcox` command in order to add new variables to the data containing the estimated cumulative baseline hazard (`basechazard`), the estimated baseline hazard contributions (`basehc`), and the estimated baseline survival function (`basesurv`). It is important to note that the baseline functions correspond to all covariates that are equal to zero in our Cox model.

Box 9.5.1 Do-file `ehi7.do`

```
version 19
set scheme sj
capture log close
set more off
log using ehi7.log, replace

use rrdat1, clear

gen des = tfin ~= ti                /*destination state*/
gen tf  = tfin - tstart + 1         /*ending time*/

gen coho2 = tb>=468 & tb<=504       /*cohort 2*/
gen coho3 = tb>=588 & tb<=624       /*cohort 3*/
gen lfx   = tstart - te             /*labor force experience*/
gen pnoj  = noj - 1                 /*previous number of jobs*/

stset tf, failure(des)

stcox edu coho2 coho3 lfx pnoj pres, nohr basesurv(s) basehc(h) basec(c)

stcurve, survival /*at mean values of all predictors*/
stcurve, cumhaz at1(coho2=0 coho3=1 edu=13 lfx=5 pnoj=1 pres=30) ///
                at2(coho2=1 coho3=0 edu=13 lfx=5 pnoj=1 pres=30) ///
                at3(coho2=0 coho3=0 edu=13 lfx=5 pnoj=1 pres=30) ///
xlabel(0(100)400) legend(row(3)) saving("Figure 9_5_1",replace)

stcurve, hazard at1(coho2=0 coho3=1 edu=13 lfx=5 pnoj=1 pres=30) ///
                at2(coho2=1 coho3=0 edu=13 lfx=5 pnoj=1 pres=30) ///
                at3(coho2=0 coho3=0 edu=13 lfx=5 pnoj=1 pres=30) ///
kernel(gauss) ///
xlabel(0(100)400) legend(row(3)) saving("Figure 9_5_2",replace)

log close
```

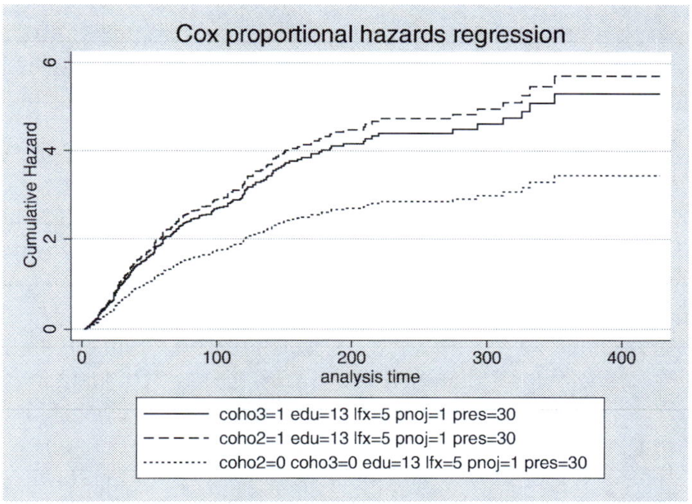

Figure 9.5.1 The plot for the cumulative hazard rates for Birth Cohorts 1, 2, and 3 was generated with do-file `ehi7.do`.

There is a simple way to graphically display estimated survival, cumulative hazard and hazard functions with Stata. If we type `stcurve, survival`, the survival function is plotted. By default, the curve is evaluated at the mean values of all the predictors, but we can also specify other values. For example, the plot can be requested for three covariate constellations (i.e., the assignment of different values to the covariates in the model). One of these covariate constellations is for Birth Cohort 1, one for Birth Cohort 2, and the last one is for Birth Cohort 3. The other covariates are fixed (`edu=13 lfx=5 pnoj=1 pres=30`; see Box 9.5.1).

Figure 9.5.1 shows a plot of the cumulative baseline rates for the three birth cohorts. Sometimes it would be helpful for interpretation if we could see a plot of the baseline rate instead of the cumulative baseline rate. Therefore, what we need is the first derivative of the cumulative rate. We can calculate this derivative by numerical differentiation. However, differentiating the cumulative rate would lead to large fluctuations. Therefore, it makes sense to smooth the cumulative rate first. By default, the hazard function is estimated by kernel smoothing of the estimated hazard contributions. Do-file `ehi7.do` shown in Box 9.5.1 can be used

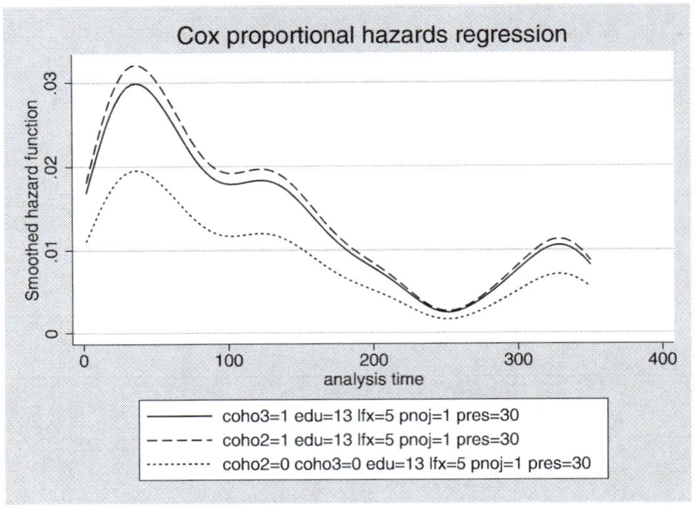

Figure 9.5.2 Smoothed baseline rates for Birth Cohorts 1, 2, and 3.

to plot the estimated hazard functions. Figure 9.5.2 shows the plot of the estimated baseline rate.

9.6 Application Example

Finally, we want to give at least one application example for the Cox model. For this purpose, we use a demographic study of divorce rates (Blossfeld, et al. 1995). From a methodological point of view, this example is particularly instructive, because it compares a piecewise constant exponential model (see chapter 5) with a Cox model. Both models are proportional hazards models. The Cox model leaves the baseline hazard rate completely unspecified, while the piecewise constant exponential model attempts to approximate the baseline rate with a series of period-specific constants. In other words, for sufficiently small time periods, a piecewise constant exponential model becomes a Cox model, which provides a direct estimation of the baseline hazard rate.

The application example comes from an international collaborative research project conducted by Alessandra De Rose (Italy) and Jan M. Hoem (Sweden) together with two authors of this book on the relationship between changes in women's educational attainment and the development of marriage disruption risks in modern societies.

The study focused mainly on two hypotheses. The first states that a higher level of education increases the risk of marital breakdown. This was to be expected, as women with higher educational attainment normally have higher abilities and are more willing to go against social norms by dissolving unhappy marriages. They are also better able to cope with the consequences, as they are better able to deal with the social or religious stigma of divorce. The second hypothesis states that the importance of women's educational attainment for disruption risks is not constant over historical time. In particular, it changes as family systems are transformed. Since the three countries under study (Italy, West Germany, and Sweden) represent very different stages in the development and differentiation of socially accepted living arrangements and levels of divorce risks, a comparison of the gradient of the effects of women's educational level on divorce risks in these countries should show that the "liberating" impact of a woman's higher educational attainment declines sharply as the family system becomes less "traditional" and as divorce habits become more permissive for all women. Specifically, the effect of women's education on divorce was expected to be stronger in Italy than in West Germany and stronger in West Germany than in Sweden.

Using event history data from Italy, West Germany, and Sweden, the authors specified women's divorce rate as a function of time-constant and time-dependent covariates using two models: an exponential model with marriage-duration-specific piecewise constant baseline rates, and a proportional hazards model that did not specify the baseline hazard rate at all.

The observation started with the time of first birth within the first marriage and ended with the event of the first date of de facto separation or, for right-censored cases, with the date of the interview, with a marriage duration of 15 years, with the date of birth of the fourth child, or with some other very specific life events (such as the death of the husband, the birth of twins, etc.)—whichever occurred first. The reason we did not go beyond a marriage duration of 15 years and did not include women with more than three children was methodological: These marriages took place under more specific conditions with very specific divorce risks.

Although the transition rate analysis focused on the effect of women's educational attainment, the authors included a series of covariables to avoid compositional effects and and to clearly highlight the direct (or

Table 9.6.1 Estimation results for a Cox model and a piecewise constant exponential model for divorce rate (results taken from Blossfeld et al. 1995).

Variable	Cox model Italy	Cox model Germany	Cox model Sweden	Piecewise constant Italy	Piecewise constant Germany	Piecewise constant Sweden
Period 0–24				-9.68	-9.58	-7.18
Period 24–48				-9.37	-7.41	-6.52
Period 48–72				-9.03	-7.48	-6.30
Period 72–96				-8.68	-7.31	-6.60
Period 96–120				-8.26	-7.07	-6.37
Period 120–144				-8.42	-7.33	-5.94
Period 144–168				-8.49	-6.65	-6.15
Period 168–				-8.66	-7.42	-5.80
Cohort 1939–43	0.36	-0.52	0.16	0.36	-0.52	0.16
	1.36	*1.54*	*0.69*	*1.36*	*1.55*	*0.71*
Cohort 1944–48	0.37	-0.10	0.49	0.38	-0.09	0.49
	1.38	*0.31*	*2.19*	*1.40*	*0.28*	*2.22*
Cohort 1949–53	0.21	0.29	0.37	0.21	0.33	0.38
	0.69	*0.88*	*1.38*	*0.70*	*0.99*	*1.41*
Cohort 1954–58	0.38	1.01	1.05	0.38	1.07	1.04
	1.08	*2.71*	*2.79*	*1.08*	*2.87*	*2.75*
Age at marriage	-0.83	-0.30	-0.56	-0.84	-0.30	-0.56
20–22	*3.82*	*1.22*	*3.55*	*3.82*	*1.21*	*3.55*
Age at marriage	-0.71	-0.65	-0.74	-0.71	-0.64	-0.74
23–25	*3.11*	*1.81*	*3.58*	*3.10*	*1.79*	*3.58*
Age at marriage	-1.21	-0.67	-0.65	-1.21	-0.66	-0.65
26–28	*3.44*	*1.18*	*2.10*	*3.44*	*1.17*	*2.10*
Age at marriage	-1.04	-1.49	-1.27	-1.03	-1.46	-1.27
28–	*2.39*	*1.43*	*2.11*	*2.38*	*1.40*	*2.11*
Second child	-1.04	-0.40	-0.75	-1.04	-0.42	-0.76
	5.32	*1.74*	*4.52*	*5.35*	*1.81*	*4.63*
Third child	-1.31	-0.66	-0.72	-1.31	-0.65	-0.74
	4.02	*1.64*	*2.96*	*4.02*	*1.63*	*3.06*
Education	0.77	0.42	0.30	0.77	0.42	0.30
middle level	*4.24*	*1.70*	*1.98*	*4.24*	*1.73*	*1.98*
Education	1.74	1.22	0.53	1.74	1.24	0.54
high level	*5.79*	*2.66*	*2.77*	*5.78*	*2.70*	*2.78*
Pregnant	0.76	0.87	0.38	0.76	0.88	0.38
at marriage	*4.12*	*4.11*	*2.78*	*4.13*	*4.13*	*2.77*
Log likelihood	-1325.3	-614.4	-1698.2	-1495.8	-759.0	-1849.5
χ^2	117.8	58.4	71.9	118.4	60.3	73.1
df	13	13	13	20	20	20

Note: Coefficients divided by their estimated standard errors are shown in *italics*.

partial) effect of educational level on the disruption risk. Most of the effect patterns for the controls were as expected and are not considered further here.

In the piecewise constant model shown in Table 9.6.1, a set of eight different time periods for marriage duration was applied. It is easy to see that the divorce rate over the entire duration of marriage is consistently lower in Italy than in Germany and consistently lower in Germany than in Sweden. There are therefore persistent differences in the risk of divorce over the duration of the marriage in each country. In all three countries, the divorce rate slightly increases with marriage duration up to a duration of about 10 years. After that, divorce rates in Sweden and West Germany become more or less stable, while in Italy a kind of trend reversal can be observed: long-term marriages become increasingly stable. To what extent this effect of marriage duration is due to selective attrition through divorce, and the amount that may be due to actual changes in propensity to divorce with duration, cannot be discussed effectively here. Nor can we answer the question of whether the reasons for divorce change with duration. Controlling for the patterns of marriage duration in the piecewise constant model and in the Cox model shows at least that the piecewise constant model provides almost identical estimates for the observed variables (cohort membership, age at marriage, children, educational level, and pregnancy at marriage) for all three countries. Thus, the results are relatively robust against the model specification. The eight period dummies in the piecewise constant model seem to yield a reasonably good approximation of the baseline hazard rate. This model, of course, has the additional advantage over the Cox model that it provides direct estimates of the baseline rate.

After controlling for important covariates for the divorce rate, it is possible to evaluate the hypothesis proposed above. Women's educational attainment was included in the form of three levels (reference category: "low educational level"). Table 9.6.1 shows that an increasing level of educational attainment does indeed have a positive effect on the rate of divorce in all three countries. Furthermore, there is indeed the expected order of these effects between the three countries. Compared to women with a lower secondary qualification, the divorce propensity of women with an upper secondary qualification is about 470 % in Italy, about 239 % in West Germany, and about 70 % in Sweden. This result means that in all three countries women's educational attainment can be used at least in part as an explanation for the high divorce rates. This is especially true in Italy, where all significant cohort effects on the divorce rate could be explained by women's increasing educational attainment. From this point of view, the initial rise in divorce rates from the mid-1960s to the mid-1980s and the following leveling off and stabilization of these rates in countries that have already reached a relatively high level of divorce rates can be understood as an expression of an initially increasing and then decreasing imbalance not only between elements of the macrostructure of society (educational expansion and structure of the family system), but also of the respective roles and expectations of the microstructure in intimate relationships.

Notes

[1]The linear combination of a row vector of possible time-dependent covariates, $A(t)$, and a column vector of associated coefficients, α, cannot contain an intercept because this is absorbed in the baseline rate, $h(t)$.

[2]The formulation implies that the baseline rates are allowed to vary between different transitions. A special type of Cox model arises if the baseline rates are constrained to be identical for all transitions (see Kalbfleisch and Prentice 1980, p. 170).

[3]We follow the formulation by Lawless (1982, p. 346). An introduction to Breslow's approximation can also be found in Namboodiri and Suchindran (1987, p. 213).

[4]Since the partial likelihood estimation is based on the order of events and censored times, the decision on a proper definition of time-dependent covariates should be based on substantive considerations, taking into account that, for a causal interpretation, the cause should precede its possible effect.

[5]Our experience with various event history models in social research suggests that changes in the specification of time-dependence can often have a strong effect on the coefficients of other time-dependent covariates, but usually have almost no effect on the coefficients of time-constant covariates.

[6]$(\exp(0.5046) - 1) \cdot 100\,\% \approx 65.6$.

[7]$(\exp(0.5046 - 0.8608) - 1) \cdot 100\,\% \approx 30\,\%$.

[8]Caution is needed here, because in this example we have not taken into account that the covariate effects are not proportional for men and women; see section 9.3.

[9]Metric covariates could be used by grouping the individual data into appropriate classes.

[10]See Kalbfleisch and Prentice (1980, p. 87), Lawless (1982, p. 365), and Blossfeld, Hamerle, and Mayer (1989).

[11]For the random effects model, see chapter 10.2.4.

Chapter 10

Problems of Model Specification

We have already emphasized that time dependence can be interpreted from different perspectives. In chapter 7, we discussed time dependence from a substantive perspective in which it was derived from a theoretically assumed underlying diffusion process or as an expression of a theoretically important latent causal factor operating in time. In this chapter, we approach time dependence from a methodological point of view and consider it as a consequence of unobserved heterogeneity. We will show that such a change of perspective means that scientists can investigate the same data using different assumptions while logically reaching valid but perhaps contradictory substantive conclusions (Manski 1993). Thus, distinguishing between competing (substantive and methodological) interpretations of the same data can be considered one of the most challenging problems in causal analysis. It is therefore not surprising that the estimation of event history models in the presence of omitted variables has become a focus of much technical work in recent years.

Section 10.1 discusses the issue of unobserved heterogeneity in general and provides some illustrative examples. Some researchers have proposed using so-called mixture models to account for unobserved heterogeneity. These approaches are demonstrated in section 10.2. In particular, we show how to estimate transition rate models where unobserved heterogeneity is assumed to follow a gamma distribution. However, our discussion is mainly critical of the usefulness of these approaches, as there is generally no way to make reliable assumptions about what is *unobserved*. Section 10.3 summarizes the discussion and stresses that unobserved heterogeneity is only *one* aspect of a broad variety of model specification problems.

10.1 Unobserved Heterogeneity

Using transition rate models, we try to find empirical evidence about how the transition rates, which describe the movements of individuals (or other units of analysis) in a given state space, depend on a set of covariates. Unfortunately, we are not always able to include all the important factors. One reason for this is the limitation of available data; we would like to include some important variables, but we simply do not have the information. Furthermore, we often do not know what is important. So what are the consequences of this situation? There are two main aspects to consider.

The first is well known from traditional regression models. Since our covariates are normally correlated, the parameter estimates depend on the specific set of covariates included in the model. Any change in this set is likely to lead to a change in the parameter estimates for the variables that were already included in previous models. Thus, in practice, the only way to proceed is to estimate a series of models with different specifications and then check whether the estimation results are stable or not. However, this sequential model specification and estimation can also be seen as a resource for theoretical development. Such a procedure can provide additional insights into the so-called context sensitivity of causal effects in the social world.

Second, a change in the covariates in a transition rate model very often also leads to a change in the time-dependent shape of the transition rate. A similar effect occurs in traditional regression models: Depending on the set of covariates, the empirical distribution of the residuals changes. But, in contrast to regression models, in which the residuals are normally only used to check the model assumptions, the residuals in transition rate models become the focus of the modeling. Indeed, when transition rate models are reformulated as regression models, the transition rate becomes a description of the residuals, and any change in the distribution of the residuals becomes a change in the time-dependent shape of the transition rate. Consequently, the empirical insight that a transition rate model provides on the time-dependent shape of the transition rate depends more or less on the set of covariates used to estimate that model.[1] Thus, the question is whether a transition rate model can provide at least some reliable insights about a time-dependent transition rate.

Before discussing this question, some examples are in order to illustrate possible consequences of "unobserved heterogeneity" (see also Vaupel and Yashin 1985; Blossfeld and Hamerle 1992). Let us assume a single transition from only one origin state to only one destination state, so that the episodes can be described by a single random variable T for the duration in the origin state. The distribution of T can be represented by a density function $f(t)$, a survivor function $G(t)$, or by a transition rate $r(t) = f(t)/G(t)$.

Let us now assume that the population consists of two groups that are distinguished by a variable x, where $x = 1$ for Group 1 and $x = 2$ for Group 2. Then we can define duration variables T_1 and T_2 separately for these two groups. And, like T, these duration variables can be described separately by density, survivor,

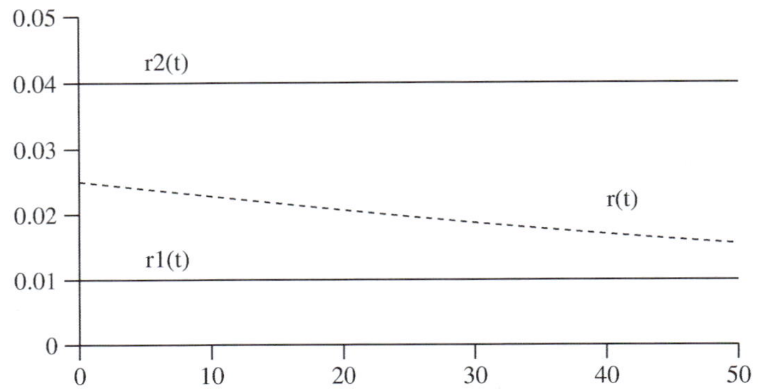

Figure 10.1.1 Mixture of two exponential transition rates.

and transition rate functions:

$$\begin{aligned} T_1 &\sim f_1(t),\ G_1(t),\ r_1(t) = f_1(t)/G_1(t), \\ T_2 &\sim f_2(t),\ G_2(t),\ r_2(t) = f_2(t)/G_2(t). \end{aligned}$$

The transition behavior in both groups can be quite different. If we are able to separate the two groups, we can gain insight into these differences by estimating $f_i(t)$, $G_i(t)$, and $r_i(t)$ separately for both groups ($i = 1, 2$). However, if we have no information about the group membership, we can only estimate the transition behavior in both groups together; that is, we can estimate only $f(t)$, $G(t)$, and $r(t)$.

The important point is that the distribution describing T can be quite different from the distributions describing T_1 and T_2; or, put in more substantive terms, the transition behavior in the whole population can be quite different from the transition behavior in the two subpopulations (groups). To describe this more formally, we can proceed as follows. First, let π_1 and π_2 denote the proportions of the two groups at the beginning of the process ($\pi_1 + \pi_2 = 1$). Then we find that

$$\begin{aligned} f(t) &= \pi_1 f_1(t) + \pi_2 f_2(t) \quad \text{and} \\ G(t) &= \pi_1 G_1(t) + \pi_2 G_2(t). \end{aligned}$$

Using the general formula $r(t) = f(t)/G(t)$, we can now derive the following expression for the transition rate in the population:

$$r(t) = r_1(t)\,\pi_1 \frac{G_1(t)}{G(t)} + r_2(t)\,\pi_2 \frac{G_2(t)}{G(t)}. \tag{10.1}$$

The transition rate $r(t)$ in the population is a weighted average of the transition rates in the two groups, $r_1(t)$ and $r_2(t)$. But the weights change during the process. In fact, the weights are just the proportions of the two groups in the "risk set" at every point in time.[2] We will give three examples:

Example 1: Suppose a sample of employees is divided into two subpopulations denoted by $x = 1$ (men) and $x = 2$ (women). The job-exit rates for these two groups are supposed to be constant:[3]

$$\begin{aligned} r_1(t) &= 0.01, \\ r_2(t) &= 0.04, \end{aligned}$$

and the proportions of the two groups in the population are assumed to be $\pi_1 = 0.5$ and $\pi_2 = 0.5$. The survivor functions are then

$$\begin{aligned} G_1(t) &= \exp(-0.01\,t), \\ G_2(t) &= \exp(-0.04\,t), \\ G(t) &= 0.5\exp(-0.01\,t) + 0.5\exp(-0.04\,t). \end{aligned}$$

Then, using (10.1), we find the transition rate

$$r(t) = \frac{0.01\exp(-0.01\,t) + 0.04\exp(-0.04\,t)}{\exp(-0.01\,t) + \exp(-0.04\,t)}.$$

Figure 10.1.1 illustrates the transition rates. It is obvious that the transition rate in the population $r(t)$ declines, although the transition rates in both subpopulations are constant. The reason for this is that the proportion of both groups in the risk set changes continuously. For instance, at the beginning of the process,

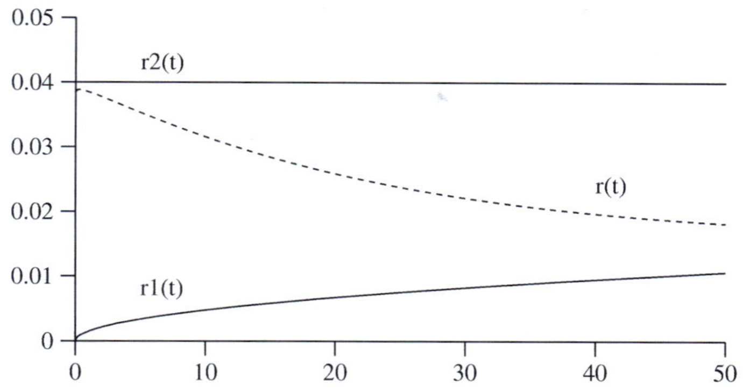

Figure 10.1.2 Mixture of an exponential and a Weibull transition rate.

there are 50 % men and 50 % women. But women, in this example, leave their jobs faster than men, which is shown by their greater transition rate. Therefore, after (say) 36 months we find

$$\begin{aligned} G_1(t) &= \exp(-0.01 \cdot 36) = 0.698, \\ G_2(t) &= \exp(-0.04 \cdot 36) = 0.237, \\ G(t) &= 0.5 \exp(-0.01 \cdot 36) + 0.5 \exp(-0.04 \cdot 36) = 0.468. \end{aligned}$$

This means that only 30.2 % of the men, but already 76.3 % of the women have left their jobs. The risk set at $t = 36$ consists of $0.5 \cdot 0.698/0.468 = 74.6\,\%$ men and $0.5 \cdot 0.237/0.468 = 25.3\,\%$ women.

Example 2: Now assume the same setup as in Example 1, but the transition rate in Group 1 is of the Weibull type, for instance (with $a = 0.01$ and $b = 1.5$):[4]

$$\begin{aligned} r_1(t) &= 1.5 \cdot 0.01^{1.5} \cdot t^{0.5}, \\ r_2(t) &= 0.04. \end{aligned}$$

The survivor functions are then

$$\begin{aligned} G_1(t) &= \exp\big(-(0.01\,t)^{1.5}\big), \\ G_2(t) &= \exp\big(-0.04\,t\big), \\ G(t) &= 0.5 \exp\big(-(0.01\,t)^{1.5}\big) + 0.5 \exp\big(-0.04\,t\big); \end{aligned}$$

and, using (10.1), we find the transition rate

$$r(t) = \frac{1.5 \cdot 0.01^{1.5} \cdot t^{0.5} \exp\big(-(0.01\,t)^{1.5}\big) + 0.04 \exp\big(-0.04\,t\big)}{\exp\big(-(0.01\,t)^{1.5}\big) + \exp\big(-0.04\,t\big)}.$$

Figure 10.1.2 illustrates the transition rates for this example. While the transition rate is increasing in Group 1 and constant in Group 2, the resulting transition rate for the population *de*creases.

Example 3: Finally, suppose that both groups follow a Weibull transition rate, for instance,

$$\begin{aligned} r_1(t) &= 1.5 \cdot 0.01^{1.5} \cdot t^{0.5}, \\ r_2(t) &= 1.2 \cdot 0.04^{1.2} \cdot t^{0.2}. \end{aligned}$$

The survivor functions are then

$$\begin{aligned} G_1(t) &= \exp\big(-(0.01\,t)^{1.5}\big), \\ G_2(t) &= \exp\big(-(0.04\,t)^{1.2}\big), \\ G(t) &= 0.5 \exp\big(-(0.01\,t)^{1.5}\big) + 0.5 \exp\big(-(0.04\,t)^{1.2}\big); \end{aligned}$$

and, using (10.1), we find the transition rate $r(t) =$

$$\frac{1.5 \cdot 0.01^{1.5} \cdot t^{0.5} \exp\big(-(0.01\,t)^{1.5}\big) + 1.2 \cdot 0.04^{1.2} \cdot t^{0.2} \exp\big(-(0.04\,t)^{1.2}\big)}{\exp\big(-(0.01\,t)^{1.5}\big) + \exp\big(-(0.04\,t)^{1.2}\big)}.$$

Figure 10.1.3 illustrates the transition rates for this example. Both groups follow an increasing transition rate, but the transition rate for the population $r(t)$ is completely different as it is bell-shaped (see also Blossfeld and Hamerle 1989a).

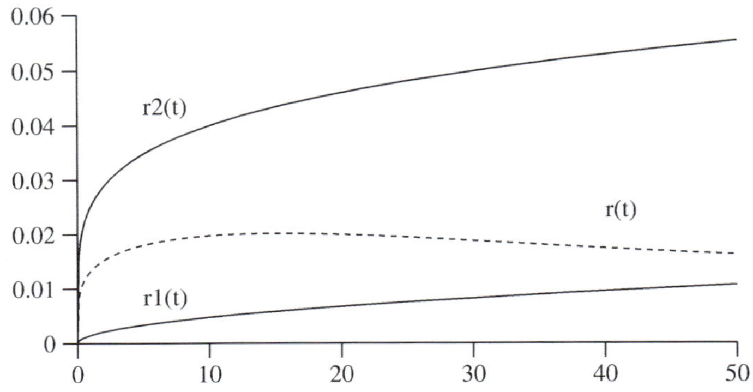

Figure 10.1.3 Mixture of two Weibull transition rates.

These examples show that the transition rate estimated for a population *can* be the result (a mixture) of quite different transition rates in the subpopulations. What are the consequences? First, this result means that an observed transition rate at the population level can be *explained* as the result of different transition rates in subpopulations. Of course, this is only a sensible strategy if we are able to identify important subpopulations. To pursue this strategy, of course, one needs *observable* criteria to divide a population into subpopulations. Merely speculating that an observed transition rate *could* be the result of quite different transition rates does not provide any further explanation. Even if there is unobserved heterogeneity (and we can usually be sure that we could not include all important covariates), more or less arbitrary distributional assumptions about unobserved heterogeneity do not lead to better models. On the contrary, the estimation results will be more dependent on assumptions than would otherwise be the case (Lieberson 1985). Therefore, we would like to emphasize our view that the most important basis for any progress in model building is sufficient and appropriate data.

There remains the problem of how to interpret a time-dependent transition rate from a causal perspective. The question is: Can time be considered as a proxy for an unmeasured variable that generates a time-dependent rate, or is it simply an expression of unobserved heterogeneity that does not allow for a substantive interpretation?[5]

10.2 Models with a Mixture Distribution

There are several suggestions on how to deal with unobserved heterogeneity in transition rate models. The basic idea underlying these proposals is to include an "error term" into the model specification. This is quite familiar from traditional regression models, but specific problems arise when trying to follow this strategy in the specification of transition rate models. These problems become visible if we begin with the following general specification of a transition rate model:

$$r(t) = f(x, \beta, \epsilon).$$

In this specification, the transition rate, $r(t)$, is a function of a vector of covariates, x, a vector of model parameters, β, to be estimated, and a stochastic variable ϵ to represent an error term. In order to make such a model estimable, additional specifications are required. First, the functional relationship $f(.)$ must be specified. There are many ways to do this, as already explained in earlier chapters of this book. But that is not enough. One also needs some assumptions about the distribution of ϵ.

This becomes clearer if we write our transition rate model as a regression model, for instance, in the following way:[6]

$$\log(T) = x\beta + \epsilon.$$

To simplify the argument, we assume that the logarithm of the duration variable, T, depends on a linear combination of covariates, $x\beta$, and, additively, on an error term, ϵ. Depending on our assumptions about the distribution of ϵ, this becomes a transition rate model. For example, we can assume that ϵ is normally distributed, $\epsilon \sim \mathcal{N}(0, \sigma^2)$—a common assumption in regression modeling. Consequently, T (conditional on the covariates) follows a log-normal distribution with the density function

$$\frac{1}{t\,\sigma}\,\phi\left(\frac{\log(t) - \beta}{\sigma}\right)$$

where ϕ is used to denote the standard normal density function. In fact, this is the lognormal transition rate model as described in section 7.5. However, there is then no longer an error term. What has happened? Our assumption about the error term in the regression model has become an assumption about the distribution of the duration variable T; so, in a sense, our transition rate model already contains an "error term"; it is, so to speak, the *distribution* of the duration variable T *conditional* on the covariates. However, this formulation is somewhat misleading, because the focus of the model has changed. In traditional regression modeling, the focus is on the *expectation* of the dependent variable, $\mathrm{E}(T)$. This focus motivates the reference to an "error term," i.e. the residuals, when attempting to explain the *expectation* of T given the covariates x. The focus in transition rate models is quite different. It is *not* the expectation of T, but the *distribution* of T (i.e., the transition rate describing how the process in a population proceeds) if time progresses. We are not interested in how the expectation of T depends on the covariates, but in how the distribution of T (i.e., the transition rate) depends on the covariates. In statistical terms, there is not much difference between these two modeling approaches; but in terms of substantive theory, the difference is quite significant. The transition rate has a substantive meaning in event history modeling as it refers to a "propensity" of individuals to change states in a state space. It is therefore not just a statistical concept used to make assumptions about an error term.

What would a transition rate model look like that contains an *additional* error term? We could split, for instance, the error term, ϵ, into two components: $\epsilon = \epsilon_1 + \epsilon_2$. Written as a regression model, we then get

$$\log(T) = x\beta + \epsilon_1 + \epsilon_2.$$

We can then make an assumption about the distribution of ϵ_1 to obtain a distributional form for T and finally arrive at a transition rate model containing an additional error term, ϵ_2. Of course, to make the model estimable, we also need a distributional assumption about ϵ_2. This is quite possible in purely formal terms, but we will obviously run into a problem. Having made distribution assumptions about ϵ_1 and ϵ_2, we have, in fact, made an assumption about $\epsilon = \epsilon_1 + \epsilon_2$. We have specified a mixture distribution for ϵ and, consequently, also for T. Thus, we finally obtain a mixture model for the duration variable, and we have lost our error term again.

The key point is that the whole procedure of splitting an error term into components is quite arbitrary. From an empirical point of view, we only have a single set of residuals *defined* by a model specification. It would be completely arbitrary to separate these residuals into two parts: one part to be interpreted as a description of the transition rate and another part to be interpreted as an error term. If there were strong theoretical and empirical reasons to justify a specific distributional assumption about the transition rate, it could make sense to include an additional error term. However, at least in sociological research, this is almost never the case. In fact, we are not aware of a single application in which the researcher had strong a priori reasons (with respect to the given data) to justify the selection of a specific type of transition rate model.

It therefore seems to make no sense to specify and estimate transition rate models on the basis of a mixture distribution—at least given the current state of sociological research. However, this result is inconclusive. The fact remains that we almost never have sufficient reasons to choose a specific transition rate model. Our estimation results will depend on the type of model. We have discussed this before and concluded that the best strategy is always to estimate a broad variety of different models in order to find robust estimation results. Thus, if one looks at the problem from this perspective, mixture models can be helpful in another way. Although these models have no utility in separating an error term from a "true" transition rate, they greatly enrich the range of models that can be used in the search for robust estimation results. Thus, if the above-mentioned strategy is followed, the use of mixture models makes perfect sense. They can be quite helpful in separating robust estimation results (i.e., estimation results that are to a large degree independent of a specific model specification) from "spurious" results, which could be defined as being highly dependent on a particular type of model.

For these reasons, we provide a brief technical discussion of mixture models in the following sections. We examine two approaches: One type of approach begins with a fully parametric specification of the distribution of the error term. In particular, the use of a gamma distribution has been proposed (Tuma 1985).[7] This approach is discussed in the following section. Another approach, proposed by Heckman and Singer (1984), is based on a discrete mixture distribution and could be described as semiparametric, since only very weak assumptions are necessary. It is not dealt with in this book.[8]

10.2.1 Models with a Gamma Mixture

In this section we follow the usual way of treating the error term in transition rate models. This is based on the assumption that there is a time-invariant unobserved constant that is specific for each individual. Furthermore, it is assumed that this unobserved constant can be represented as the realization of a random variable that is identically distributed for all individuals and independent of observed covariates (Davies and Crouchley 1985; Crouchley and Pickles 1987). This error term is then multiplicatively linked to the transition rate. The idea is that this random variable captures both individual heterogeneity and deviations from the supposed shape of duration dependence (Galler and Pötter 1990).

More precisely, the transition rate $r(t \mid x, v)$ is assumed to depend on a vector, x, of observed covariates and on a scalar stochastic term, v, that is not observed. Our question is how the variables in x influence the transition rate, but this cannot be observed directly. In any observed sample of episodes, the transition rate also depends on the values of v.[9]

The same applies to the associated density and survivor functions. The relationship between transition rate, density, and survivor function must therefore be written as follows

$$r(t \mid x, v) = \frac{f(t \mid x, v)}{G(t \mid x, v)}, \tag{10.2}$$

$$G(t \mid x, v) = \exp\left(-\int_0^t r(\tau \mid x, v)\, d\tau\right). \tag{10.3}$$

In order to find an estimation approach, we make some simplifying assumptions. First, that the transition rate can be expressed as

$$r(t \mid x, v) = r^u(t \mid x)\, v \qquad v \geq 0. \tag{10.4}$$

We assume that the component $r^u(t \mid x)$ of this expression, called the *underlying* rate, is given parametrically according to one of the standard parametric transition rate models discussed in chapters 4 and 7. The cumulative transition rate can then be written as

$$H(t \mid x, v) = v\, H^u(t \mid x) = v \int_0^t r^u(\tau \mid x)\, d\tau. \tag{10.5}$$

The second basic assumption is that the stochastic term, v, follows a gamma distribution with expectation $\mathrm{E}(v) = 1$. This assumption implies that the density function of v can be written as

$$f_v(v) = \frac{\kappa^\kappa\, v^{\kappa-1}}{\Gamma(\kappa)} \exp(-\kappa\, v) \qquad \kappa > 0.$$

The variance of this distribution is $\mathrm{Var}(v) = 1/\kappa$. The next step is to calculate the resulting mixture distribution for the observed durations. First, we obtain for the density and survivor functions

$$f(t \mid x) = \int_0^\infty f(t \mid x, v)\, f_v(v)\, dv,$$

$$G(t \mid x) = \int_0^\infty G(t \mid x, v)\, f_v(v)\, dv.$$

These mixtures are expectations according to the distribution of v. The calculation is easy, because the gamma distribution implies the basic equality

$$\int_0^\infty \exp(-v\, s(t))\, f_v(v)\, dv = \left[1 + \frac{1}{\kappa}\, s(t)\right]^{-\kappa}$$

that holds for any real valued function $s(t)$.[10] Using (10.3) and (10.5), the unconditional survivor function is therefore

$$G(t \mid x) = \int_0^\infty \exp\left(-v\, H^u(t \mid x)\right) f_v(v)\, dv = \left[1 + \frac{1}{\kappa} H^u(t \mid x)\right]^{-\kappa}. \tag{10.6}$$

The unconditional density function can be determined by differentiating the negative value of the unconditional survivor function, resulting in

$$f(t \mid x) = r^u(t \mid x) \left[1 + \frac{1}{\kappa} H^u(t \mid x)\right]^{-\kappa-1}. \tag{10.7}$$

Finally, the unconditional transition rate is

$$r(t \mid x) = \frac{f(t \mid x)}{G(t \mid x)} = r^u(t \mid x) \left[1 + \frac{1}{\kappa} H^u(t \mid x)\right]^{-1}. \tag{10.8}$$

Several Destination States. So far, we have derived a mixture model for episodes with one origin and one destination state. However, the extension to a situation with several destinations is simple. For the transition to the destination state k, the model specification is

$$r_k(t \mid x, v_k) = r_k^u(t \mid x)\, v_k \qquad v_k \geq 0$$

where $r_k^u(t \mid x)$ is the underlying rate used for model formulation. The error term v_k is assumed to be specific for each transition and gamma-distributed with unit expectation. Finally, in the same way as previously shown for the single transition case, we can derive expressions for the observed transition rates that can be used for model estimations.

Maximum-Likelihood Estimation. The approach to deriving transition rate models with an additional gamma-distributed error term can be summarized as follows: We start with a parametrically given transition rate, as in (10.4), and assume that the unobserved heterogeneity is gamma-distributed with unit mean and variance $1/\kappa$. We then derive the observed mixture distribution described by $f(t \mid x)$, $G(t \mid x)$, and $r(t \mid x)$, given in (10.6), (10.7), and (10.8), respectively. This applies to the case of a single transition, but an extension to the case to several transitions is completely analogous. It follows that the log likelihood can be set up in the usual way. We only give the formula for the single transition case. With $\mathcal{N}$ and $\mathcal{E}$ denoting the sets of all episodes and of episodes with an event, respectively, the log likelihood can be written as

$$\ell = \sum_{i \in \mathcal{E}} \log(r(t_i)) + \sum_{i \in \mathcal{N}} \log(G(t_i)) =$$
$$\sum_{i \in \mathcal{E}} \log\left(r^u(t_i)\right) - \log\left(1 + \frac{1}{\kappa} H^u(t_i)\right) - \sum_{i \in \mathcal{N}} \kappa \log\left(1 + \frac{1}{\kappa} H^u(t_i)\right).$$

Unfortunately, the usual way of applying the episode-splitting method is no longer possible. Let us assume an episode $(0, t)$, starting at time zero and ending at time t, and let us assume that there is a covariate that changes its value at t_x $(0 < t_x < t)$. Then it would be perfectly possible to include this information in the calculation of the cumulated rate $H^u(t \mid x)$, that is needed to calculate the density and survivor functions of the unconditional distribution. However, the episode-splitting method is only applicable in its usual form if the calculation can be separated into two steps. The first step has the information about the $(0, t_x)$ split, and the second has information about the (t_x, t) split. Howerver, it is obvious that the unconditional density and survivor functions cannot be calculated in these two distinct steps. Therefore, we use split episodes in our examples.

Gamma Mixture Models in Stata. In principle, the approach of adding a gamma-distributed error term can be used with any parametric transition rate model. To request the estimation of the model with an additional gamma-distributed error term, you only have to select the option `frailty(gamma)`. The estimation results will show you the estimated parameters, an estimate of the variance of the frailties, and a likelihood-ratio test of the null hypothesis that this variance is zero. You can also choose the inverse-Gaussian distribution by simply typing `frailty(invgauss)`.

10.2.2 Exponential Models with a Gamma Mixture

To illustrate the estimation of gamma mixture models with Stata, we first use an exponential model. The transition rate, which now depends on v, can be written as

$$r(t \mid v) = a\, v \qquad a,\, v \geq 0.$$

The underlying model, which has already been described in chapter 4, is the exponential model, which means that $r^u(t) = a$ and $H^u(t) = a\, t$. The observed mixture distribution is described by

$$\begin{aligned} G(t) &= (1 + d\, a\, t)^{-\frac{1}{d}}, \\ f(t) &= a\, (1 + d\, a\, t)^{-\frac{1}{d}-1}, \\ r(t) &= a\, (1 + d\, a\, t)^{-1} \end{aligned}$$

where d is the variance of the gamma distributed error term. If $d > 0$, there is of course a negative time dependence, although the underlying rate is a time-independent constant.

The model has two parameters: a, the constant transition rate of the underlying exponential model, and d, the variance of the mixing gamma distribution. The parameterization is as follows:

$$a = \exp(A\alpha) \quad \text{and} \quad d = \exp(\delta).$$

It is assumed that the first component of the covariate (row) vector, A, is a constant equal to one. The associated coefficient vector, α, together with δ, are the model parameters to be estimated. The estimate of the variance of the mixing gamma distribution is $\exp(\delta)$.

Box 10.2.1 Estimation results of **ehd2.do** with a gamma mixture

```
         failure _d:  des
   analysis time _t:  tf

Fitting exponential model:

Fitting constant-only model:
Iteration 0:   log likelihood = -903.19585
Iteration 1:   log likelihood = -902.99555
Iteration 2:   log likelihood = -902.99548

Fitting full model:
Iteration 0:   log likelihood = -912.23868
Iteration 1:   log likelihood = -870.86706
Iteration 2:   log likelihood = -866.14271
Iteration 3:   log likelihood = -866.07659
Iteration 4:   log likelihood =  -866.0763
Iteration 5:   log likelihood =  -866.0763

Exponential PH regression
Gamma frailty

No. of subjects =          600       Number of obs    =        600
No. of failures =          458
Time at risk    =        40782
                                     LR chi2(6)       =      73.84
Log likelihood  =    -866.0763       Prob > chi2      =     0.0000

------------------------------------------------------------------------
      _t |      Coef.   Std. Err.      z    P>|z|     [95% Conf. Interval]
---------+--------------------------------------------------------------
     edu |   .0872679   .0310473     2.81   0.005     .0264163    .1481195
   coho2 |   .5706931   .1444685     3.95   0.000       .28754    .8538462
   coho3 |    .416298   .1512207     2.75   0.006     .1199109    .7126851
     lfx |  -.0046331   .0010997    -4.21   0.000    -.0067884   -.0024778
    pnoj |   .0850445    .054537     1.56   0.119    -.0218461    .1919351
    pres |  -.0320635   .0066215    -4.84   0.000    -.0450415   -.0190855
   _cons |  -3.986876   .3430481   -11.62   0.000    -4.659238   -3.314515
---------+--------------------------------------------------------------
/lntheta |  -.8019158   .1836758    -4.37   0.000    -1.161914    -.441918
---------+--------------------------------------------------------------
   theta |   .4484689   .0823729                      .3128868    .6428024
------------------------------------------------------------------------
Likelihood-ratio test of theta=0: chibar2(01)    = 47.72
                                  Prob>=chibar2 = 0.000
```

To demonstrate the model estimation, we use do-file **ehd2.do** which is shown in Box 4.1.3. The estimation results for an exponential model without a gamma-distributed error term are shown in Box 4.1.4. In order to estimate a mixture model, we type

```
streg edu coho2 coho3 lfx pnoj pres, dist(e) nohr fr(gamma)
```

Box 10.2.1 presents estimates of the parameter vector α (see `_t`) and an estimate of δ (see **/lntheta**). The comparison of these estimates with the results of the simple exponential model in Box 4.1.4 shows that there are no substantial changes in the estimated parameters. In addition, we obtain a quite significant estimate for the variance of the mixing gamma distribution: $\hat{d} = \exp(-0.8019) = 0.448$. This is supported by using the likelihood-ratio test statistic used to compare the two models. This is 47.7 with one degree of freedom. The result indicates that we have a decreasing transition rate absorbed by the error term in relation to the assumption of a constant transition rate for the underlying model (see Figure 10.2.1). Of course, this does not prove that all deviations from a constant transition rate are due to "unobserved heterogeneity." As discussed at the beginning of this section, the benefit of estimating a mixture model is not to separate an error term from a "true" transition rate, but to provide some means for assessing the robustness of parameter estimates for *observed* covariates.

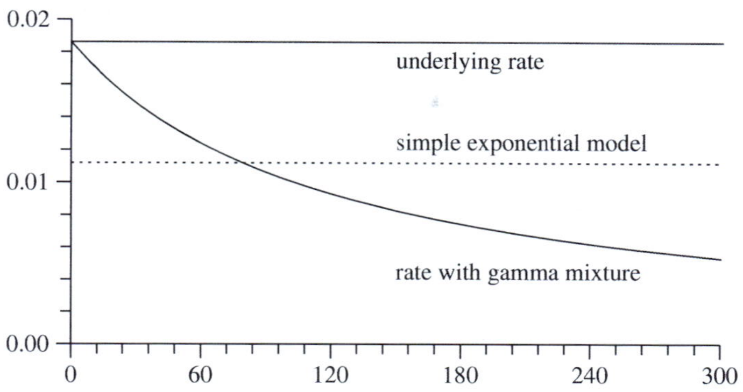

Figure 10.2.1 Transition rates estimated with an exponential model without (dotted) and with (solid) a gamma mixture distribution. Model specification according to Box 4.1.3; all covariate values set to zero.

10.2.3 Weibull Models with a Gamma Mixture

For our second illustration, we use the Weibull model (see section 7.3). The transition rate, now conditional on v, is $r(t \mid v) = r^u(t)\, v$ with $v \geq 0$. The underlying transition rate and the cumulative rate are given by

$$r^u(t) = b\, a^b\, t^{b-1} \quad \text{and} \quad H^u(t) = (a\, t)^b$$

according to the standard Weibull model. The observed mixture distribution is described by

$$\begin{aligned} G(t) &= \left(1 + d\,(a\,t)^b\right)^{-\frac{1}{d}}, \\ f(t) &= b\, a^b\, t^{b-1} \left(1 + d\,(a\,t)^b\right)^{-\frac{1}{d}-1}, \\ r(t) &= b\, a^b\, t^{b-1} \left(1 + d\,(a\,t)^b\right)^{-1}. \end{aligned}$$

The model has three parameters: a and b are the parameters of the underlying Weibull distribution, and d represents the variance of the mixing gamma distribution. The model is parameterized as follows:

$$a = \exp(-A\alpha), \quad b = \exp(B\beta), \quad d = \exp(\delta).$$

It is assumed that the first component of each of the covariate (row) vectors, A and B, is a constant equal to one. The corresponding coefficient vectors, α and β, together with δ, are the model parameters to be estimated.

To illustrate the model estimation, we use do-file `ehg6.do` shown in Box 7.3.3 (see section 7.3 on Weibull models). The estimation results for a Weibull model without a gamma-distributed error term are shown in Box 7.3.4. To estimate a mixture model, we specify the `frailty(gamma)` option. Estimation results are shown in Box 10.2.2.

Box 10.2.2 shows estimates of the parameter vector α (see `_t`) and estimates of β_0 (see `/ln_p`) and δ (see `/lntheta`). Comparing the estimates for α with the corresponding estimates in Box 7.3.4 (i.e., the simple Weibull model without a gamma mixture distribution) shows that there are no substantial changes.[11] However, not only do we obtain a fairly significant estimate for the variance of the mixing gamma distribution, $\hat{d} = \exp(\hat{\delta}) = 2.189$. The shape parameter of the underlying Weibull distribution has also changed considerably. In the simple Weibull model, it was $\hat{b} = 0.914$; now it is $\hat{b} = \exp(0.6693) = 1.953$. As shown in Figure 10.2.2, the transition rate has changed quite dramatically. Although the Weibull model only allows for monotonically increasing or decreasing transition rates, mixing the Weibull distribution with a gamma distribution allows for bell-shaped transition rates; and since this seems most appropriate for our data, such a rate can eventually be estimated. The underlying Weibull transition rate is just an arbitrary part of the mixture distribution, or something like a residual to make the mixing distribution fit the data. This again demonstrates that one should be very careful when interpreting the various parts of a mixture distribution in theoretical terms.

These examples show that using a mixture model can significantly improve a model's ability to fit a given set of data with one of its possible time-dependent transition rates. As a by-product, this could lead to more reliable parameter estimates for the covariates.

As a final example, Figure 10.2.3 shows the transition rates for a log-normal model with and without a mixing gamma distribution.[12] Since we already know that the log-normal distribution is appropriate for our data, we do not expect much improvement by using a mixture distribution. However, as can be seen in Figure 10.2.3, the final mixing distribution shows that the (population level) transition rate is somewhat steeper than that which can be expressed with the simple log-normal model.

Box 10.2.2 Estimation results of `ehg6.do` with a gamma mixture distribution

```
       failure _d:  des
 analysis time _t:  tf

Fitting Weibull model:

Fitting constant-only model:

Iteration 0:   log likelihood = -1010.7268
Iteration 1:   log likelihood = -875.38172
Iteration 2:   log likelihood =  -871.8774
Iteration 3:   log likelihood = -871.85633
Iteration 4:   log likelihood = -871.85632

Fitting full model:

Iteration 0:   log likelihood = -896.80052
Iteration 1:   log likelihood = -838.04895
Iteration 2:   log likelihood = -836.07108
Iteration 3:   log likelihood = -834.66076
Iteration 4:   log likelihood = -834.66053
Iteration 5:   log likelihood = -834.66053

Weibull AFT regression
Gamma frailty

No. of subjects =          600      Number of obs    =        600
No. of failures =          458
Time at risk    =        40782
                                    LR chi2(6)       =      74.39
Log likelihood  =   -834.66053      Prob > chi2      =     0.0000

----------------------------------------------------------------------
      _t |     Coef.   Std. Err.      z    P>|z|     [95% Conf. Interval]
---------+------------------------------------------------------------
     edu | -.0719787   .0265541    -2.71   0.007    -.1240239   -.0199335
   coho2 | -.4640506   .1205172    -3.85   0.000      -.70026   -.2278411
   coho3 | -.4014352    .128362    -3.13   0.002      -.65302   -.1498504
     lfx |  .0041267   .0008468     4.87   0.000      .002467    .0057865
    pnoj | -.1143679   .0469785    -2.43   0.015     -.206444   -.0222918
    pres |  .0260338   .0053382     4.88   0.000     .0155711    .0364966
   _cons |  3.492617    .295304    11.83   0.000     2.913832    4.071402
---------+------------------------------------------------------------
   /ln_p |  .6692704   .0879964     7.61   0.000     .4968007    .8417401
/lntheta |  .7833431    .171606     4.56   0.000     .4470016    1.119685
--------+-------------------------------------------------------------
       p |  1.952812   .1718404                      1.643455    2.320401
     1/p |   .512082   .0450614                      .4309599    .6084742
   theta |  2.188777   .3756073                      1.563617    3.063888
----------------------------------------------------------------------
Likelihood-ratio test of theta=0: chibar2(01)   = 104.09
                                  Prob>=chibar2 = 0.000
```

Specification of the Error Term

The main disadvantage of mixture models is that parameter estimates *can* be highly sensitive to the assumed parametric form of the error term. For example, Heckman and Singer (1982), who based the systematic part of the model on the same assumptions, estimated four different unobserved heterogeneity models: one with a normal, one with a log-normal, and one with a gamma distribution of the error term, as well as a model with a nonparametric specification of the disturbance. They found that the estimates of these models were surprisingly different. In other words, the identification problem might be shifted only to another level: Misspecification of the transition rate by neglecting the error term might be replaced by misspecification of the parametric distribution of the error term. Thus, a powerful theory for the distribution of the stochastic error term would be required. However, given the various sources of misspecification described later, it is very unlikely that any theory can provide solid guidance for a specific parametric distribution of the error term. In other words, misspecification is likely.

This makes a semiparametric specification of the error term attractive. Heckman and Singer (1982, 1984) proposed to use a discrete distribution with a varying number of mass points to approximate the unknown distribution of the error term. Unfortunately, this solution also only shifts the identification problem to an-

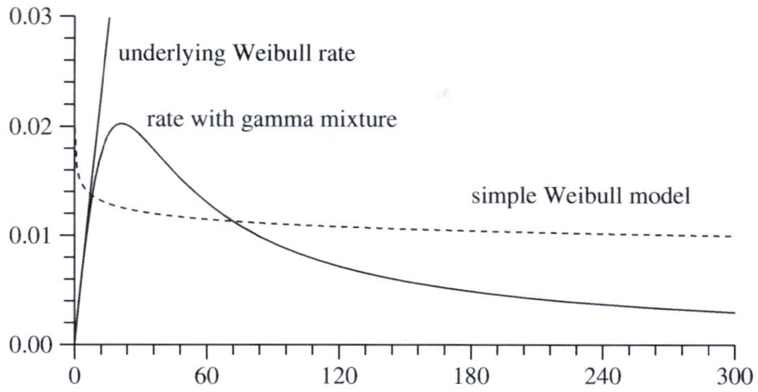

Figure 10.2.2 Transition rates estimated with a Weibull model without (dotted) and with (solid) a gamma mixture distribution. Model specification according to Box 7.3.3; all covariate values set to zero.

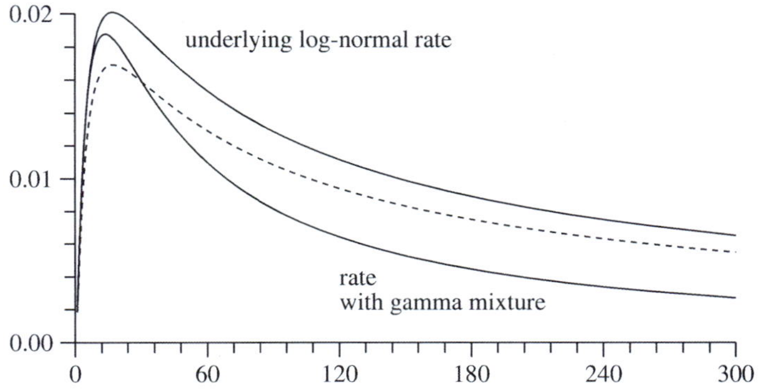

Figure 10.2.3 Transition rates estimated with a log-normal model without (dotted) and with (solid) a gamma mixture distribution. Model specification according to Box 7.5.3; all covariate values set to zero.

other level. This is because Trussell and Richards (1985) reported a strong dependence of the semiparametric estimator on the choice of the duration dependence in transition rate models. Once more, a powerful argument would be needed to justify a particular parametric form of duration dependence. However, as discussed in section 7.1, this will rarely be the case in empirical applications (see also Heckman and Walker 1987).

A further step towards solving the identification problem would be to adopt a nonparametric specification of both the duration dependence and the unobserved heterogeneity term. However, it appears that these models are only identifiable in the case of complete survivor functions for all values of the covariates (Elbers and Ridder 1982). They are also very sensitive to small perturbations in the values of the estimated survivor function (Galler and Pötter 1990). Thus, given the common limitations of event history data in practical research (censored information, measurement error), it seems impossible to estimate both duration dependence and unobserved heterogeneity nonparametrically in one model (Galler and Pötter 1990).[13]

10.2.4 Random Effects for Multiepisode Data

In chapter 9.4, we have already presented a method for dealing with multie-pisode data in the framework of Cox models. The method presented there is to assume a specific hazard rate for each individual. As usual in semiparametric regression models, the hazard rates are not estimated, but controlled for. Here, we want to provide some information on fully parametric alternatives implemented in Stata.

In section 10.2, we started from the accelerated failure time model and split the error term into two parts, ε_1 and ε_2. Now we consider that different persons j might have several spells i. The error term is split into one part, $\varepsilon_{\mathrm{ij}}$, which is specific to the spells of persons and describes the shape of T, and a second part, which is person-specific, υ_{j}. This results in

$$log(T_{\mathrm{ij}}) = x_{\mathrm{ij}}\ \beta_{\mathrm{ij}} + \varepsilon_{\mathrm{ij}} + \upsilon_{\mathrm{j}}.$$

By transforming this formula, it also covers proportional hazard models:

$$r(t_{\mathrm{ij}}) = h_0(t)\ exp(x_{\mathrm{ij}}\beta_{\mathrm{ij}} + \upsilon_{\mathrm{j}}),$$

with the baseline rate $\mathrm{h}_0(\mathrm{t})$.

One method that can be used to estimate such a model is the shared-frailty model. This model is a generalization of frailty models. To fit such a model, which assumes that the frailty is group-specific, you can use the option `shared(varname)`. Inserting the person identifier `id` in this option leads to a person-specific ν_j. By default, Stata assumes that the error term ν is gamma-distributed, as described in the previous section. Since the syntax is straightforward, we do not present it here.

Stata also offers other methods for estimating random-effects survival-time models. These are based on the assumption that the person-specific errors, which are unobservable, follow a normal distribution with $N(0,\sigma^2_\nu)$.[14] The syntaxes are rooted in different traditions: one in the tradition of panel data and one in the tradition of multilevel analysis.

In the field of panel data analysis, Stata uses the prefix `xt`. `xt` is used to set the data, describe data and variables, and perform the model estimation. In addition to `stset`, we have to declare the data as panel data using the command `xtset panelvar`, where `panelvar` refers to the (group or) person identifier. It does not matter whether `stset` or `xtset` comes first. In the next step, the `streg` commands described in chapters 4 to 7 are given the prefix `xt`. The semi-parametric Cox model cannot be estimated. A syntax for an exponential transition rate model with a random intercept for persons is:

```
stset tf, failure(des)          /*define single episode data*/
xtset id
xtstreg edu coho2 coho3 lfx pnoj pres, dist(e) nohr
```

According to the output in Box 10.2.3, persons have between 1 and 9 jobs with an average of 3.0 (see `Obs per group` and the following lines). The lower part of Box 10.2.4 contains information on the person-specific variance `/sigma2_u`, which is estimated as `.496`. In addition, Stata automatically performs a likelihood-ratio test to determine whether the random coefficient model provides a better fit compared to the ordinary exponential model. Since the test statistic is almost 41 with one degree of freedom, this model fits the data much better.

If we compare the results with those in Box 4.1.4, it can be seen that the current model places more emphasis on the previous number of jobs and less on general labor force experience. According to Box 10.2.3, jobs with a higher ordinal number are more stable (there is no significant effect in Box 4.1.4), and general labor force experience has no influence on job exits (there is a highly significant and negative coefficient in Box 4.1.4). The current model places more emphasis on differences within persons. Persons with many jobs might have especially short ones at the beginning of their career and more stable ones later on. In a purely single-episode model, these effects might be set off by some people with only one permanent, lifelong job. This might cause the differences in results between Boxes 4.1.4 and 10.2.3.

The other convention for syntax originates from multilevel analysis, where terms such as Level 1 observations (spells) are nested in Level 2 (persons). In the simplest version, which is called a random-intercept model, we have to declare the data as survival time data. Then we add the prefix `me` to the estimation command and extend the command with `|| lev2:` before the comma. `lev2` is a variable in the data set, e.g. a person identifier. In our example, the syntax is:

```
stset tf, failure(des)     /*define single episode data*/
mestreg edu coho2 coho3 lfx pnoj pres || id:, dist(exp) nohr
```

The output is placed in Box 10.2.4. There is hardly any difference to the previous results. The small deviations are due to the different numbers of integration points. By default, `xtstreg` uses 12; `mestreg`, 7 points. In both commands, the integration points can be specified with the option `intpoints(#)`. The use of the same numbers of integration points leads to identical estimation results.

Commands with the prefix `me` can be extended in several ways: First, researchers can specify random coefficients for some covariates, often referred to as random slopes. If a researcher has good reasons to assume that the effect of a covariate `x1` on the transition intensity varies over persons, a random effect can be specified for this covariate. To do this, the covariate x1 must be placed after the colon `|| lev2: x1`. In addition, the `mestreg` command allows the specification of a third level `|| lev3: || lev2:` . The Stata reference manual illustrates a three-level model: namely, job spells (Level 1) nested in persons (Level 2) and persons nested in birth cohorts (Level 3) with a slightly modified version of our data set `rrdat1.dta` (StataCorp 2017b, pp. 458–461).

Box 10.2.3 Estimation results using do-file **ehd2.do** with a random intercept specified with Stata's syntax on panel data (**xtset, xtstreg**)

```
Random-effects exponential PH regression        Number of obs      =        600
Group variable:                    id           Number of groups   =        201

                                                Obs per group:
                                                              min =          1
                                                              avg =        3.0
                                                              max =          9

Integration method: mvaghermite                 Integration pts.   =         12

                                                Wald chi2(6)       =      88.62
Log likelihood = -2445.5658                     Prob > chi2        =     0.0000
---------------------------------------------------------------------------
        _t |      Coef.   Std. Err.      z    P>|z|     [95% Conf. Interval]
-----------+---------------------------------------------------------------
       edu |   .0844583   .0344922     2.45   0.014      .0168549    .1520617
     coho2 |   .6205112    .175156     3.54   0.000      .2772118    .9638106
     coho3 |   .5474561   .1744061     3.14   0.002      .2056265    .8892858
       lfx |  -.0014849   .0011192    -1.33   0.185     -.0036784    .0007087
      pnoj |  -.1963633   .0680519    -2.89   0.004     -.3297426   -.062984
      pres |  -.0315989   .0066048    -4.78   0.000      -.044544   -.0186538
     _cons |  -4.285118   .3922293   -10.93   0.000     -5.053873   -3.516363
-----------+---------------------------------------------------------------
/sigma2_u |   .4960335   .1214078                       .3070247    .8013989
---------------------------------------------------------------------------
LR test vs. exponential model: chibar2(01) = 40.84 Prob >= chibar2 = 0.0000
```

Box 10.2.4 Estimation results using do-file **ehd2.do** with a person-specific random intercept estimated by a multilevel mixed effects parametric survival model (**mestreg**)

```
Mixed-effects exponential PH regression             Number of obs      =    600
Group variable:                    id               Number of groups   =    201

                                                    Obs per group:
                                                                  min =      1
                                                                  avg =    3.0
                                                                  max =      9

Integration method: mvaghermite                     Integration pts.   =      7

                                                    Wald chi2(6)       =  88.56
Log likelihood = -2445.5646                         Prob > chi2        = 0.0000
---------------------------------------------------------------------------
        _t |      Coef.   Std. Err.      z    P>|z|     [95% Conf. Interval]
-----------+---------------------------------------------------------------
       edu |   .0844504   .0345023     2.45   0.014      .0168272    .1520736
     coho2 |   .6205506   .1751808     3.54   0.000      .2772025    .9638987
     coho3 |   .5474475   .1744568     3.14   0.002      .2055184    .8893766
       lfx |  -.0014809   .0011203    -1.32   0.186     -.0036766    .0007149
      pres |  -.0315992   .0066052    -4.78   0.000     -.0445452   -.0186532
     _cons |  -4.284749   .3923536   -10.92   0.000     -5.053748    -3.51575
-----------+---------------------------------------------------------------
id         |
var(_cons)|   .4973179   .1227258                       .3066045    .8066584
---------------------------------------------------------------------------
LR test vs. exponential model: chibar2(01) = 40.85 Prob >= chibar2 = 0.0000
```

10.3 Discussion

So far, we have discussed the issue of "unobserved heterogeneity." We have tried to show that the neglect of important covariates is indeed a serious source of model misspecification. However, we have also tried to show that there are no simple solutions to this problem. In particular, transition rate models with a mixture distribution do not provide a simple solution to the problems posed by unobserved heterogeneity, although they can be quite helpful in the search for robust parameter estimates. This is simply because these models provide a broader range of possible shapes for the transition rate, so they can better fit a given set of data. Nonetheless, the most promising strategy for coping with unobserved heterogeneity is to look for more appropriate data.

One should, however, not forget that unobserved heterogeneity is only one source of problems in our search for useful and robust models. In fact, there are many other problems in model specification that are no less important. It therefore seems sensible to conclude this chapter with a brief discussion of the most important problems of causal analysis with event history data.

(1) The most significant aspect of model specification is to find an appropriate definition of the state space. For example, consider an investigation of women's transition rates between the states "employed,"

"housewife status," and "unemployed" over the life course. If women in part-time and full-time employment behave differently and/or if covariates have distinct effects that are dependent on the origin and destination states "full-time" and "part-time employment," then a study neglecting the full-time and part-time divide in the state space will produce omitted differences.[15] Another instructive example of misspecification of the state space was given in chapter 4, where we demonstrated that estimating the effect of education on the job-exit rate is problematic because education has a positive effect on upward moves and a negative effect on downward moves. Thus, not taking differences in destination states into account leads to a confounding of the contradictory effects of education.

(2) It may not be possible to include all important variables in the model specification. As already stressed earlier, the resulting parameter estimates are more or less dependent on the set of covariates actually used, and omitting some important covariates is therefore an important source of misspecification.[16] For instance, we have shown in some of our examples that the job-exit behavior of men and women is quite different in some respects; but in many of our analyses we have not explicitly considered gender differences.[17] Therefore, this might be an important source of misspecification. In a serious study, one would include covariates for gender differences and investigate the resulting changes in parameter estimates.

(3) Another important type of misspecification occurs when the initial conditions at entry into the episode (in single and, much more importantly, in multiepisode models) are not taken into account. Differences at the beginning of the spells are usually the consequence of previous selection processes. As mentioned above, most sociological research is based on nonexperimental observations of highly selective processes, which as such may influence the estimated transition rate. Let us take the job-exit example again. It is very likely that (a) there is self-selection among individual employees depending in part on their preferences for particular jobs, (b) there is selection by employers due to their hiring and personnel policies, and (c) there is selectivity due to forces exogenous to the covariates included in the model (e.g., socioeconomic background, ethnicity, etc.). In short, various sorting processes have already taken place when individuals enter into a new job episode, and they may massively influence the results of estimated transition rates. This is particularly true in models with multiple episodes, where one should try to additionally control for the dependencies between successive episodes for each of the individuals (see section 4.3). Although event history models allow for the inclusion of information about process history much better than cross-sectional or panel models, few researchers can claim that their models have included all relevant variables on the history of the process.

(4) Misspecification can result when important time-dependent covariates are not taken into account or when time-dependent covariates are modeled as time-constant covariates and measured only once.[18] In chapter 6, we demonstrated that time-dependent covariates are observations of the sample path of parallel processes (at the individual level, the level of small groups, the intermediate level of organizations or households, or at the macrolevel of a society). Omitting these important processes, or including them only as time-constant covariates, is, of course, a source of misspecification. Although we have argued that the real strength of event history models, at least compared to cross-sectional and panel models, is their ability to include these parallel processes appropriately, it can easily happen that important time-dependent covariates are neglected or have to be neglected due to missing data. The effect of time-dependent covariates can also be very sensitive to the points in time at which they should change their value. This was demonstrated in Example 3 of section 6.6, which analyzed the effect of pregnancy on entry into marriage for couples living in consensual unions.

(5) Possible misspecification can occur if interaction effects are not taken into account. This is a serious issue in many social science applications, as the relationship between the transition rate and any given (time-constant or time-dependent) explanatory variable is often not the same across all values of the remaining covariates. If a transition rate model includes more than two covariates, many interaction terms (even of higher order) may be required. An instructive example of incorrect estimation results due to an omitted interaction term was given in chapter 9, where we included marriage as a time-dependent covariate in the job-exit rate model without taking into account that it has opposite effects on the job-exit rate for men and women.

(6) Another problem that might result in misleading estimation results is measurement errors. In transition rate models, this problem concerns not only covariates but also the information on the starting and ending times of episodes (see the discussion in Sandefur and Tuma 1987).

(7) Another, but no less important, source of misspecification is the model type. As we have discussed in previous chapters, there is a broad range of transition rate models that can be used for practical research. The use of different model types is likely to lead to different estimates of the effects of covariates on the transition rate. As discussed in chapter 8, there are some methods to investigate whether a model fits the data, but these methods are quite limited. Based on maximum likelihood estimation, there are generally no

conclusive statistical criteria that can be used for model selection. Thus, in practice, the most useful strategy is to estimate and compare a variety of different models and to find out to what extent the estimation results are robust (i.e., do not depend on the selected model; see also Galler and Pötter 1992).

(8) Finally, the functional relationship between the transition rate and the various types of covariates just discussed (as well as duration dependence and possible interaction terms) could also be a source of misspecification. Typically, linear or log-linear functions are used to link model parameters to covariates. There are, however, many other possibilities, and the estimation results may depend on the type of link function used to specify a model.

To summarize, specification bias is pervasive in empirical social research. What can be recommended in such a situation—a situation that does not provide a satisfactory technical solution to the problem of model specification? First, it is important to remember that unobserved heterogeneity means that important information is either omitted or not observable in the model. Thus, if we have a clear hypothesis about what it is or could be that is not in the model, then we can offer the following advice: Try to find better data that allows a representation of the important factors in the model. If, however, we have no idea about what is missing from the model, it would certainly be wrong to assume away the problem after diligently attempting to include a large number of variables in the covariate vector as controls. As Lieberson (1985) has convincingly argued, this control variable approach not only leads to an underestimation or overestimation of the true relationship. In certain circumstances, the true relationship is even approximated before rather than after controls. A similar situation can occur when a stochastic error term is assumed to capture the unobserved heterogeneity. Since, by definition, we have no information about what is not observed, we are not able to assess assumptions about its distribution.

In particular, we do not believe that all the various sources of misspecification described earlier can be integrated into just one stochastic error term. For example, including an error term into a job-exit rate model does not bring us closer to a solution of the problem that education has a positive effect on upward moves and a negative effect on downward moves. Of course, such a conclusion can only be drawn on the basis of theoretical considerations. In fact, our brief discussion of possible sources of misspecification and the various examples in this book demonstrate that misspecification can at least be greatly reduced if event history model selection is theoretically guided. In our job-exit model, for example, an improvement in the model specification could be achieved simply by estimating it with regard to upward and downward exits, differentiating between men and women, including parallel processes (e.g., family events, events at the level of the firm, changes in the macrostructure), as well as more appropriate information on the history of the process. The choice of a model is therefore not simply a technical decision that is made at the beginning of an analysis and can then be forgotten. Rather, the results depend critically on a theoretically guided model choice, and any differences among models can only be reconciled if one is aware of the type of model used, and of the fact that interpretation may be contingent on the model chosen.

As stated by Manski (1993), the identification of causal factors is not an all-or-nothing proposition. Very often, we have only proxy variables or insufficient data to estimate the value of a parameter; but we may still be able to bound the parameter, identify its direction of influence, or its statistical significance. Sociological theory, at least at its current stage of development, rarely predicts a specific value of the coefficients. Most hypotheses we have seen in empirical social science applications specify the direction of influence of the causal variables, or whether they have any effect at all. Thus, it seems that the fixation of sociologists on point estimation has inhibited appreciation of the usefulness of statements about bounds (Manski 1993). All in all, we need to develop a greater tolerance for ambiguity. As Manski (1993, p. 51) puts it, "Social scientists (...) often seem driven to draw sharp conclusions, even when these can be generated only by imposing much stronger assumptions than can be plausibly defended. Scientific reporting and policy analysis would be improved if we would face up to the fact that we cannot answer all of the questions that we ask."

At the end of this book, we describe in the next chapter the method of sequence analysis, which is often a very useful method for describing event history data.

Notes

[1]It also depends on the type of model. As shown in chapter 7, all parametric models are only capable of capturing a limited range of shapes of a time-dependent transition rate.

[2]The "risk set" at time t is the set of individuals who have not yet had a transition before t. Therefore, it is the number of individuals at $t = 0$ multiplied by the survivor function $G(t)$.

[3]That is, we assume an exponential model for both groups; see chapter 4.

[4]See section 7.3 for a definition of the Weibull model.

[5]There is, however, another aspect to this problem. From a purely descriptive point of view, a time-dependent transition rate provides some empirically relevant information about the population itself. Some researchers seem to assume that this is *not* the case. The argument is as follows: Since a time-dependent transition rate could be the result of quite different transition rates in unobserved subpopulations, the transition rate for the population could only be "apparent." We agree with

this reasoning insofar as there is always the possibility that the observed transition rate for a population could be the result of quite different transition rates in certain subpopulations. This delivers an opportunity to *explain* the transition rate at the population level by referring to the various transition rates in subpopulations. However, we do not agree with the conclusion that the observed transition rate at the population level is only "apparent." It remains an empirical fact. Otherwise, it would make no sense to explain the transition rate for the population. The conclusion should be quite different. Instead of using the examples given previously to support speculation about "true" versus only "apparent" time dependencies in the transition rate, statements about time-dependent transition rates should be made explicitly dependent on specific populations. Then it is not contradictory to claim, for example, that there are different transition rates for both men and women and that a completely different transition rate applies to a population consisting of men and women. If, to stay with this example, there are constant, but different transition rates for both men and women, the transition rate for the population consisting of men and women will decrease with time. Thus, the statement that there is a decreasing transition rate for the population consisting of men and women will not become false if we learn that this population consists of men and women and that transition rates for men and women are time-constant but different. The statement about the transition rate for the population remains true and provides a more or less valuable *empirical* insight.

[6]Many authors have discussed the formal similarities between regression and transition rate models; see, for instance, Lawless (1982) and Petersen (1990, 1991). See also the discussion of SAS models in Blossfeld, Hamerle, and Mayer (1989).

[7]In principle, any other distributional assumption resulting in correspondingly defined mixture models is possible. The gamma distribution is just a particularly convenient choice because the resulting mixture distribution is easily tractable.

[8]Trussell and Richards (1985) discuss their experience with this approach.

[9]This stochastic variable v corresponds to the error term ϵ_2, which was previously introduced in the regression framework. Since the regression model was formulated in terms of $\log(T)$, the error term now becomes $\exp(\epsilon_2)$ and is multiplicatively linked to the transition rate.

[10]See, for instance, Lancaster (1990, p. 328).

[11]Again, note that the signs of the estimates of the coefficients in parameter vector α have to be reversed in order to compare these estimates with the other examples in this book.

[12] The model is specified as shown in Box 7.5.3 (do-file `ehg11.do`); see section 7.5 on log-normal models. The gamma mixture model is estimated using the do-file `ehg11.do` with the additional option `frailty(gamma)`.

[13] A potentially successful compromise between a fully nonparametric and a parametric specification of the duration dependence could be a piecewise constant exponential model, as shown in chapter 5. This model could be mixed with a nonparametric specification of the unobserved error term. However, if a larger number of dummy variables is used, the identification of this model, while theoretically possible, would require a large amount of exact data (see Galler and Pötter 1990).

[14]For the likelihood function and estimation methods, see Stata's reference manual (StataCorp 2017a, 2017b).

[15]However, to some extent one could also include the differences of full-time and part-time employment by using a dummy variable and some of its interaction effects with other covariates.

[16]We do not speak of "biased" parameter estimates because the concept of unbiased estimation has no clear meaning outside of descriptively defined estimation problems. The problem is that our estimates are more or less dependent on model specification, so the question is how to find an appropriate model that may, in turn, depend heavily on its intended use.

[17]We did this because our example data set is small and we only wanted to use one basic model throughout the book for didactic reasons.

[18]A serious source of misspecification is often a "one time" measurement of explanatory variables (usually at the time of a retrospective interview), sometimes even after the event under study has occurred (see Sandefur and Tuma 1987).

Chapter 11

Sequence Analysis

BRENDAN HALPIN

11.1 What is Sequence Analysis?

Sequence analysis can be defined as an approach to the analysis of longitudinal data (such as life-course histories) that treats whole runs of such data as units ("sequences") and defines measures of similarity or difference between pairs of sequences. It uses this pairwise measure for a variety of purposes, including (1) generating data-driven classifications of sequences, (2) assigning sequences to *a priori* empirical or theoretical ideal types, (3) examining how diversity of sequences differs across groups (such as cohorts or social classes), and (4) examining the determinants of dyadic sequence similarity (e.g., between spouses). Sequence analysis also defines a number of unary measures, such as sequence complexity.

As a form of longitudinal data analysis sequence analysis complements event history analysis, by focusing on time series such as life-course careers as wholes rather than on their specific features such as transition rates. By focusing on wholes, it can gain an overview of the patterns of the trajectories in a way that more focused techniques might not achieve. It has descriptive and exploratory strengths, but is less adapted to testing well-specified statistical hypotheses.

Sequence analysis was introduced into sociology by Andrew Abbott in a series of articles arguing the importance of taking time into account (see expecially Abott and Tsay, 2000). Abbott was inspired by Sankoff and Kruskal (1983) who, *inter alia*, introduced to a wider audience the optimal matching algorithm that efficiently and flexibly calculates distances between pairs of sequences. As a result, optimal matching analysis (OMA) and sequence analysis have come to be almost synonymous in the social sciences. Thanks in part to OMA's early implementation in the software package TDA (Rohwer and Pötter 2007) it also became popular in Europe (see Brüderl and Scherer 2004); and with the development of the R package TraMineR (Gabadinho et al. 2009) has become even more widely established in the social sciences (see also Blanchard et al. 2014). Whereas OMA is very popular, a number of other algorithms for defining inter-sequence distance are in common use and these will be discussed below.

11.1.1 Sequence Data

Sequence analysis can be applied to any sort of ordered categorical data. Categorical life histories are an important example. The most common applications of the computer science techniques behind sequence analysis are in text processing and molecular biology. Neither macromolecules such as DNA nor textual data are, in a simple sense, temporally longitudinal, but they are ordered. Applications in the social sciences include the sequences of pages visited in an interaction with a web site (Wang and Zaïane 2002), steps in a dance (Abbott and Forrest 1986), the rhetorical structure of journal articles (Abbott and Barman 1997), or the steps a survey agency takes in order to contact respondents (Kreuter and Kohler 2009). It is also frequently applied to life-course or labor-market histories such as the transition from school to work (McVicar and Anyadike-Danes 2002; Bras et al. 2010; Anders and Dorsett 2017), the transition to retirement (Han and Moen 1999; Fasang 2012; McDonough et al. 2017), or trajectories in multiple domains (Pollock 2007; Wiggins et al. 2007). See Liao et al. (2022) for a more extensive overview also Halpin (2013).

The basic nature of a sequence is an ordered set of observations in a categorical space. Sequences can be of uniform or varying length. We can consider the essential task of sequence analysis to be the mapping of information on differences within this categorical state space to differences between trajectories that move within that space. Along with this key focus on pairwise similarity or distance, sequence analysis also has some unary measures that summarize characteristics of individual sequences such as their complexity.

11.1.2 The Value of a Holistic View

Time series such as life-course sequences or labor-market histories can be seen as epiphenomenal outcomes of more fundamental underlying processes that drive transitions between states. The whole history is then an accumulation of these processes over time (with a substantial element of chance). So why is it worth focusing on the traces, rather than on the underlying data generation process?

Whereas the underlying transition patterns may be apprehended more directly by methods such as event history analysis or Markov modeling, the processes are potentially complex and difficult to grasp in their entirety. The generating processes are complex:

- Individuals bring different characteristics from the beginning
- Individual history matters via both duration dependence and path dependence (individuals accumulate characteristics)
- Time matters:
  - Calendar time (e.g., economic cycles): The state distribution may change dramatically
  - Developmental time, or maturation: The rate and nature of transitions may change dramatically during the life course
  - Transitions are affected by and themselves affect processes in other life-course domains (e.g., fertility affects labor market participation, and *vice versa*).

Clearly, it would be impractical to model these processes in their full complexity using stochastic approaches such as event history analysis. This would require too many parameters. In order to remain tractable, most conventional analysis narrows the focus in some manner by, for example, looking only at single spells or durations to specific events. It is thus hard to gain an overview of the whole life course with conventional approaches, and there is also the possibility of *emergent* features: features that emerge out of the complexity of the multiple constitutive processes in a way that is difficult to predict.

Sequence analysis thus offers a capacity to provide overviews of long trajectory data. In contrast, however, it is somewhat weaker when it comes to statistically testing well-specified hypotheses. Thus, it is best seen as a complement to event history analysis that offers a descriptive and exploratory capacity.

11.2 Defining Distances

Defining distance between sequences is the key to sequence analysis. We can think of this operation as a mapping of distances between states in a state space to distances between sequences moving through that state space. A variety of means exist to carry out this mapping.

11.2.1 Hamming Distance

If the state space is real, we could imagine defining intersequence distance as the sum of the absolute (or squared) distance at each time point in a pair of sequences. If the state space is categorical, simple Hamming distance can be defined as the count of mismatches that is, points at which the sequences are not in the same state. If it is possible to define distances between the states (other than 0/1 for match/mismatch), we can calculate a more refined Hamming distance that is the sum of distances between the sequences at each time point. Hamming distance can be thought of as capturing full or partial similarity (or difference) at the same time.

11.2.2 Optimal Matching Distance

Hamming distance picks up similarity at the same time, but will not be sensitive to similarity that is displaced in time. Thus, `ABABA` and `BABAB` will count as completely dissimilar, despite having the subsequence `ABAB` in common. The optimal matching algorithm (OMA) calculates the Needleman-Wunsch distance (Needleman and Wunsch 1970) between two sequences, adding the operations of insertion and deletion to the direct state comparison carried out by the Hamming algorithm (referred to in this context as substitution). In this framework, we can delete the last `A` of `ABABA` and insert an initial `B` to obtain resulting sequences that match exactly. Whereas the Hamming distance is $5 \times d(A, B)$ the Needleman-Wunsch or OMA distance is twice the cost of the indel operation (since insertions and deletions are interchangeable—an insertion in Sequence 1 is effectively identical to a deletion in Sequence 2—we refer to them as *indels*). More generally, the OMA distance is the cost of the cheapest set of indels and substitutions that change one sequence into the other. The "optimal" of "optimal matching algorithm" derives from the fact that the algorithm determines the cost of the cheapest set of operations efficiently (determining the actual set of operations required is slightly more complicated, not least because more than one set can have the lowest cost).

The OMA distance can thus be considered as an extension of the Hamming distance that captures full or partial similarity (or difference) while allowing for some displacement in time (or "alignment"). Note that displacement is not costless: Similarity at similar time points will count as having a smaller distance than similarity at more different time points.

In effect, the substitution operation is exactly the same as the Hamming element-wise distance calculation, except that it can be displaced in time. For sequences of equal length, indels will be used only to improve the match; and the extent to which this takes place is affected by how large the indel cost is relative to the substitution costs. The lowest the indel cost can effectively be is half the largest substitution cost. If lower than this, the largest substitution cost will never be used, because an insertion plus a deletion will be cheaper. At the extreme, if the indel cost is less than or equal to half the *lowest* substitution cost, only indel operations will take place—in this case, the OMA distance is equal to the length of the longest common subsequence (LCS), which is in itself an important distance measure.

In the other direction, if the indel cost is sufficiently large, only substitution operations will be used; in which case, OMA returns the Hamming distance (for sequences of equal length; if sequence length differs, indels are always necessary). For life-course data such as that treated in this book with spells in states typically rather longer than the time unit used, indels will be used relatively infrequently, meaning that the OMA distance will not differ sharply from the Hamming distance. However, where it does differ, it usually picks up meaningful similarity that Hamming fails to detect.

Note that for life-course and similar data, it is generally not desirable to compare sequences of different length, partly because we are no longer comparing like with like, and partly because the difference in length of the pair of sequences may have a disproportionate effect relative to their content. It is best to compare such sequences over a fixed span, such as age range or total duration.[1] For sequences in other domains, such as, say, interviewer contact sequences in surveys, complete sequences may naturally have differing lengths, and comparing such sequences therefore makes sense.

11.2.3 Other Distances

A number of other distances are available. "Dynamic Hamming" is an algorithm that determines the distances between states on the basis of the transition rates between them, in which the transition rates are allowed to vary as time progresses. Dynamic Hamming distance is often used with time-diary sequence data, arguing that at times of the day when transitions are high (e.g., at the start of the typical working day), differences between states are less important than at times when transitions are low (Lesnard 2008). It is more appropriate for applications in which time has a "ruler" such as hours of the day, days of the week, or months of the year; and less appropriate for those in which time is more developmental and potentially elastic, such as the transition from school to work.

Time warp edit distance (TWED) works like OMA, but instead of insertion and deletion, it locally expands and compresses the time dimension. This is arguably better suited to use with spell data (Halpin 2014).

A number of sequence-analytic measures have been developed by Elzinga that focus on order and the concept of subsequence—that is, an ordered subset of a sequence (e.g., `ACE` is a subsequence of `ABCDE`). One measure is the number of matching subsequences, determined by enumerating all the subsequences in two sequences and counting those that match. A version of this measure allows the elements of sequences to be weighted by duration, so that each token can represent a spell in a particular state with a particular duration (Elzinga 2003; Elzinga et al. 2008).

11.2.4 Determining State Distances

Distances between the categories of the state space may be determined in a number of ways. It may be done on an *a priori* basis by the analyst using existing theory or empirical knowledge to relate the states. In the examples below, we assign distances in this manner—this is akin to, for example, putting linear scores on an ordinal variable. Empirical information can also be used: For example, one could score states by their mean value on observed variables (e.g., score occupational groups by average income, or average years of education) and use Euclidean distances based on the standardized scores.

A common practice is to use transition rates between the states as an index of their closeness. While this is intuitively appealing and in certain circumstances may be a good idea, it is by no means guaranteed that there will be a simple relationship between true distance and transition rates. Moreover, there is no guarantee that differences based on transition rates will be metric (i.e., will have the required properties of a distance measure). Matrices of state distances (substitution matrices) and transition matrices have the same form, but substitutions and transitions are very different phenomena: Transitions are changes that occur in time, whereas substitutions are cross-sequence comparisons of elements. Nonetheless, it may be a useful strategy.

See `help sdmksubs` for more information on generating transition-based substitution matrices in Stata. This command offers a number of ways of generating substitution costs from transition rates, including the conventional $2 - p_{ij} - p_{ji}$, a measure that uses all outflow proportions in rows i and j, measures based on residuals, and chi-square distance. Simulations suggest that measures that use more information than the basic $2 - p_{ij} - p_{ji}$ perform better (e.g., the "`dtype(pct)`" or "`dtype(pearson)`" options).

Topological models of the pattern of transitions may offer a useful alternative (Halpin and Chan 1998; Bukodi et al. 2016).

A final note: for distance measures that use the state distance, the values in the substitution matrix often matter more than the choice of algorithm (e.g., Hamming vs. OMA vs. TWED). The recommendation is to experiment with a number of settings to understand how they affect findings.

11.3 Doing Sequence Analysis in Stata

Sequence analysis in Stata is provided via an add-on, SADI (Halpin 2017). Installation of SADI, plus the required `moremata` package, is as follows (see the do-file `ehk1.do`):

Installation of both packages, plus the required `moremata` package, is as follows (see the do-file `ehk1.do`):

```
net from http://teaching.sociology.ul.ie/sadi
net install sadi
ssc install moremata
```

Note that SADI can also be installed from SSC via `ssc install sadi`, but it is less frequently updated there.

11.3.1 Example Data

In this chapter, we will use a data set from NEPS as used in the Appendix, but considered as sequence data. We will look at careers from age 16–35, creating a categorical state variable that is occupation in three categories (low, medium, and high on the International Socio-Economic Index [ISEI] (Ganzeboom and Treiman, 2003)) by full-time/part-time with not in the labor market as a separate category (resulting in a seven-state space).

We convert the spell format data to calendar format (from one record per spell to one record per individual with one state variable per time unit). We reduce the monthly data to 80 quarters, from age 16 to age 35, by taking the state in the first month of the quarter, because 240 months makes the sequences too long. This makes a very small (but detectable) difference to the resulting distances, but speeds up processing by a factor of at least 3^2 (the time to compare two sequences is approximately proportional to the product of their lengths; and because every pair is compared, involving up to $\frac{N(N-1)}{2}$ comparisons, this can be important). (See do-file `ehk2.do` for the code to carry out this data manipulation.)

11.3.2 A First Look at the Data

The principal visualisation commands provided are `sdchronoplot` and `sdindexplot`. These are based on Ben Jann's `heatplot` program (Jann, 2019) , which must first be installed (do "`ssc install heatplot`", "`ssc install palettes`" and "`ssc install colrspace`" to install in Stata). Most of the `heatplot` command's options can be used with `sdchronoplot` and `sdindexplot`.

The output of `sdchronoplot` is a *chronogram*, simply the time-dependent distribution of states, which gives a useful overview. Here we break it down by gender (see do-file `ehk3.do` for this section).

```
use rrcalendarq
sdchronoplot state*, by(sex) legend(subtitle("Status"))
```

As Figure 11.3.1 shows, nonemployment dominates at the beginning for both genders, but then declines. Almost all males slowly end up in full-time work, whereas women enter the labor market more quickly but retain a higher proportion of nonemployment. After about age 25, women spend more time in part-time work. Women are more present in medium ISEI occupations, with more men in both high and low ISEI occupations.

However, while this graphic device summarizes the evolution of labor market participation reasonably well, it loses all sight of longitudinal continuity and presents just period-specific summaries. If we want to view data that preserve the individual perspective, we need the graphic called the indexplot. This shows each individual's trajectory as a horizontal line. However, unless order is imposed, the graphic is not very

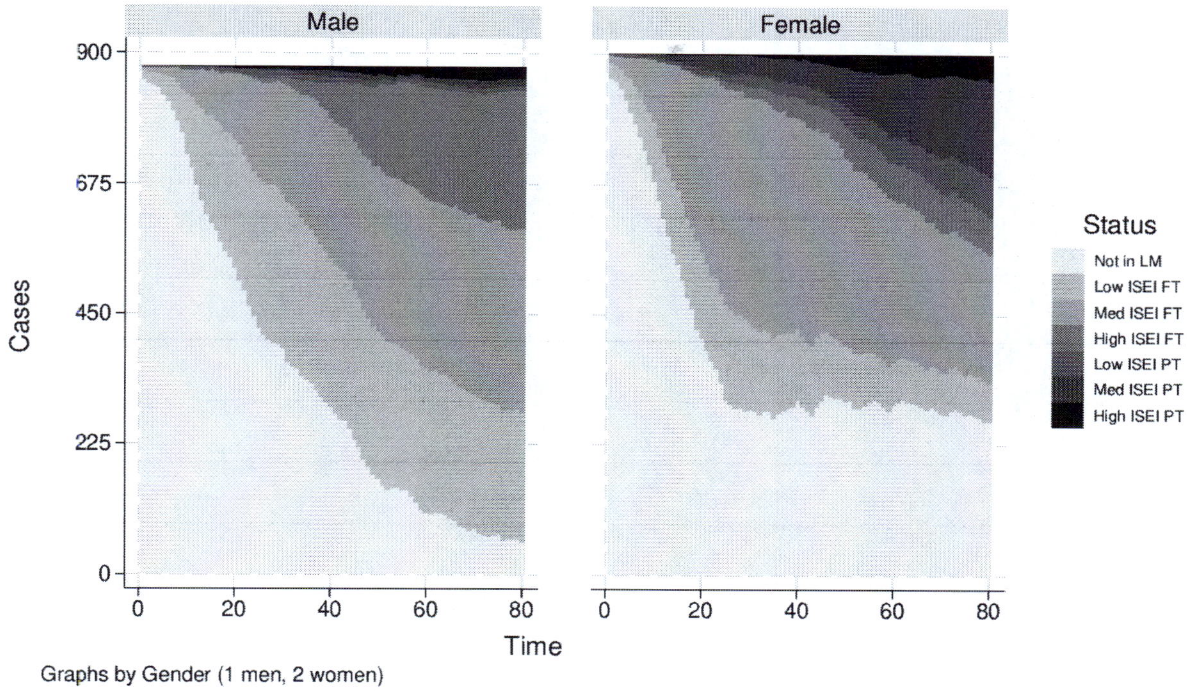

Figure 11.3.1 State distribution over time, by gender.

informative (a default strategy is "lexical" order, i.e., sorting by the states from left to right, which is achieved by the `order(state*)` option used in the example below).

The `sdindexplot` command draws indexplots.

```
sdindexplot state*, by(sex) order(state*) ///
legend(subtitle("Status"))
```

As can be seen in Figure 11.3.2, `sdindexplot` orders the data "lexically" (i.e., sorted as if alphabetically, thanks to the `order(state*)` option). As a result, the pattern until the first transition is clear, but is afterwards rather chaotic. To exploit the indexplot properly, we need to find a way of ordering the cases so that it becomes readable. This is one of the things that sequence analysis achieves.

11.4 Unary Summaries

Univariate summaries of sequences are also useful. Some that are obviously interesting are the number of spells and the cumulated duration in each state. We can look at Shannon entropy as an index of complexity (which does not take order into account, but only the distribution of durations across the several states), or the spell-wise measures, the number of distinct subsequences, and Elzinga's turbulence measure (Elzinga 2010), that weights the number of distinct subsequences inversely with the variance of spell durations (see also do-file `ehk3.do` for this section).

The `sdentropy` command calculates both the cumulative duration (as a proportion of total duration) and the Shannon entropy of the distribution of states (see Box 11.4.1 and Figure 11.4.1). However, the entropy measure is blind to order. To take order into account, we look at the number of distinct subsequences (NDS), and Elzinga's turbulence. This requires that we first generate a spell-wise version of the calendar data. The command `cal2spell` in Box 11.4.2 creates pairs of state and duration variables that represent the calendar as runs in the same state. The commands `ndsub` and `turbulence` use these pairs to calculate the number of distinct subsequences and the turbulence coefficient. Sequences with many spells in different states will have high numbers of distinct subsequences (e.g.,, ABC has eight distinct subsequences, "" [i.e., the null or empty sequence], A, B, C, AB, AC, BC, and ABC, whereas ABB has only six,"", A, B, AB, BB, and ABB). The turbulence index weights this inversely with the variation in duration of the spells (specifically, by the theoretical maximum variance divided by the observed variance).

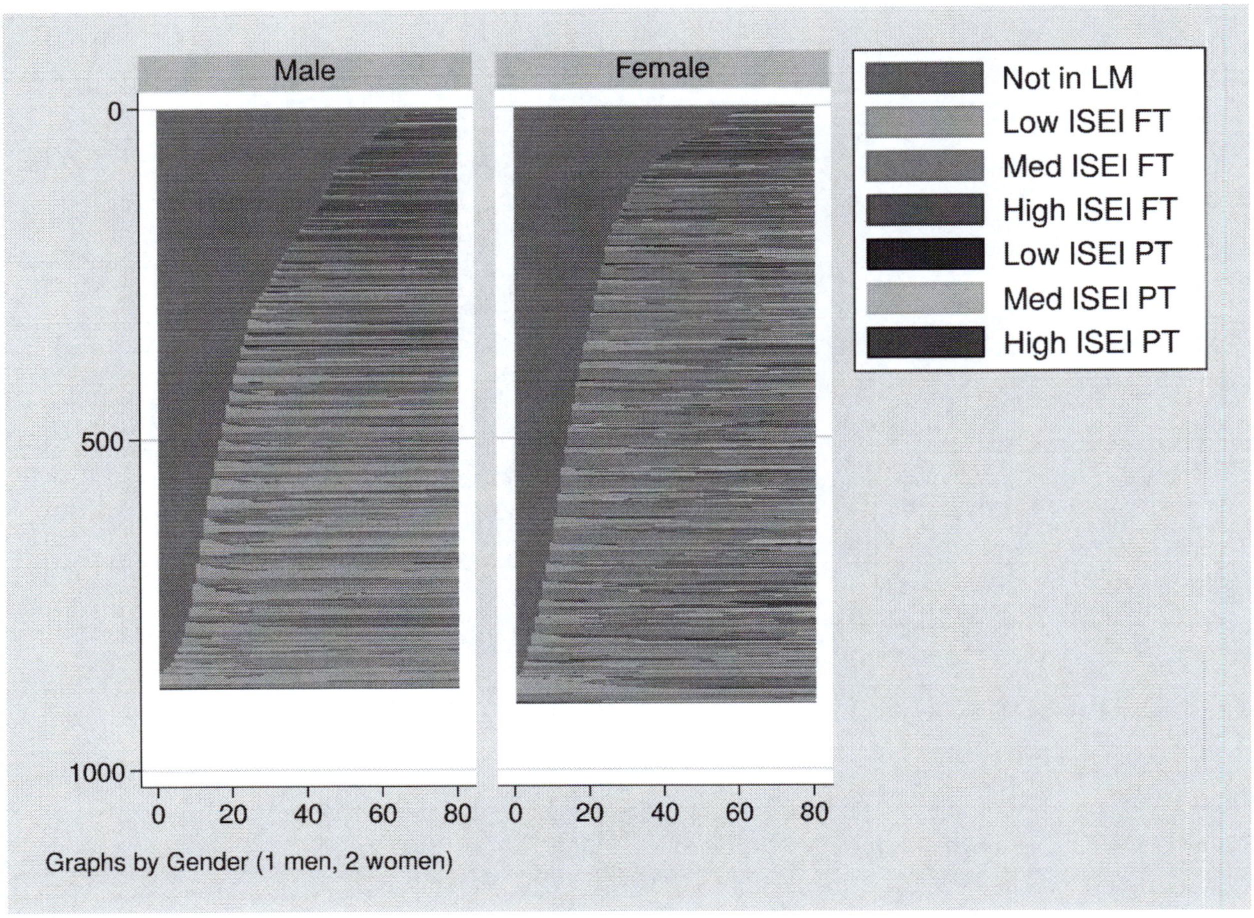

Figure 11.3.2 An indexplot, lexically ordered, by gender.

Box 11.4.1 Calculating cumulated duration and entropy (see do-file **ehk3.do**)

```
. sdentropy state*, gen(entropy) cdstub(dur) nstates(7)
. su dur* entropy

 Variable |       Obs        Mean    Std. Dev.       Min        Max
----------+---------------------------------------------------------
     dur1 |     1,778    .4047174    .2319978          0          1
     dur2 |     1,778    .1801884    .2876401          0          1
     dur3 |     1,778    .2687641    .3033331          0          1
     dur4 |     1,778     .071991    .1564796          0       .975
     dur5 |     1,778    .0209224    .0880239          0          1
----------+---------------------------------------------------------
     dur6 |     1,778    .0384139     .123682          0       .975
     dur7 |     1,778    .0150028    .0743906          0         .9
  entropy |     1,778    1.024739    .3676291          0   2.324768

. hist entropy
(bin=32, start=0, width=.07264901)
```

From the correlations, we see that turbulence correlates quite highly with entropy (despite entropy not taking order into account) and with the number of spells.

11.5 Intersequence Distance

Sequence analysis operates by defining (and then analyzing) intersequence distance or dissimilarity. Pairs of sequences can be judged (by a variety of algorithms) to be more or less similar, and this information can be used for various purposes including data-driven classification of sequences, comparison to ideal types, and assessment of differing variability of sequences across categories such as cohort, as well as for dyadic approaches that focus on pairs such as spouses, siblings, or mothers and daughters.

The simplest sort of distance or dissimilarity applies to sequences of equal length, and involves counting the number of time points at which they do not match. If the state variable is a scale variable, we could

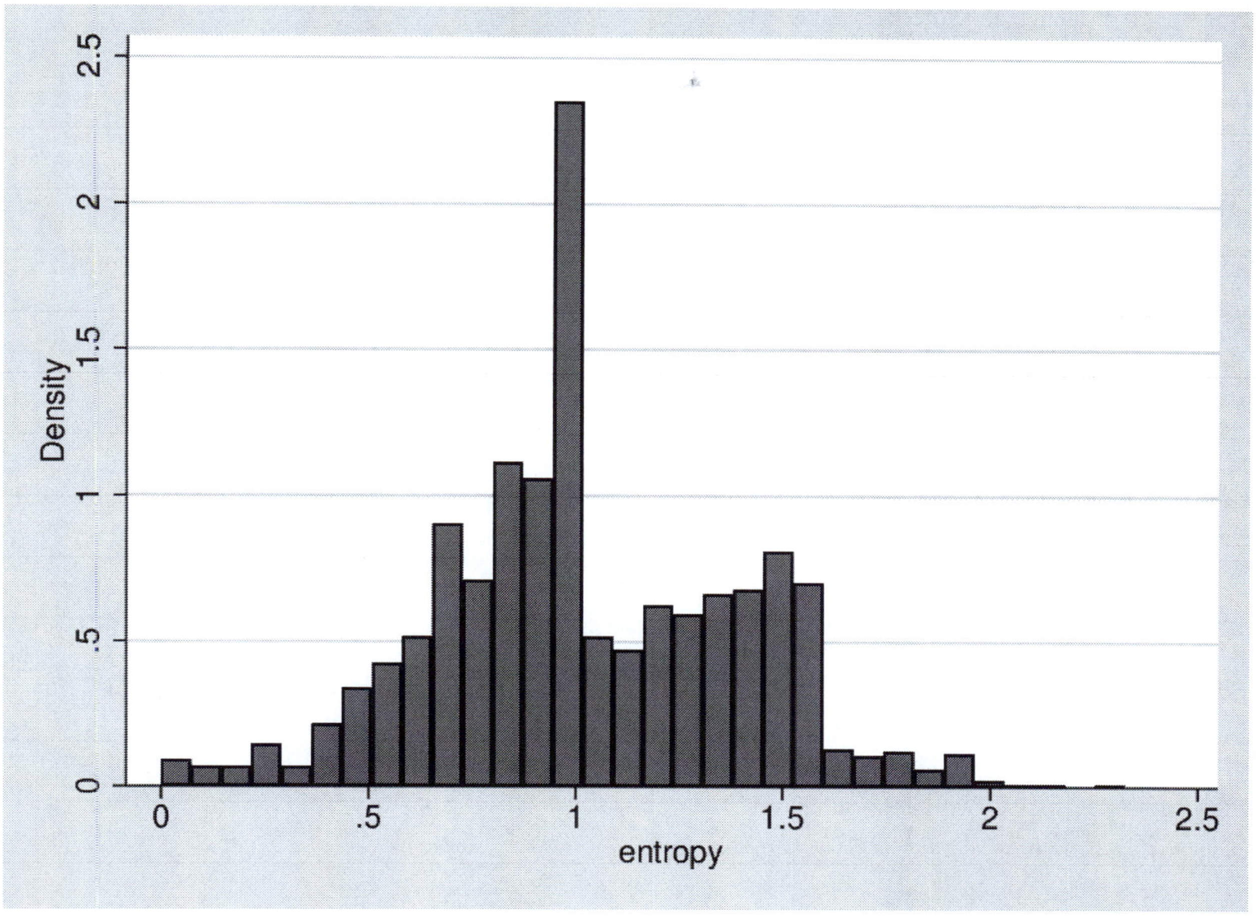

Figure 11.4.1 The distribution of entropy.

instead sum the absolute difference at each time point. If it is a categorical variable and it is possible to define distances between the states of the variable, we can sum the differences at each time point. This is known as the Hamming distance, and can be said to pick up *full or partial similarity at the same time.*

Defining State-Space Distances

In this case, we have a seven-category state space. Insofar as sequence analysis involves mapping distances between the states in the state space to distances between the sequences, we first need to define distances

Box 11.4.2 Number of distinct subsequences and turbulence (see do-file `ehk3.do`)

```
. cal2spell, state(state) spellvar(sp) length(len) nspells(nsp)

. ndsub, statevars(sp*) nsp(nsp) nstates(7) gen(phi)

. turbulence, gen(turb) sta(sp*) len(len*) nsp(nsp) nst(7)

. su entropy nsp phi turb

    Variable |       Obs        Mean    Std. Dev.       Min        Max
-------------+--------------------------------------------------------
     entropy |     1,778    1.024739    .3676291          0   2.324768
         nsp |     1,778    4.140045    2.099941          1         16
         phi |     1,778    50.65129    386.0347          2      14122
        turb |     1,778    6.305856    2.361259          1   17.10609

. corr entropy nsp phi turb
(obs=1,778)

             |  entropy      nsp      phi     turb
-------------+------------------------------------
     entropy |   1.0000
         nsp |   0.5170   1.0000
         phi |   0.1443   0.3657   1.0000
        turb |   0.6994   0.7701   0.2967   1.0000
```

Box 11.5.1 Hamming distance (see do-file ehk4.do)

```
. matrix subs = (0,   3,   3,   3,   3,   3,   3 \ ///
>              3,   0,   1,   2, 0.5, 1.5, 2.5 \ ///
>              3,   1,   0,   1, 1.5, 0.5, 1.5 \ ///
>              3,   2,   1,   0, 2.5, 1.5, 0.5 \ ///
>              3, 0.5, 1.5, 2.5,   0,   1,   2 \ ///
>              3, 1.5, 0.5, 1.5,   1,   0,   1 \ ///
>              3, 2.5, 1.5, 0.5,   2,   1,   0)

.
. sdhamming state1-state80, subs(subs) pwd(ham)
Not normalising distances with respect to length

. matlist ham[1..6,1..6]

             |       c1          c2          c3          c4          c5          c6
   ----------+------------------------------------------------------------------
          r1 |        0
          r2 |      163           0
          r3 |       68         103           0
          r4 |      143          74          88           0
          r5 |      148          81          95          54           0
          r6 |       31         157          67         112         119           0
```

between these seven states. One way to do this is to say that the three occupational groups are on a single scale (the distance from low to medium is 1, and from low to high is 2), and that there is an additional difference of 0.5 for jobs that differ in being full-time vs. part-time. We can then set the nonemployment state to be equally distant from all, at 3, giving the following state-distance matrix:

	1	2	3	4	5	6	7
1 Not in LM	0	3	3	3	3	3	3
2 Low ISEI FT	3	0	1	2	.5	1.5	2.5
3 Med ISEI FT	3	1	0	1	1.5	.5	1.5
4 High ISEI FT	3	2	1	0	2.5	1.5	.5
5 Low ISEI PT	3	.5	1.5	2.5	0	1	2
6 Med ISEI PT	3	1.5	.5	1.5	1	0	1
7 High ISEI PT	3	2.5	1.5	.5	2	1	0

Note that this intentionally sets "not in LM [labor market]" apart, with a greater distance to any employed state than the maximum distance between employed states. Setting distances is a tricky aspect of sequence analysis, because it has a big effect on the resulting sequence distances (examples in this section are in do-file ehk4.do).

To calculate Hamming distance, we need to create a (symmetric, square) matrix containing the state-space distances, and enter the sdhamming command. Remember that calculating pairwise distances for all pairs is a slow process, because $\frac{N(N-1)}{2}$ comparisons need to be made (see Box 11.5.1, in which distances between the first six sequences are shown). The maximum possible distance is 80×3—the length of the sequence multiplied by the highest state-space distance; and the minimum is zero—between identical sequences.

11.6 What to Do with Sequence Distances?

The immediately obvious thing to do with a matrix of all pairwise inter-sequence distances is to perform a cluster analysis; in other words, to use the distances to generate a data-driven classification of the sequences. Are the sequences distributed in the multidimensional space implied by their pairwise distance in such a way that we can capture a good deal of their variation by grouping them?

```
capture drop g6
clustermat wards ham, add
cluster generate g6 = groups(6)
sdchronogram state*, by(g6)
```

We cluster using Ward's distance. This is a hierarchical algorithm, and tends to create clusters of approximately equal size. We pick six clusters here simply for ease of presentation,[2] but nonetheless the chronogram in Figure 11.6.1 shows that the six groups have very different patterns (see do-file ehk4.do). However, summarizing across individuals here probably hides a good deal of heterogeneity in the clusters. Let us look at the indexplot again:

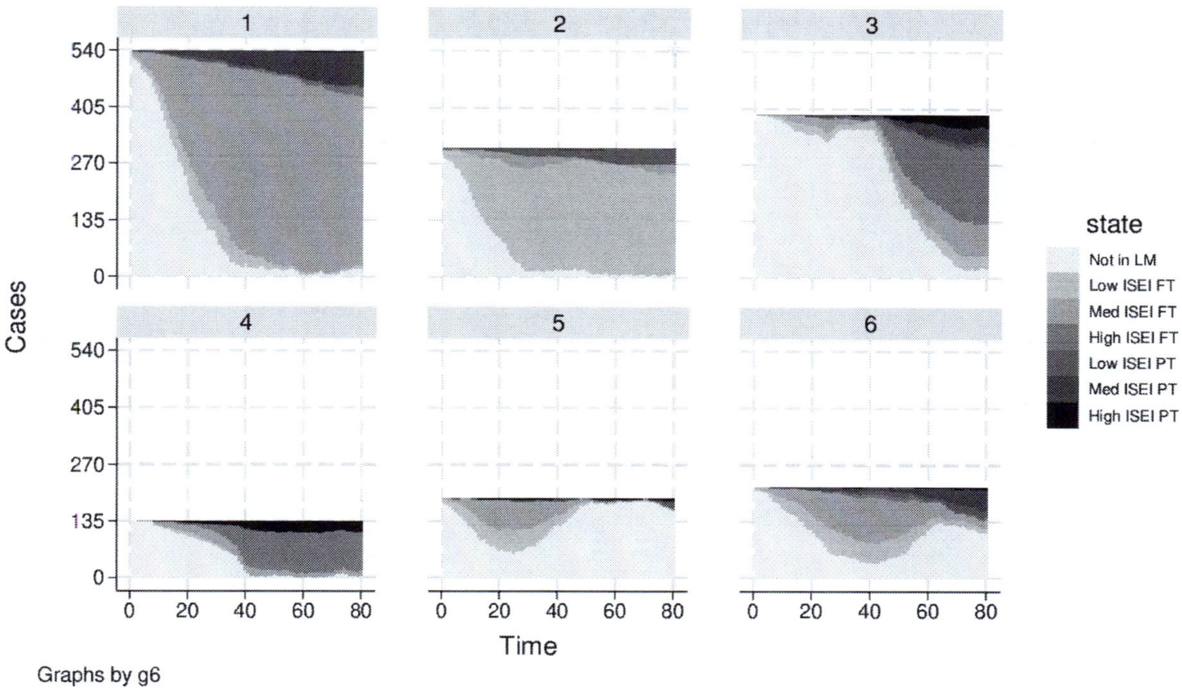

Figure 11.6.1 State distribution over time by cluster.

```
capture drop g999
cluster gen g999 = groups(9999), ties(fewer)
sqindexplot, by(g6) order(g999)
```

The g999 variable contains the cluster solution with the maximum possible number of clusters (effectively as many clusters as there are distinct sequences; 9999 can be replaced by any number greater than or equal to the number of distinct sequences). Because in hierarchical clustering such as Ward's, all simpler cluster solutions are made by successively combining cases with cases into groups and then groups with groups, its order is the dendrogram order. Using this variable in the `order()` option puts the cases in dendrogram order, and this makes the indexplot much more legible when using a hierarchical clustering algorithm. It effectively renders the structure of subclusters visible within clusters.

Figure 11.6.2 demonstrates a lot of the features of sequence analysis and cluster analysis. First, sequences can be similar without being identical (e.g., having good matches early and poor matches late or vice versa). Second, at their edges, clusters will include cases that are not very close to the cluster core. Cluster 3 is a case in point. Although it has a consistent theme of transition into the labor market in the late 20s, the destinations are various and some people have early labor market experience. Some of this can be solved by using more disaggregated cluster solutions, but some element of diversity within clusters must just be accepted.

A note of caution: Unless the data have very strong structure, the nature of the clusters will be somewhat arbitrary; and, in particular, the assignment of cases to clusters will be somewhat unstable. If clusters are to be used as dependent or explanatory variables in further analysis, this needs to be borne in mind. If you want to compare two cluster solutions, the `ari` and `permtab` commands are worth exploring.

Unary Summaries of Clusters

The unary summaries outlined above (cumulated duration, number of spells, entropy, number of distinct subsequences, turbulence, etc.) can be informative with respect to clustering. Box 11.6.1 shows how the clusters differ according to these measures.

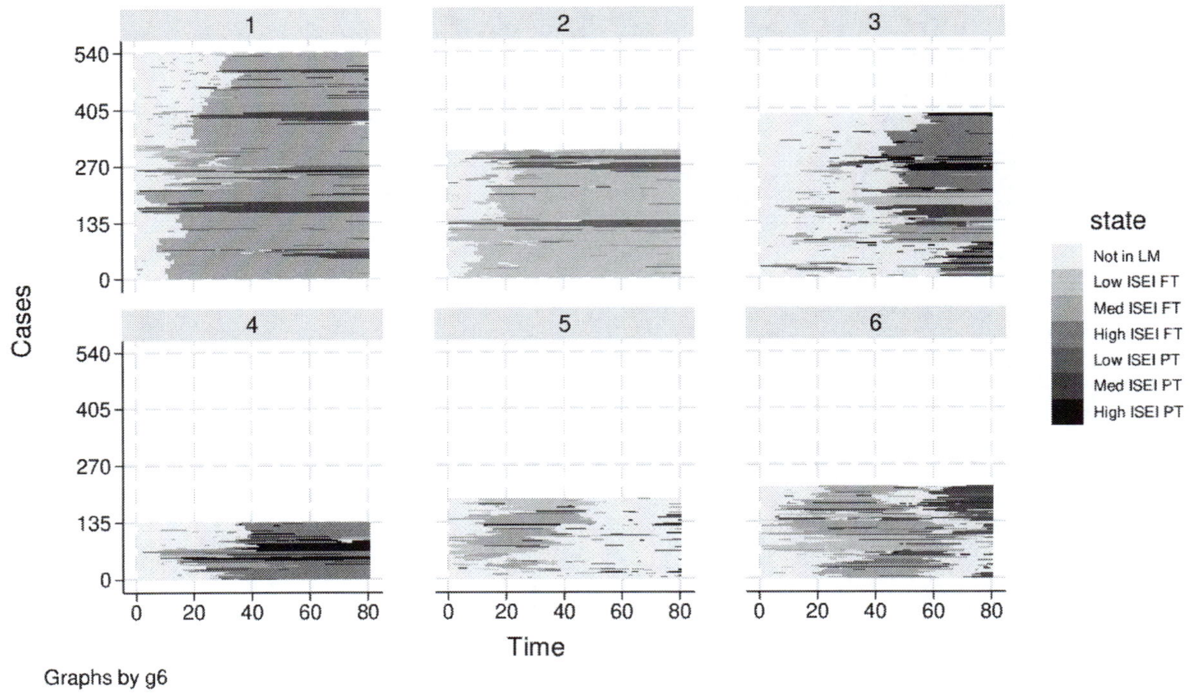

Figure 11.6.2 Indexplot by cluster.

Box 11.6.1 Summarizing the clustering with unary measures

```
. table g6, c(mean dur1 mean dur2 mean dur3 mean dur4 ) format(%5.2f)

---+-----------------------------------------------------------
g6 | mean(dur1)  mean(dur2)  mean(dur3)  mean(dur4)
---+-----------------------------------------------------------
 1 |      0.24        0.05        0.62        0.01
 2 |      0.20        0.69        0.04        0.00
 3 |      0.62        0.06        0.09        0.16
 4 |      0.37        0.02        0.10        0.40
 5 |      0.72        0.12        0.13        0.01
 6 |      0.48        0.15        0.25        0.02
---+-----------------------------------------------------------

. table g6, c(mean dur5 mean dur6 mean dur7 ) format(\%5.2f)

---+-----------------------------------------------
g6 | mean(dur5)  mean(dur6)  mean(dur7)
---+-----------------------------------------------
 1 |      0.01        0.07        0.00
 2 |      0.06        0.00        0.00
 3 |      0.01        0.03        0.03
 4 |      0.00        0.00        0.11
 5 |      0.01        0.01        0.00
 6 |      0.03        0.06        0.00
---+-----------------------------------------------

. table g6, c(mean entropy mean nsp mean phi mean turb) format(\%5.2f)

---+-----------------------------------------------------------------------
g6 | mean(entropy) mean(nsp) mean(phi) mean(turb)
---+-----------------------------------------------------------------------
 1 |         1.00         3.71        27.93          5.57
 2 |         0.88         3.82        29.11          5.26
 3 |         1.07         4.08        34.59          6.64
 4 |         1.27         4.11        31.16          7.21
 5 |         0.85         4.48        38.80          6.47
 6 |         1.22         5.52       188.94          8.31
---+-----------------------------------------------------------------------
```

Box 11.7.1 Calculating optimal matching distances

```
. oma state1-state80, len(80) subs(subs) indel(1.5) pwd(oma)
Not normalising distances with respect to length
(0 observations deleted)
1471 unique observations

. corrsqm ham oma, nodiag
VECH correlation between ham and oma: 0.9260
Diagonal suppressed
```

11.7 Optimal Matching Distance

Hamming distance picks up full and partial similarity at the same time, but optimal matching (OMA) distance can recognize similarity that is displaced in time by virtue of adding the *indel* operation to the state-comparison operation of Hamming. The **oma** command is similar to the **sdhamming** command, but requires two extra options: **len()**, because it is capable of comparing sequences of different lengths; and **indel()**, to set the index cost. For sequences of fixed length, **len()** can be set to a constant.

Box 11.7.1 shows how to calculate OMA distances (see do-file **ehk4.do**) and compare them to the Hamming distances already calculated (again, remember that calculating all pairwise distances can be slow). The **corrsqm** command calculates a correlation of 0.926 between the Hamming and OMA distance matrices, which means they are really quite similar but not identical.

We can learn a little about how OMA differs from Hamming by looking at pairs of cases. Looking around in the current data, the distances between Cases 3 and 8 differ fairly sharply, with a Hamming distance of 85 but an OMA distance of only 37. The **sdstripe** command creates text representations of the sequences for easy examination, and helps us to explore the source of the difference:

```
sdstripe state*, sym("_LMHlmh") gen(seqstr)
list seqstr if inlist(_n,3,8)
```

```
3.   _______________LLLLLLLLLLLL______________MMMMMMMMMMMMMMMMMMMMMMMMMMMMMMMMMMMMMMM
8.   ________________M_____________MMMMMMMMMMMMMMMMMMMMMMMMMMMMMMMMMMMMMMMMMMM_____MM
```

The Hamming comparison looks as follows, with the sum of point-wise differences coming to 85.

```
3.   _______________LLLLLLLLLLLL______________MMMMMMMMMMMMMMMMMMMMMMMMMMMMMMMMMMMMMMM
8.   ________________M_____________MMMMMMMMMMMMMMMMMMMMMMMMMMMMMMMMMMMMMMMMMMM_____MM
     ===============313333333333===33333333333================================33333==
```

The OMA comparison looks like this: First delete 11 elements in each sequence, at a cost of 22 × 1.5:

```
3.   _______________LL..........______________MMMMMMMMMMMMMMMMMMMMMMMMMMMMMMMMMMMMMMM
8.   ________________M_____________......MMMMMMMMMMMMMMMMMMMMMMMMMMMMMMMMMMMMM.....MM
```

Then do an element-wise comparison of the resulting, quite similar, sequences. The total "cost" or dissimilarity is 33 points due to indels and 4 points due to substitution (or element-wise comparison).

```
3.   _______________LL_____________MMMMMMMMMMMMMMMMMMMMMMMMMMMMMMMMMMMMMMM
8.   ________________M_____________MMMMMMMMMMMMMMMMMMMMMMMMMMMMMMMMMMMMMMM
     ===============31====================================================
```

Substantively, these sequences share the feature of late entry to medium ISEI/FT work, with a short earlier spell breaking the long spell of nonemployment. However, from a Hamming point of view, there are large spells in which one sequence is in work (low or medium) and the other is in nonemployment, thereby yielding a high difference score (the fact that the substitution matrix makes nonemployment the most distinct state has a big influence here). OMA sees this similarity and gives a lower difference, but still penalizes the fact that the similarity is not at the same time. Thus, Hamming focuses on full or partial similarity at the same time, whereas OMA focuses on full or partial similarity at the same time or displaced in time, but penalizes displacement.

A key question for the analyst is whether the smaller OMA difference of 37 (in which the difference reflects the fact that while similar things happen, they happen at different times) is more meaningful than the Hamming difference of 85 (which reflects the fact that while some things are the same or similar at the same times, there are many time points with differences).

See page 193 for discussion of how to set the indel cost for OMA.

Box 11.8.1 Further distance measures (see do-file ehk5.do)

```
. matrix flat = J(7,7,1) - I(7)

. sdhamming state1-state80, subs(flat) pwd(hmf)
Not normalising distances with respect to length

. oma state1-state80, len(80) subs(flat) indel(0.5) pwd(omf)
Not normalising distances with respect to length
1471 unique observations

.
. capture drop dur1-dur7 entropy
. sdentropy state*, gen(entropy) cdstub(dur) nstates(7) shortlab
. cal2spell, state(state) spellvar(sp) length(len) nspells(nsp)

. dynhamming  state1-state80, pwd(dyn)
1471 unique observations

. twed state1-state80, subsmat(subs) pwd(twd) length(72) lambda(0.5) nu(0.04)
Not normalising distances with respect to length
1471 unique observations

. twed state1-state80, subsmat(flat) pwd(twf) length(72) lambda(0.5) nu(0.04)
Not normalising distances with respect to length
1471 unique observations

. combinadd sp1-len16, pwsim(xts) nspells(nsp) nstates(7) rtype(d)
Running spell-wise Elzinga sequence comparisons
Reading data and enumerating tuples
Pairwise processing
```

Box 11.8.2 Correlations between distance measures

```
. matlist distcorr, format(%6.3f)

             |     c1      c2      c3      c4      c5      c6      c7      c8
-------------+----------------------------------------------------------------
     Hamming |  1.000
         OMA |  0.926   1.000
 Dyn Hamming |  0.753   0.757   1.000
        TWED |  0.766   0.758   0.952   1.000
         X/t |  0.042   0.009   0.033   0.149   1.000
Hamming flat |  0.761   0.763   1.000   0.954   0.033   1.000
    OMA flat |  0.637   0.767   0.929   0.875  -0.005   0.927   1.000
   TWED flat |  0.748   0.760   0.985   0.977   0.059   0.985   0.920   1.000
```

11.8 Special Topics

11.8.1 Other Distance Measures

OMA and Hamming distance are the most commonly used measures in sequence analysis. SADI offers a number of other measures, including dynamic Hamming, TWED, and a version of Elzinga's combinatorial measures. Dynamic Hamming distance calculates the time-dependent pattern of transition rates to calculate a changing pattern of distances between the states, and uses this in a Hamming-style comparison (i.e., it looks for similarity at the same time and does not compare across time). TWED uses a similar logic to OMA, but instead of deleting and inserting tokens, it locally stretches and compresses the time dimension, using a stiffness parameter (nu) and a gap penalty (lambda) instead of an indel cost (it uses a substitution matrix exactly as OMA does). Elzinga's methods focus on counting the number of subsequences common to both sequences, and the particular method most appropriate to life-course data is probably the "X/t" measure that works on sequences of spells weighted by duration (rather than sequences of uniform time units). This measure, calculated by `combinadd`, uses the spell variables calculated by `cal2spell` above. See Box 11.8.1, in which these measures are calculated, along with OMA and Hamming with a "flat" state-distance matrix in which all states are set equally distant from each other. Box 11.8.2 shows the correlations between the various distances. Note, in particular, the impact of changing the substitution matrix, and the fact that the X/t duration-weighted spell measure (which focuses on order much more than on when things happen) correlates very weakly with other measures. See Halpin (2014) and Studer and Ritschard (2015) for further discussion of these and other distance measures (see do-file `ehk5.do`).

11.8.2 Ideal Types

The examples above focus on using sequence analysis to generate a data-driven classification using the matrix of all pairwise distances. Sequence analysis can also be helpful in assigning cases to pre-determined ideal sequences. It can be difficult to determine rules to allocate messy sequence data to categories, but with

Box 11.8.3 OMA distance to ideal-type sequences (see do-file `ehk6.do`)

```
// Create 4 ideal typical sequences
//  1: Slightly delayed entry, spell in medium, exit
//  2: Slightly delayed entry, spell in medium until end
//  3: Very delayed entry, spell in high until end
//  4: Earlier entry, spell in low until end
input state1-state80
1 1 1 1 1 1 1 1 1 1 1 1 1 1 1 1 3 3 3 3 3 3 3 3 3 3 3 3 3 3 ///
3 3 1 1 1 1 1 1 1 1 1 1 1 1 1 1 1 1 1 1 1 1 1 1 1 1 1 1 1 ///
1 1 1 1 1 1 1 1 1 1 1 1 1 1 1 1 1 1 1 1 1
1 1 1 1 1 1 1 1 1 1 1 1 1 1 1 1 3 3 3 3 3 3 3 3 3 3 3 3 3 3 ///
3 3 3 3 3 3 3 3 3 3 3 3 3 3 3 3 3 3 3 3 3 3 3 3 3 3 3 3 3 ///
3 3 3 3 3 3 3 3 3 3 3 3 3 3 3 3 3 3 3 3 3
1 1 1 1 1 1 1 1 1 1 1 1 1 1 1 1 1 1 1 1 1 1 1 1 1 1 1 1 1 1 ///
1 1 1 1 1 1 1 1 1 1 1 1 1 1 1 1 1 1 4 4 4 4 4 4 4 4 4 4 4 ///
4 4 4 4 4 4 4 4 4 4 4 4 4 4 4 4 4 4 4 4 4
1 1 1 1 1 1 1 1 1 1 1 1 2 2 2 2 2 2 2 2 2 2 2 2 2 2 2 2 2 2 ///
2 2 2 2 2 2 2 2 2 2 2 2 2 2 2 2 2 2 2 2 2 2 2 2 2 2 2 2 2 ///
2 2 2 2 2 2 2 2 2 2 2 2 2 2 2 2 2 2 2 2 2
end

// Visually examine them
sdstripe state*, sym("_LMHlmh") gen(seqstr)
list seqstr
drop seqstr

// Append the real data:
// the ideal type sequences will be the first four cases
append using rrcalendarq

// Define substitution costs
matrix subs = (0,   3,   3,   3,   3,   3,   3 \ ///
3,   0,   1,   2, 0.5, 1.5, 2.5 \ ///
3,   1,   0,   1, 1.5, 0.5, 1.5 \ ///
3,   2,   1,   0, 2.5, 1.5, 0.5 \ ///
3, 0.5, 1.5, 2.5,   0,   1,   2 \ ///
3, 1.5, 0.5, 1.5,   1,   0,   1 \ ///
3, 2.5, 1.5, 0.5,   2,   1,   0)

// Geneate OM distances to the ideal types, in an Nx4 matrix
oma state1-state80, subsmat(subs) pwd(idtyp) ref(4) length(80) indel(1.5)

// Turn matrix into variables idtyp1-idtyp4 and
// assign each case to the nearest type
svmat idtyp
egen minidt = rowmin(idtyp*)
gen g4 = 0
forvalues x = 1/4 {
 replace g4 = `x' if idtyp`x'==minidt
}

tab g4
```

Box 11.8.4 The four reference sequences

```
  . sdstripe state*, sym("_LMHlmh") gen(seqstr)
  Creating long string representation

  . list seqstr

  -------------------------------------------------------------------------------------
  |                                                                            seqstr |
  -------------------------------------------------------------------------------------
  |                                                                                   |
1.| ________________MMMMMMMMMMMMMMMM________________________________________________ |
2.| ________________MMMMMMMMMMMMMMMMMMMMMMMMMMMMMMMMMMMMMMMMMMMMMMMMMMMMMMMMMMMMMMMM |
3.| ________________________________________________HHHHHHHHHHHHHHHHHHHHHHHHHHHHHHHH |
4.| ____________LLLLLLLLLLLLLLLLLLLLLLLLLLLLLLLLLLLLLLLLLLLLLLLLLLLLLLLLLLLLLLLLLLLL |
  |                                                                                   |
  -------------------------------------------------------------------------------------
```

sequence analysis, we can calculate distance to the reference sequences and allocate cases to the group of the closest. These reference sequences may be defined theoretically or drawn from earlier analyses. In the following example, we generate four sequences that characterize very different patterns and then calculate all cases' distances to these four cases. Note that calculating distances to a small number of reference sequences is much faster than calculating all pairs (see do-file `ehk6.do`).

In the code displayed in Box 11.8.3, four reference sequences are generated and placed at the start of the data file (Box 11.8.4 displays the sequences). The `oma` command with the `ref(4)` option then calculates the distance from each data sequence to each of the first four sequences. The remaining code maps the distance matrix to variables, `idtyp1` to `idtyp4`, and generates the variable `g4`, assigning each case to a group based

Table 11.8.1 Multi-domain block-structured substitution matrix.

	c1	c2	c3	c4	c5	c6	c7	c1	c2	c3	c4	c5	c6	c7
r1	0	3	3	3	3	3	3	1	4	4	4	4	4	4
r2	3	0	1	2	.5	1.5	2.5	4	1	2	3	1.5	2.5	3.5
r3	3	1	0	1	1.5	.5	1.5	4	2	1	2	2.5	1.5	2.5
r4	3	2	1	0	2.5	1.5	.5	4	3	2	1	3.5	2.5	1.5
r5	3	.5	1.5	2.5	0	1	2	4	1.5	2.5	3.5	1	2	3
r6	3	1.5	.5	1.5	1	0	1	4	2.5	1.5	2.5	2	1	2
r7	3	2.5	1.5	.5	2	1	0	4	3.5	2.5	1.5	3	2	1
r1	1	4	4	4	4	4	4	0	3	3	3	3	3	3
r2	4	1	2	3	1.5	2.5	3.5	3	0	1	2	.5	1.5	2.5
r3	4	2	1	2	2.5	1.5	2.5	3	1	0	1	1.5	.5	1.5
r4	4	3	2	1	3.5	2.5	1.5	3	2	1	0	2.5	1.5	.5
r5	4	1.5	2.5	3.5	1	2	3	3	.5	1.5	2.5	0	1	2
r6	4	2.5	1.5	2.5	2	1	2	3	1.5	.5	1.5	1	0	1
r7	4	3.5	2.5	1.5	3	2	1	3	2.5	1.5	.5	2	1	0

on the nearest reference sequence.

This can also be a useful strategy when working with either very large data sets such as those derived from register data or very large surveys. We start by doing a cluster analysis of a random sample of a manageable size, extract medoid sequences, and use these as reference sequences in an analysis of the whole data set. Cases in the wider data set can be assigned to the cluster of the closest reference sequence. While this will not necessarily give identical results to those from an analysis of the whole data set (if it were feasible), it should be quite close, particularly if using a partitioning aroung medoids PAM cluster algorithm (i.e., a partitioning algorithm rather than an agglomerative algorithm such as Ward's).[3]

11.8.3 Multichannel Sequence Analysis

Multichannel sequence analysis (MCSA) is a sequence analysis that deals with sequences in multiple parallel domains. Examples in the literature often combine domains such as residential, labor market, partnership formation and fertility histories (e.g., Pollock 2007).

As a simple example, here we add marital status (available in this data set only as never vs. ever married). First, we create the state variable (see do-file `ehk7.do`):

```
use rrcalendarq, clear
reshape long state, i(id) j(date)
gen marstate = 1
replace marstate = 2 if date>marq
```

In order to do sequence analysis on multiple domains, we combine these into a single variable (States 1 to 7 are the seven labor-market states for those never married, 8 to 14 for ever married):

```
egen mc = group(marstate state)
reshape wide state marstate mc, i(id) j(date)
order id state* marstate* mc*
```

The challenge is to determine the new substitution costs. The simplest is to add costs for the two domains. If, for instance, we allow a difference of 1 between never and ever married, and a difference of 2.5 between low ISEI/PT and high ISEI/FT, then the difference between low ISEI/PT, never married and high ISEI/FT, ever-married is $2.5 + 1 = 3.5$.

We can create a new substitution matrix as follows:

```
matrix sub1 = subs + J(7,7,1)
matrix mcsub = subs, sub1  sub1, subs
```

This yields the block-structured 14-by-14 matrix shown in Table 11.8.1. This is a simple example, because the additional state space has only two categories. For more categories, this strategy can be extended in a straightforward fashion.

Box 11.8.5 demonstrates how to calculate and cluster the distances and produce indexplots of both domains. Figure 11.8.1 presents indexplots of the resulting clusters showing labor market (left) and marital

Box 11.8.5 OMA on multiple domains (see do-file `ehk7.do`)

```
oma mc*, subs(mcsub) len(80) indel(2) pwd(mcpwd)
clustermat wards mcpwd, add
cluster gen m=groups(4/12 16 32 9999), ties(fewer)
sdindexplot state*, by(m6, legend(off) title("Job history")) ///
 order(m9999) name(gr1, replace) legend(subtitle("Status"))
sdindexplot marstate*, by(m6, legend(off) title("Marital history")) ///
 order(m9999) name(gr2, replace) legend(subtitle("Status"))
graph combine gr1 gr2
```

status trajectories (right). We can see how both domains affect the clustering. For instance, from the point of view of labor-market histories, Clusters 1 and 2 are very similar, typified by an entry to medium-ISEI full-time work from not in the labor market at a relatively young age. However, they are markedly different in their marriage history, with Cluster 1 marrying very late, if at all.

11.8.4 Dyadic Analysis

Dyadic analysis involves looking at paired sequences such as spouses' labor market history, mother–daughter fertility histories, or sibling careers (Han and Moen 2001; Raab et al. 2014; Karhula et al. 2019). Sequence analysis generates a single measure of distance per pair that is then usually the subject of further analysis. For smaller datasets, dyadic distances can be extracted from the pairwise distance matrices as discussed above, but the dedicated `sdlongdyad` command calculates dyadic distances (for OMA and TWED) without calculating all pairwise distances, and can therefore work with large numbers of dyads and very long sequences. However, calculating all pairwise distances has some advantages:

- It allows comparison of the dyadic distance (ego to alter) with the distances of ego to all other individuals (see e.g., Liao, 2021)
- It allows use of all distance measures (not just OMA and TWED)
- It allows comparison of dyads where individuals have sequences of different lengths.

Because the example data sets in this book do not contain dyads, we create a simulated couple dataset of 300 couples, each individual with a 40-unit sequence that is more or less similar to her partner's. (See do-file `ehk8.do`.)

The `sdlongdyad` command needs data in long format, with one row per dyad per time unit, and two state variables, one for each individual in the dyad. If we start with data in the usual wide format, with

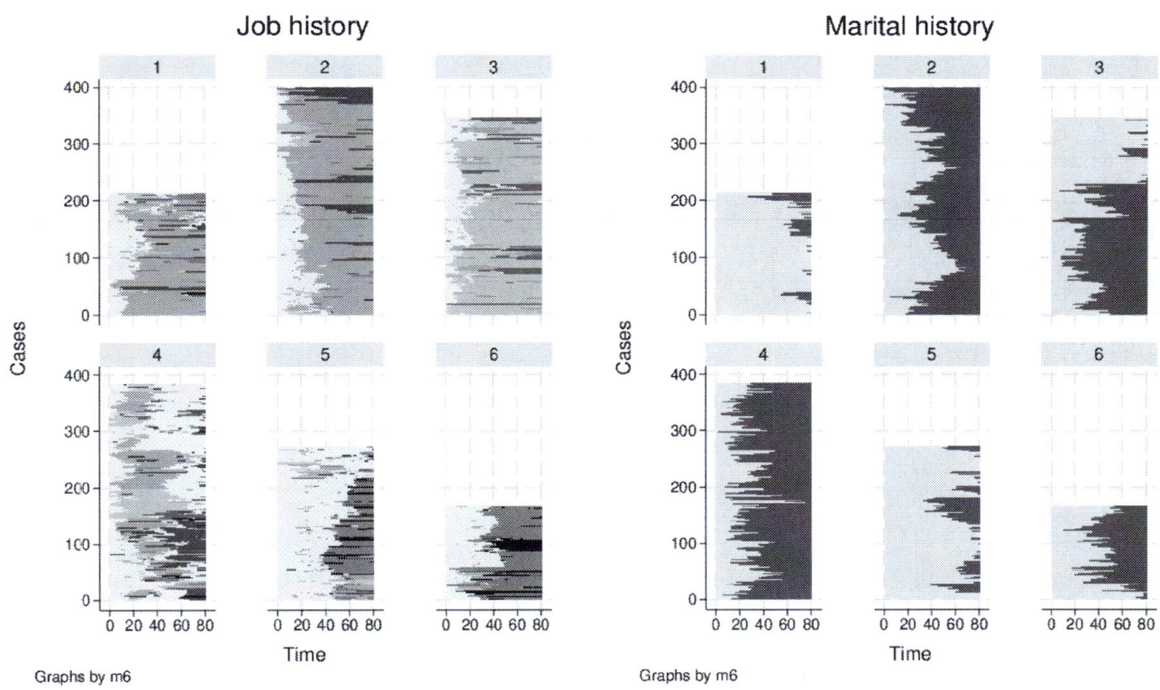

Figure 11.8.1 Multichannel indexplots by cluster.

one row per individual containing state variables `s1-s40`, a dyad-level ID `pairid`, and a within-dyad ID `indtype`, we need to restructure it as follows:

```
use dyadsim
// dyadsim.dta is created by ehk8.do
// sort pairid indtype
// reshape long s, i(pairid indtype) j(t)
// reshape wide s, i(pairid t) j(indtype)
```

The two `reshape` commands first put the data in long format by individual and then into long format by dyad. The resulting structure has one row per time unit per dyad, with one variable each for ego and alter.

The dyadic distances are calculated as follows (using a default substitution cost of 1.0):

```
sdmksubs, dtype(flat) subsmat(sm) nstates(4)
sdlongdyad s1 s2, dyadid(pairid) subsmat(sm) ///
indel(0.5) dist(dd) mtype(oma)
```

The resulting `dd` matrix has three columns, the dyad ID, the length of the dyad's common sequences, and the dyadic distance.

Generating all pairwise distances, on the other hand, requires the data to be in the standard format of one row per individual. The complication lies in extracting the dyad distances from the resulting pairwise distance matrix. One way to do this is to sort such that all egos come first, then all alters, with sorting by dyad-id within the groups. Thus, if there are 300 pairs, ego i is in row i and alter i is in row $300 + i$. The pairwise matrix generated by running `oma` in the usual fashion has the dyadic distances in two locations: in one minor diagonal in the top right and one in the lower left. Table 11.8.2 shows how this would look for a small example with five pairs. We can use Mata to extract these into variables (see Box 11.8.6 and do-file `ehk9.do`).

Box 11.8.6 Calculating dyadic distances (see do-file `ehk9.do`)

```
use dyadsim, clear

// Sort in "individual type" blocks
sort indtype pairid

// Get distance matrix for all pairs
sdmksubs, dtype(flat) subsmat(sm) nstates(4)
oma s*, subsmat(sm) indel(0.5) pwd(dyads) len(40)

mata:
// Copy the Stata matrix into Mata
d = st_matrix("dyads")

// Extract one submatrix (upper right) and then its diagonal
d1 = d[1..rows(d)/2,rows(d)/2+1..rows(d)]
dd1 = diagonal(d1)

// Double up to match the data
dd  = dd1 \ dd1

// Push it from Mata into a Stata variable, "dist".
newvars = st_addvar("double", "dist")
st_store(., newvars, dd)
end

su dist
```

While generating all pairwise distances is inefficient if all we want is the dyads' distances, it allows us to look at the distribution of distances from each ego to all alters and not just to its partner. These distances pertain to what Karhula et al. (2019) refer to as "unrelated dyads" (see also Liao 2019). If the true dyads are more similar, their distance will be systematically less than the mean across all dyads, related and unrelated. The code in Box 11.8.7 generates the mean distance of each individual to all corresponding others (i.e., each mother to all daughters, each daughter to all mothers) and its standard error, and returns these as variables. Figure 11.8.2 shows the distribution of the dyad distance less the distance to all others.

Table 11.8.2 Dyad distances for small example (5 dyads), showing the dyadic distances in two minor diagonals.

	Ego 1	Ego 2	Ego 3	Ego 4	Ego 5	Alter 1	Alter 2	Alter 3	Alter 4	Alter 5
Ego 1						d(1)				
Ego 2							d(2)			
Ego 3								d(3)		
Ego 4									d(4)	
Ego 5										d(5)
Alter 1	d(1)									
Alter 2		d(2)								
Alter 3			d(3)							
Alter 4				d(4)						
Alter 5					d(5)					

Box 11.8.7 Extracting mean distance to all others

```
mata
// Mean distance (and SE) to all alters (rows/cols as appropriate)
dm1 = mean(d1)'
dm2 = mean(d1')'

dse1 = J(rows(d)/2, 1, .)
dse2 = J(rows(d)/2, 1, .)
for (i=1; i<= rows(d)/2; i++) {
 dse1[i] = sqrt(variance(d1[,i])) :/ sqrt(rows(dse1)-1)
 dse2[i] = sqrt(variance(d1[i,]')) :/ sqrt(rows(dse2)-1)
}

// Double up to match the data frame
dm  = dm1 \ dm2
dse = dse1 \ dse2

ddata = dm, dse

newvars = st_addvar("double", ("dmean", "dse"))
st_store(., newvars, ddata)
end

su dist dmean dse
gen delta = dist - dmean
kdensity delta, xline(0) bw(8) ///
 title("Density of difference from mean distance")
```

11.9 Conclusion

Sequence analysis offers an exploratory and descriptive approach to the analysis of life-course data that is complementary to event history analysis. By treating careers or trajectories as holistic units, it gives us an alternative to event history analysis's focus on the determinants of hazard rates. It can thus yield an overview of general patterns that is difficult to achieve with event history analysis's concern with instantaneous change. There is a trade-off: Event history analysis can test formal hypotheses about the processes generating the data, while sequence analysis looks for patterns, making numerous assumptions on the way.

The main function of sequence analysis is to translate information (or theoretical judgements) about distances between categorical life-course states into distances between the trajectories of individuals moving through these states. As we have seen, these distances can be then used for generating data-driven classifications, to assign cases to *a priori* theoretical or empirical ideal types, to assess how diversity of experience differs across characteristics such as cohort or social class, or to examine the determinants of similarity between dyads such as couples. It can be particularly helpful when we want to understand how processes in different life domains are synchronized, as with multi-channel sequence analysis. There are two critical sets of decisions to make when conducting sequence analysis. First, how to turn state distances into sequence distances, and second, how to determine state distances. There are many ways to translate state distances into sequence distances, ranging from simple (such as Hamming distance) to more complex (such as Elzinga's subsequence counting). A key difference between the algorithms is how they recognize similarity displaced in time. Hamming distance (and its dynamic version) simply do not recognize displaced similarity. Such algorithms are therefore more appropriate where the time dimension has a clock or a calendar, for example, where the trajectories represent daily, weekly or annual time use, or for dyadic analysis where both trajectories in the dyad are simultaneous (e.g., looking at synchronization of couples' labor market histories, which are simultaneous, but not similarities between mother-daughter fertility histories, which are not). Where time is more developmental than calendar driven, such as in the transition from school to work, we want to detect similarity between sequences with a similar order of events, even where one individual may progress more quickly than another. For such problems, the optimal matching distance or time warp edit distance

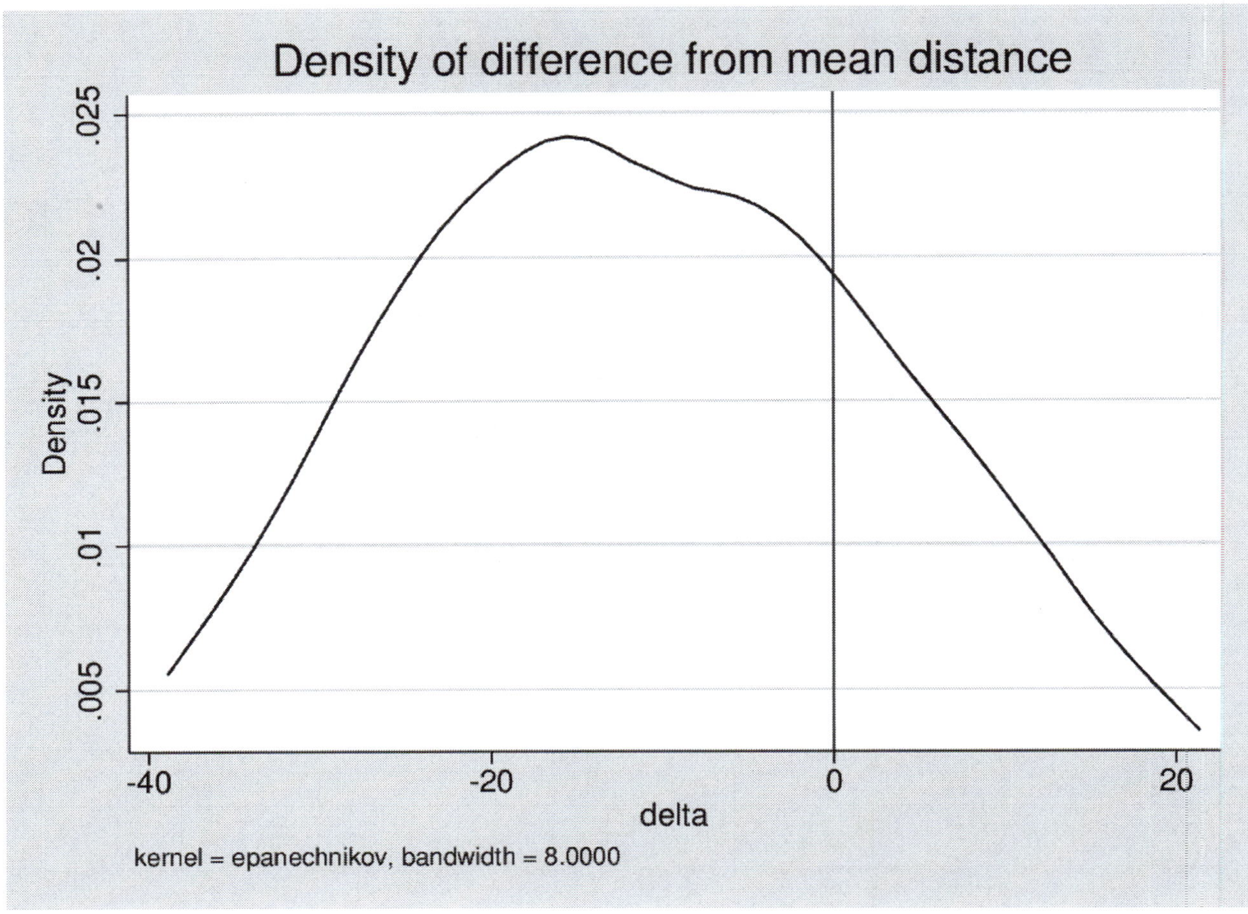

Figure 11.8.2 Dyad distance compared to mean distance to all others.

are more attractive, as these recognize displaced similarity. Algorithms such as Elzinga's duration-weighted number of common subsequences have even more flexibility regarding the time dimension, focusing on shared order of states to a greater degree than their strict timing.

For all sequence analysis algorithms (and also, it must be noted, for event history analysis), a decision has to be made about the state space, i.e., which states to combine and which to keep distinct. For instance, do we combine "never married" and "divorced" into "single"? Do we distinguish employers with more than twenty employees from those with fewer? For sequence analysis algorithms that take account of state distance (e.g., Hamming, OMA and TWED, but not dynamic Hamming or subsequence approaches), the analyst has the further problem of deciding and justifying the state distances used. As we have seen, this is both difficult and important. The definition of the state space and the choice of distances will have at least as big an effect as choice of algorithm.

The choices of algorithm and parameterization, including the choice of state distances, make sequence analysis quite flexible, but also mean that it has lots of "researcher degrees of freedom". That is, the outputs can depend very much on the choices the analyst makes, which means that the results can be seen as lacking objective validity. This must be mitigated as far as possible by being very clear about the logic of the choice of parameterization (especially state distances) and the consequences of the algorithm used, as well as robustness checking (varying parameterization and algorithm). But however clear the logic of the choices made, sequence analysis will never yield results as convincing as those from conventional statistical models testing well-specified (if often necessarily narrow) hypotheses, and its value lies in the descriptive and exploratory capacities it offers us, complementary to conventional approaches.

Notes

[1]If the sequences are truncated, due, for instance, to late entry or early exit (e.g., to retirement), it may make sense to pad the sequences with a special state that is neutral in terms of its distances to other states (Bukodi et al. 2016, pp. 838-839).

[2]Note that if we want to choose the number of clusters using formal criteria, Stata's `clustermat stop` functions do not work as expected with distance matrices. Enter `search calinksi` and `search dudahart` to find alternatives (see Halpin 2016).

[3]This is not directly possible with existing Stata cluster analysis tools. Enter `net from http://teaching.sociology.ul.ie/statacode` and then `net describe clutils` for tools that allow you to extract medoids, and to do PAM clustering on distance matrices.

Appendix: Exercises

As mentioned in chapter 2.1, we offer exercises that Thorsten Schneider developed for the second edition of this book. These examples are based on a data set provided by the German National Educational Panel Study (NEPS) (Blossfeld and Roßbach 2019).[1] This data set `rrdat2.dta` is a small sample of the so-called Starting Cohort 6, version D_7-0-0, that has been anonymized in several ways. It is restricted to persons born and living in West Germany who entered the labor market at least once. The data file `rrdat2.dta` can be easily downloaded at `https://doi.org/10.5157/NEPS:Campus:1.0.0`.

The exercises follow the main chapters and start with chapter 2. The do-file for retrieving the data set is `a_ehb0.do` (see Box A.1). The prefix `a_` refers to Appendix; the rest follows a structure similar to the files in the main chapters. The Stata description of the data set can be found in Box A.2.[2]

Box A.1 Do-file `a_ehb0.do`

```
version 19

set more off

capture log close
log using a_ehb0.log, replace

use rrdat2, clear

describe

label list en1957    /* reports all possible value labels of variable edu_cas */

tab edu_cas edu_yoe if noj == 1
```

Box A.2 Variables in data file `rrdat2.dta` (produced by the do-file `a_ehb0.do`)

```
. describe

 Contains data from rrdat2.dta
  obs:          7,392                  NEPS Campus File, doi:10.5157/NEPS:Campus:1.0.0;
                                                             derived from SC6 D_7-0-0

 vars:             17
 size:        192,192
--------------------------------------------------------------------------------------
 variable   storage  display   value        variable
 name       type     format    label        label
--------------------------------------------------------------------------------------
 id         int      %9.0g                  ID of individual
 noj        byte     %2.0f                  Serial number of job
 start_y    int      %39.0f    enmiss       Year job started
 start_m    byte     %28.0g    en1937       Month job started
 fin_y      int      %17.0f    enext2       Year job ended
 fin_m      byte     %28.0g    en1937ext1   Month job ended
 sex        byte     %9.0g     en157        Gender (1 men, 2 women)
 i_y        int      %4.0f                  Year last interview
 i_m        byte     %10.0g    en1874       Month last interview
 b_y        int      %4.0f                  Year of birth
 b_m        byte     %10.0g    en1874       Month of birth
 mar_y      int      %39.0g    enmiss       Year of first marriage
                                               (-2 if no marriage)
 mar_m      byte     %17.0g    en1874       Month of first marriage
                                               (-2 if no marriage)
 isei08     float    %37.2f    enext4       Socio-economic status (ISEI-08) of job i
 ftpt       byte     %17.0g    en1496       Job volume at begin of job i
                                               (1 full-, 2 part-time)
 edu_cas    byte     %142.0g   en1957       educational attainment, recent, CASMIN
 edu_yoe    byte     %25.0f    enext3       educational attainment, recent, [in years]
--------------------------------------------------------------------------------------
 Sorted by: id  noj

 . label list en1957    /* reports all possible value labels of variable edu_cas */
 en1957:
 -55 Not determinable
   0 [1a] No school leaving qualification
   1 [1a, 1b, 2b] No school leaving qualification or lower General education
   2 [1c] Basic vocational training above and beyond compulsory schooling
   3 [2b] Intermediate general education
   4 [2a] Intermediate vocational qualification, or secondary programmes in which
     general intermediate schooling is combined by vocational training
   5 [2c_gen] General maturity: Full maturity certificates (e.g. the Abitur, A-levels)
   6 [2c_voc] Vocational maturity: Full maturity certificates including vocationally
     specific schooling or training
   7 [3a] Lower tertiary education: Lower level tertiary degrees, generally of shorter
     duration and with a vocational orientation
   8 [3b] Higher tertiary education: The completion of a traditional, academically
     orientated university education
```

Box A.2 (cont.) Variables in data file `rrdat2.dta` (produced by the do-file `a_ehb0.do`)

```
. tab edu_cas edu_yoe if noj == 1

educational          |
attainment, recent,  |        educational attainment, recent, [in years]
CASMIN               |     9        12        13        15        16        18 |     Total
---------------------+------------------------------------------------------------+----------
[1a, 1b, 2b] No schoo |    32         0         0         0         0         0 |        32
[1c] Basic vocational |     0       495         0         0         0         0 |       495
[2a] Intermediate voc |     0         0       666         0         0         0 |       666
[2c_gen] General matu |     0         0         4         0         0         0 |         4
[2c_voc] Vocational m |     0         0         0       294         0         0 |       294
[3a] Lower tertiary e |     0         0         0         0       107         0 |       107
[3b] Higher tertiary  |     0         0         0         0         0       337 |       337
---------------------+------------------------------------------------------------+----------
Total                 |    32       495       670       294       107       337 |     1,935
```

The variables are:

`id`	Individual identifier, values arbitrarily chosen.
`noj`	The serial number of a job. The first job of an individual always has the value 1.
`start_y`	Calendar year of job start.
`start_m`	Calendar month of job start (from 1 for January to 12 for December).
`fin_y`	Calendar year of job end.
`fin_m`	Calendar month of job end (from 1 for January to 12 for December).
`sex`	Gender of individual: 1 for men, 2 for women.
`i_y`	Calendar year of last interview.
`i_m`	Calendar month of last interview (from 1 for January to 12 for December).
`b_y`	Calendar year of birth.
`b_m`	Calendar month of birth (from 1 for January to 12 for December).
`mar_y`	Calendar year of marriage; if not yet married, -2.
`mar_m`	Calendar month of marriage; if not yet married, -2.
`isei08`	Status of current job measured by the International Socio-Economic Index (ISEI), version 2008; unidimensional scale with low values for low status and high values for high status.
`ftpt`	Job volume at the start of the current job; 1 refers to full-time, 2 refers to part-time employment (note: the data set used here contains neither changes in the volume within one job spell nor marginal employment).
`edu_cas`	Education of individual; last recorded information in the survey. The coding scheme follows the classification developed in the project Comparative Analysis of Social Mobility in Industrial Nations (CASMIN) (Braun and Müller 1997). The different categories are displayed in Box A.2. Due to low case numbers and data protection, CASMIN 1a, 1b, and 2b are aggregated into one category here.
`edu_yoe`	Education of individual; last recorded information in the survey. The coding assumes 9 years of school for those with a lower secondary level qualification without further training and up to 18 years for those with a university degree. For more information on how different educational groups are assigned to this variable, see the cross-tabulation of `edu_cas` and `edu_yoe` in Box A.2 and the official NEPS documentation by Zielonka and Pelz (2015).

Chapter 2

Familiarize yourself with the data set

a) Open data set rrdat2.dta. Describe the life course of the persons with the `id 1002, 1005,` and `1006` (note: you can use `browse` to open the data set in an editor or you can use the `list` command).

- When was the person born?
- Is the person a man or a woman?
- What is the education level?
- Describe the job history in detail.
- Has the person ever married? If yes, when did the person enter marriage for the first time?

b) How many job episodes and how many persons are in the data set? How many women are in the data set?

c) How many persons have exactly three job episodes (no less than three, but not more than three job episodes)?

d) Generate a new variable cohort and categorize the year of birth into cohorts, each cohort covering 5 years (e.g. 1944–1948, 1949–1953, etc.). An exception is the youngest cohort that spans only 3 birth years. (Note: after `generate newvar = oldvar`, you can change values in newvar with `replace` or with `recode`; for details, type `help replace` or `help recode` in Stata's command line)

- How many persons are in each cohort?
- What is the average job number in each cohort?
(Note: you can use `bysort id: generate last = noj if _n == _N` and `table cohort if last == noj, c(mean noj) f(%3.1f)`, or some other commands.)

Define episode data on job exits

e) Transform the start and the end of a job as well as the date of the interview into century months (CM).

f) Calculate the duration of a job (add 1 month to guarantee that a job lasts at least 1 month). Define the destination states. All jobs ending before the last interview have an exit. Declare your data set as survival time. Use the concept of single episode data. What is the total analysis time at risk? (The result is `576,475`.)

Advanced: Define episode data on entry into first marriage

g) Before looking at the data, first think about the following questions: What is the origin state in this case? What are the possible destination states? When do episodes start? This means, from which age onward can people get married (i.e., are they at risk)? When do episodes end?

h) Hints and suggestions on data preparation:

- The data contain multiple rows or jobs for most individuals. Entry into first marriage is a single process. Restrict the data or the episode definition to the first observation of each person (otherwise you will be counting persons as often as their number of jobs).
- In Germany, legal marriage is allowed from age 16 onward (although only with parental admission if under 18). Persons are not at risk if they are small children. Because of this, start the process with the 16th birthday (16 years * 12 months/year + 1 month).
- Add one month to the duration (time till marriage) to make sure that those marrying immediately after turning 16 have no episode duration of 0.
- Be sure that you treat those who have not married yet as right-censored at the time of the interview.

i) Declare your data set as survival time. (Note: 1 person married 3 months before turning 16 and after `stset`, Stata says: '`2 observations end on or before enter()`'). What is the total analysis time at risk? (The result is `335,786`.)

Chapter 3

Use `rrdat2.dta`, define single episode data on job exits, and declare the data to be survival time data (see above 2.f).

a) Estimate separate survivor functions on job exits for the first and the second job. Use the Kaplan-Meier method (`sts list`).

b) How large is the proportion of first jobs still held after 12, 24, 60, and 120 months?

c) What is the median time till job exit in the first and in the second job (median time: 50% are still in the current job, 50% already left)?

d) Produce one figure containing the survivor functions on job exits for the first and the second job.

Advanced: Survivor functions on entry into first marriage
Use the definition of episode data on entry into first marriage developed above (see above 2.i).

e) Estimate separate survivor functions (`sts list`) for men and women on entry into first marriage.

f) How large is the proportion of men who have already married before the age of 20, 25, or 30 years?

g) What is the median age of marriage for women and men (median age: 50% are still single, 50% already married)?

h) Produce one figure containing the survivor functions on entry into first marriage for both women and men.

i) Describe the survivor functions. What are the main differences between men and women?

j) Conduct two tests of whether differences in the survivor functions for men and women are statistically significant. Use the log-rank and the Wilcoxon (Breslow) test. Give an interpretation of the test results.

Chapter 4

Use `rrdat2.dta`, define single episode data on job exits, and declare the data to be survival time data (see above 2.f).

a) Estimate an exponential model on job exit (a transition rate model with the assumption that the duration variable T follows an exponential distribution) without covariates.

b) How large is the estimated hazard rate?

c) Based on the parametric estimation result, when do you observe the median?

d) Based on the parametric estimation result, how large is the proportion of jobs lasting longer than 12 months?

e) Estimate an exponential model with those covariates used in Box 4.1.3 in the main text. The variable on years of education is slightly different in `rrdta2.dta` than in `rrdat1.dta`. In addition, there is a variable on job status in `rrdat2.dta`, whereas there is a variable on job prestige in `rrdat1.dta`. To generate the cohort dummies, use the categorization described above in 2.d. You will find that the prefix `by` and `egen`-function with option `min(exp)` are useful for generating the variable on labor force eperience `lfx` (see Box 2.2.3 in chapter 2 or type `help egen` in Stata's command line). The covariates are: years of education (not CASMIN!), cohort dummies, labor force experience, previous number of jobs, and status of current job (ISEI).

f) Give an interpretation of the results.

g) Based on this estimation, what is the ratio of job exit rates for the second oldest versus the oldest cohort (all other things being equal)?

h) Based on the estimation results of 4.e, how much larger or smaller is the job exit rate for persons with a higher tertiary degree compared to persons with university entrance qualification without a tertiary or vocational degree? Report the percentage change, the so called alpha effect.

i) Extend the model by using working part-time as a further covariate (code part-time as 1, full-time as 0).

j) Does this enlarged model lead to a better fit compared to the model estimated under 4.e? Conduct a likelihood ratio test. Do not forget to specify the correct degrees of freedom.

k) Give an interpretation on the working hours (part-time vs. full-time).

l) Competing risks: Make a distinction between job exits followed immediately or very soon by a new job and job exits followed by a pause (= more than 3 months between end of current job and start of the next job or no subsequent job during the observation period). See Box 2.2.3 in chapter 2, if you need help on how to approach information from a subsequent job spell.

m) Estimate two competing risk models with all previously used covariates (years of education, cohort dummies, labor force experience, previous number of jobs, ISEI, part-/full-time). How do the results differ between both estimations?

Advanced: Transition rate models on entry into first marriage
Use the definition of episode data on entry into first marriage developed above (see above 2.i).

n) Estimate an exponential model on entry into first marriage (a transition rate model with the assumption that the duration variable T follows an exponential distribution) without any covariate.

o) How large is the estimated hazard rate?

p) Based on the parametric estimation result, at which age do we observe the median?

q) Based on the parametric estimation result, what are the percentages of persons already married when reaching the ages of 18 or 38?

r) Estimate an exponential model on entry into first marriage with gender as covariate (1 for women, 0 for men).

s) Give an interpretation for the coefficient for gender.

t) Based on this estimation, at which age are 50% of men and 50% of women married (median age at marriage by gender)?

u) Based on this estimation, display one survivor function for women and one for men, both in the same figure.

Chapter 5

Use `rrdat2.dta`, define single episode data on job exits, and declare the data to be survival time data (see above 2.f).

a) Estimate a piecewise constant exponential (PCE) model on job exit without any covariate. Use time periods spanning over 6 months for the first 3 years in a job and from the 3rd to the 20th year in a job for time periods of 12 months (i.e., (0,6], (6,12], (12,18], (18,24], (24,30], (30,36], (36,48], (48,60], and so on, with the last interval being $(240,+\infty)$).

b) Predict the hazard rate and plot it.

c) Describe the hazard rate.

d) Reestimate the previous model with the restriction that there is a time-constant rate after 12 years in a job. (Hint: One way to do this is to simply add up all single dummies on being more than 12 years in a job. This new variable can be included in the model instead of the single dummies on longer job durations.)

e) Does this restricted model fit the data less well than the previous one? Conduct a likelihood ratio test. Do not forget to specify the correct degrees of freedom.

f) Based on the last model, predict values for the survivor function. Plot and then describe the survivor function.

Advanced: Transition rate models on entry into first marriage
Use the definition of episode data on entry into first marriage developed above (see above 2.i).

g) Estimate a piecewise constant exponential (PCE) model on entry into first marriage without any covariate. Please use the following time periods: age 16–20 years (equivalent to the process time 0 to 48 months), 20–24, 24–28, 28–32, 32–36, 36–40, 40–44, and 44 years or above.

h) Describe the shape of the hazard rate. At which ages do people have a high propensity to marry; at which ages a low one?

i) Plot the hazard rate and compare the plot with your previous description (5h).

j) Reestimate the PCE model with gender as a covariate (1 for women, 0 for men).

k) Give an interpretation for the coefficient on gender. How many times larger or smaller is the transition rate for women than for men?

l) Plot the hazard rate for women and men into one graph (Hint: to put different lines into one graph, you can use syntaxes such as `line VAR2 VAR1, sort || line VAR3 VAR1, sort` or `twoway line VAR3 VAR2 VAR1, sort`).

m) Reestimate the PCE model with period-specific effects for gender.

n) Does this enlarged model lead to a better fit compared to the model estimated under 5.j? Conduct a likelihood ratio test. Do not forget to specify the correct degrees of freedom.

o) Plot the hazard rate for women and men based on the estimation results from 5.m into one graph. How do the transition rates differ between men and women? Are the gender-specific differences in this graph the same as those in the previous one (see 5.i)?

Chapter 6

Use `rrdat2.dta`, define single episode data on job exits, and declare the data to be survival time data (see above 2.f).

a) Estimate an exponential model on job exit with gender (1 for women, 0 for men), general labor force experience (`lfx`), and job-specific labor force experience (`lfx24`) as covariates. For job-specific labor force experience, split job duration every 24 months and put the new time varying covariate (`lfx24`) into the estimation.

b) Enlarge the previous model with a time-varying covariate on marriage (1 for married, else 0). (Note: you have to split episodes). Does this model lead to a better model fit? Conduct a likelihood ratio test. Do not forget to specify the correct degrees of freedom.

c) Estimate an exponential model on job exit for women with only general labor force experience (`lfx`) and job-specific labor force experience (`lfx24`, see above). Take marriage history into account by differentiating between the first 18 months in a marriage (=1, else 0) and married 18 months or longer (=1, else 0) (neglecting divorce or widowhood). (Note: you have to split episodes again.) Interpret the outcome.

Advanced: Transition rate models on entry into first marriage
Use the definition of episode data on entry into first marriage developed above (see 2.i).

d) Reestimate the piecewise constant exponential (PCE) model on entry into first marriage with a covariate on gender (1 for women, 0 for men) as already described in 5.j. Use the same time periods as described above: age 16–20 years (equivalent to the process time 0 to 48 months), 20–24, 24–28, 28–32, 32–36, 36–40, 40–44, and 44 years or more.

e) To test whether transition to marriage becomes more likely after labor market entry, split the episodes on entry into first marriage at labor market entry. Then define a binary variable with value 1 if a labor market entry has already taken place, else 0. Enlarge the previous model with the new time-varying covariate.

f) Interpret the outcome.

g) Is the effect of labor market entry more important for women or for men? Insert an interaction term into the model that is 1 for women after labor market entry (otherwise 0).

h) Split the episodes on a monthly base. Estimate a transition rate model with covariates on gender and months since 16th birthday (= process time) in linear and quadratic form. Take into account interaction terms between gender and process time in both linear and quadratic form.

i) Based on this estimation, at which age do men and at which age do women have their largest intensity to marry? (Hint: use the first derivation and set it to zero.)

j) Based on the estimation specified in 6.h, plot the survivor function for men and the survivor function for women into one figure.

Chapter 7

Use `rrdat2.dta`, define single episode data on job exits, and declare the data to be survival time data (see above 2.f).

a) Estimate a Gompertz model with the following covariates linked to the b parameter: years of education, cohort dummies, labor force experience, previous number of jobs, and status of current job (ISEI) (for details on covariates see above 4.e).

b) Plot two hazard rates in one figure. Assume for both hazard rates that persons are in their first job and born between 1964 and 1968. In the first case, persons hold a university degree (`edu_yoe = 18`) and a job status of 65; in the second case, they have a "basic vocational training above and beyond compulsory schooling plus vocational training" (`edu_yoe = 12`) and a job status of 40. (Hint: A convenient way is to use `stcurve` here; for more information, type `help stcurve` in Stata's command line.)

c) Estimate a Gompertz model with years of education linked to b and c parameters. To keep it simple, do not link any other covariate apart from years of education to the parameters. Plot the hazard rate for persons with a university degree and for persons with basic vocational training above and beyond compulsory schooling plus vocational training. (Note: stcurve does not work, when covariates are linked to the c parameter. You can follow strategies and syntaxes already used in the exercises for chapter 5.) Interpret the outcome.

Advanced: Transition rate models on entry into first marriage
Use the definition of episode data on entry into first marriage developed above (see 2.i).

d) Specify a log-logistic model with gender linked to the a parameter as a covariate (1 for women, 0 for men).

e) Plot the hazard function for women and men in one figure.

f) At which age ("t_{max}") do women and men have their highest propensity to marry? And how large is the rate ("r_{max}") for women and for men at these ages (see chapter 7.4 for details)?

g) Specify a log-logistic model with gender linked as a covariate to the a as well as to the b parameter (Note: to link gender to the b parameter, use the ancillary option described for Weibull models in chapter 7.3).

h) Plot the hazard function for women and men in one figure. Compare this plot to the previous one, 7.e.

Chapter 8

No specific exercises for chapter 8.

Chapter 9

Use `rrdat2.dta`, define single episode data on job exits, and declare the data to be survival time data (see above 2.f).

a) Estimate a Cox model with the following covariates: years of education, cohort dummies, labor force experience, previous number of jobs, and status of current job (ISEI) (for details on covariates see above 4.f).

b) Is the proportionality assumption violated in the previous model (9.a)? Test this for cohort membership. Use one model only for testing all cohort dummies and use the logarithm of job duration in the option `texp( )`. Give an interpretation of the results.

c) Restrict your sample to persons with at least three job episodes (`bysort id: gen max_noj = _N`).

d) Estimate a Cox model with previous number of jobs and status of current job (ISEI) as covariates.

e) Reestimate the previous model with fixed effects; this means, stratification by `id`.

f) How do the results differ between the models specified in 9.d and 9.e?

Advanced: Parametric and semiparametric transition rates on entry into first marriage
Use the definition of episode data on entry into first marriage developed above (see 2.i).

g) Estimate a Cox as well as an exponential model with coviariates on gender (1 for women, 0 for men), cohort dummies, and the time-varying covariate on labor market entry (1 after, 0 before entry; see above 5.e).

h) Drop the two cases in which marriage took place before the 16th birthday. Also drop the time-varying covariates on labor market entry and undo previous episode splitting (you might use `stjoin`). Define time until marriage right from birth and declare data to be survival time data.

i) Estimate again a Cox and an exponential model with covariates on gender, cohort membership, and labor market entry. The results of both Cox models are identical. Why? In contrast, the estimated coefficients differ between the two exponential models. Why?

j) Proportionality assumption: Test time dependence for the covariate on labor market entry in the Cox model.

k) In chapter 9.1, it is written that the Breslow method "is sufficient if there are only 'relatively few' ties." Check how common multiple ties are in the data set.

l) Because of many ties, estimate a Cox model with the exact marginal likelihood method (`exactm`). Use gender, cohort membership, and labor market entry as covariates. Do the results differ from those for the previous Cox model?

Chapter 10

Use `rrdat2.dta`, define single episode data on job exits, and declare the data to be survival time data (see above 2.f).

a) Use the following covariates: years of education, cohort dummies, labor force experience, previous number of jobs, and status of current job (ISEI) (for details on covariates see above 4.f) and estimate exponential models

- with shared frailty for job episodes belonging to one person, and
- with a random intercept for persons (`xtstreg`).

Compare the results of both models.

b) Use the same covariates and estimate a piecewise constant exponential model with the time periods (0,12], (12,24], (24,36], (36,48], (48,60], (60,72], (72,196], (96,120], and $(120,+\infty)$

- with shared frailty for job episodes belonging to one person, and
- with a random intercept for persons (`xtstreg`. Note: values reported for `Obs per group: min = avg = max =` are rather useless after splitting.)

Compare the `LR test of theta=0:` after model estimation in 10.a with those in 10.b. What might be the reasons for the large differences?

End of exercises.

Notes

[1] "From 2008 to 2013, NEPS data were collected as part of the Framework Programme for the Promotion of Empirical Educational Research funded by the German Federal Ministry of Education and Research (BMBF). Since 2014, the NEPS survey has been conducted by the Leibniz Institute for Educational Trajectories (LIfBi) at the University of Bamberg in cooperation with a nationwide network" (`https://www.neps-data.de/Mainpage`). For terms and conditions for accessing the NEPS Scientific Use Files see `https://www.neps-data.de/tabid/296/language/en-US`

[2] As mentioned in chapter 2.2, do not forget to change the working directory to the correct drive and directory by using the command `cd drive:directory_name`. The path of your current working directory is displayed in the bottom of your Stata window and will also be shown in the Results Window if you type `cd`.

References

Aalen, O. O. (1978). Nonparametric inference for a family of counting process. *Annals of Statistics, 6*, 701–726.

Aalen, O. O. (1987). Dynamic modelling and causality. *Scand. Actuarial Journal, 12*, 177–190.

Abbott, A. (1995). Sequence analysis: new methods for old ideas. *Annual Review of Sociology, 21*, 93–113.

Abbott, A., & Barman, E. (1997). Sequence comparison via alignment and Gibbs sampling. *Sociological Methodology, 27*, 47–87.

Abbott, A., & Forrest, J. (1986). Optimal matching methods for historical sequences.*Journal of Interdisciplinary History, XVI*, 471–494.

Abbott, A., & Tsay, A. (2000). Sequence analysis and optimal matching methods in sociology.*Sociological Methods and Research, 29*, 3–33.

Agarwal, S., & Kapoor, R. (2023). Value creation tradeoff in business ecosystems: Leveraging complementarities while managing interdependencies. *Organization Science, 34(3)*, 1216-1242.

Aitchison, J., & Brown, J. A. C. (1973). *The lognormal distribution with special reference to its uses in economics*. Cambridge: Cambridge University Press.

Allison, P. D. (1982). Discrete time methods for the analysis of event histories. In S. Leinhardt (Ed.), *Sociological Methodology* (pp. 61–98). San Francisco: Jossey-Bass.

Allison, P. D. (1996). Fixed-effect partial likelihood for repeated events. *Sociological Methods & Research, 25*, 207–222.

Allison, P. D. (2014). *Event history and survival analysis*. 2nd ed. Thousand Oaks, CA: Sage.

Allmendinger, J. (1989a). *Career mobility dynamics: A comparative analysis of the United States, Norway, and West Germany*. Studien und Berichte no. 49, Max-Planck-Institut für Bildungsforschung, Berlin.

Allmendinger, J. (1989b). Educational systems and labor market outcomes. *European Sociological Review, 5*, 231–250.

Allmendinger, J. (1994). *Lebensverlauf und Sozialpolitik: Die Ungleichheit von Frau und Mann und ihr öffentlicher Ertrag*. Frankfurt: Campus Verlag.

Allmendinger, J. et al. (2019). Adult education and lifelong learning. In H.-P. Blossfeld & H.-G. Roßbach (Eds.), Education as a Lifelong Process – The German National Educational Panel Study (NEPS). *Zeitschrift für Erziehungswissenschaft, vol. 3*, Wiesbaden: VS Verlag, 325–346.

Anders, J., & Dorsett, R. (2017). What young English people do once they reach schoolleaving age: a cross-cohort comparison for the last 30 years. *Longitudinal and Life Course Studies, 8*, 75–103.

Andersen, P. K., Borgan, O., Gill, R. D., & Keiding, N. (1993). *Statistical models based on counting processes*. New York: Springer-Verlag.

Andreß, H.-J. (1985). *Multivariate Analyse von Verlaufsdaten*. Mannheim: ZUMA.

Andreß, H.-J. (1989). Recurrent unemployment—the West German experience: an exploratory analysis using count data models with panel data. *European Sociological Review, 5*, 275–297.

Andreß, H.-J. (1992). *Einführung in die Verlaufsdatenanalyse*. Köln: Zentrum für historische Sozialforschung.

Andreß, H.-J., Golsch, K., & Schmidt, A. W. (2013). *Applied panel data analysis for economic and social surveys*. Springer Science & Business Media.

Arminger, G. (1990). Testing against misspecification in parametric rate models. In K. U. Mayer & N. B. Tuma (Eds.), *Event history analysis in life course research* (pp. 253–268). Madison: University of Wisconsin Press.

Arminger, G., Clogg, C. C., & Sobel, M. E. (Eds.). (1995). *Handbook of statistical modeling for the social and behavioral sciences*. New York: Plenum.

Arminger, G., & Müller, F. (1990). *Lineare Modelle zur Analyse von Paneldaten*. Opladen: Westdeutscher Verlag.

Arrow, K. (1973). Higher education as a filter. *Journal of Public Economics, 2*, 193–216.

Baccaïni, B., & Courgeau, D. (1996). The spatial mobility of two generations of young adults in Norway. *International Journal of Population Geography, 2*, 333–359.

Balbo N., & Barban, N. (2014). Does Fertility Behavior Spread among Friends? *American Sociological Review, 79*, 412–431.

Baltagi, B. H. (2015). *The Oxford handbook of panel data* Oxford University Press.

Barron, D. N. (1993). The analysis of count data: overdispersion and autocorrelation. In P. Marsden (Ed.), *Sociological Methodology* (pp. 179–220). San Francisco: Jossey-Bass.

Becker, G. S. (1975). *Human capital*. New York: Columbia University Press.

Becker, G. S. (1981). *A treatise on the family*. Cambridge: Harvard University Press.

Becker, R. (1993). *Staatsexpansion und Karrierechancen. Berufsverläufe im öffentlichen Dienst und in der Privatwirtschaft*. Frankfurt: Campus.

Becker, R., & Blossfeld, H.-P. (1991). Cohort-specific effects of the expansion of the welfare state on job opportunities: a longitudinal analysis of three birth cohorts in the FRG. *Sociologische Gids, 4*, 261–284.

Becker, R., Blossfeld, H.-P., & Mayer, K. U (2024). Socio-economic change and intergenerational class mobility: A dynamic analysis of the experiences of West Germans born between 1929 and 1971. *Research in Social Stratification and Mobility 92*, 100956.

Begall, K., & Mills, M. C. (2013). The Influence of Educational Field, Occupation, and Occupational Sex Segregation on Fertility in the Netherlands. *European Sociological Review, 29*, 720–742.

Bennett, D. S. (1997). Measuring rivalry termination, 1816–1992. *Journal of Conflict Resolution, 41*, 2227–2254.

Bernardi, F. (1999a). Does the husband matter? Married women and employment in Italy. *European Sociological Review, 15*, 285–300.

Bernardi, F. (1999b). *Donne fra famiglia e carriera. Strategie di coppia e vincoli sociali (Women between family and career. The strategy of couples and social constraints).* Milan: Franco Angeli.

Bernardi, F. (2001). Is it a timing or a probability effect? Four simulations and an application of transition rate models to the analysis of unemployment exit. *Quality and Quantity, 35*, 231–252.

Bernardi, L., Huinink, J., & Settersten Jr, R. A. (2019). The life course cube: A tool for studying lives. *Advances in life course research, 41*, 100258.

Bernardi, F., Layte, R., Schizzerotto, A., & Jacobs, S. (2000). Who exits unemployment? Institutional features, individual characteristics and chances of getting a job: a comparison of Great Britain and Italy. In D. Gallie & S. Paugam (Eds.), *Welfare regimes and the experience of unemployment*. Oxford University Press.

Bernardi, F., & Martínez-Pastor, J.-I. (2011). Female Education and Marriage Dissolution: Is it a Selection Effect? *European Sociological Review, 27*, 693–707.

Bernasco, W. (1994). *Coupled careers. The effects of spouse's resources on success at work*. Amsterdam: Thesis.

Bihagen, E., Nermo, M., Stern, C., & Åberg, Y. (2017). Elite mobility among college graduated men in Sweden: skills, personality and family ties. *Acta Sociologica, 60*, 291–308.

Bilsborrow, R. E., & Akin, J. S. (1982). Data availability versus data needs for analyzing the determinants and consequences of internal migration: an evaluation of U.S. survey data. *Review of Public Data Use, 10*, 261–284.

Blanchard, P., Bühlmann, F., & Gauthier, J.-A (Eds.) 2014. *Advances in Sequence Analysis: theory, Method, Applications.* Berlin: Springer.

Blau, D. M. (1994). Labor force dynamics of older men. *Econometrica, 62*, 117–156.

Blau, D. M., & Riphahn, R. T. (1999). Labor force transitions of older married couples in Germany. *Labour Economics, 6*, 229–251.

Blau, P. M., & Duncan, O. D. (1967). *The American occupational structure*. New York: Wiley.

Bloom, D. E. (1982). What's happening to the age at first birth in the United States? A study of recent cohorts. *Demography, 19*, 351–370.

Blossfeld, G. J. (2022). *Education, Work, and Family Events in Women?s Lives: Long-Term Developments and Recent Trends in East and West Germany*. Frankfurt and New York: Budrich Academic Press.

Blossfeld, G. J. (2026). *Comparison Between Continuous-Time and Discrete-Time Event History Analysis Using Stata*. New York and Abington: Routledge.

Blossfeld, H.-P. (1985). *Bildungsexpansion und Berufschancen*. Frankfurt and New York: Campus.

Blossfeld, H.-P. (1986). Career opportunities in the Federal Republic of Germany: a dynamic approach to the study of life-course, cohort, and period effects. *European Sociological Review, 2*, 208–225.

Blossfeld, H.-P. (1987a). Labor market entry and the sexual segregation of careers in the FRG. *American Journal of Sociology, 93*, 83–118.

Blossfeld, H.-P. (1987b). Zur Repräsentativität der Sfb-3-Lebensverlaufsstudie. Ein Vergleich mit Daten aus der amtlichen Statistik. *Allgemeines Statistisches Archiv, 71*, 126–144.

Blossfeld, H.-P. (1989). *Kohortendifferenzierung und Karriereprozeß. Eine Längsschnittstudie über die Veränderung der Bildungs- und Berufschancen im Lebenslauf*. Frankfurt: Campus.

Blossfeld, H.-P. (1990). Changes in educational careers in the Federal Republic of Germany. *Sociology of Education, 63*, 165–177.

Blossfeld, H.-P. (1992). Is the German dual system a model for a modern vocational training system? *International Journal of Comparative Sociology, 33*, 168–181.

Blossfeld, H.-P. (1994). *Causal modeling in event history analysis*. Paper prepared for the XIII World Congress of Sociology, Bielefeld. University of Bremen: mimeo.

Blossfeld, H.-P. (Ed.) (1995). *The new role of women. Family formation in modern societies*. Boulder: Westview Press.

Blossfeld, H.-P. (2017a). Evidence of causation—The contribution of life course research, Part I: Dominant models of causal inference and their limitations in life course research. In In R. Scott & M. Buchmann (General Editors) with S. Kosslyn (Consulting Editor), *Emerging trends in the social and behavioral sciences*, (pp. 1–20). New York City, US: Wiley & Sons.

Blossfeld, H.-P. (2017b). Evidence of causation—The contribution of life course research, Part II: Causation as generative process. In R. Scott & M. Buchmann (General Editors) with S. Kosslyn (Consulting Editor), *Emerging trends in the social and behavioral sciences*, (pp. 1–20). New York City, US: Wiley & Sons.

Blossfeld, H.-P., & Becker, R.(2017). Entry of men into the labor market in West Germany and their career mobility (1945-2008). A long-term longitudinal analysis identifying cohort, period, and life-course effects. *Journal for Labour Market Research, 50*, 113–130.

Blossfeld, H.-P., & Blossfeld, G.J. (2015a). Life Course and Event History Analysis. In James D. Wright (Ed.), *International Encyclopedia of the Social & Behavioral Sciences*, 2nd edition, Vol 14., (pp. 51–58) Oxford: Elsevier.

Blossfeld, H.-P., & Blossfeld, G.J. (2015b). Event History Analysis. In Henning Best and Christof Wolf (Eds.), *The Sage Handbook of Regression Analysis and Causal Infernece*, 2nd edition, Vol 14., (pp. 359–385) Los Angeles (CA): Elsevier.

Blossfeld, H.-P., Buchholz, S., & Hofäcker, D. (2006). *Globalization, uncertainty and late careers in society.* London: Routledge.

Blossfeld, H.-P., Buchholz, S., Skopek, J., & Triventi, M. (Eds.) (2016b). *Secondary education models and social inequality. An international comparison.* Cheltenham, UK, and Northampton, MA, USA: Edward Elgar.

Blossfeld, H.-P., & De Rose, A. (1992). Educational expansion and changes in entry into marriage and motherhood: the experience of Italian women. *Genus, 3–4*, 73–91.

Blossfeld, H.-P., De Rose, A., Hoem, J., & Rohwer, G. (1995). Education, modernization, and the risk of marriage disruption: differences in the effect of women's educational attainment in Sweden, West-Germany, and Italy. Stockholm Research Reports in Demography, no. 76, Stockholm. In K. Oppenheim Mason & A.-Magritt Jensen (Eds.), *Gender and family change in industrialized countries* (pp. 200–222). Oxford: Clarendon.

Blossfeld, H.-P., & Drobnič, S. (2001). *Careers of couples in contemporary societies: from male breadwinner to dual-earner families.* Oxford: Oxford University Press.

Blossfeld, H.-P., Drobnič, S., & Rohwer, G. (1998). Les trajectoires professionnelles des couples mariés en Allemagne. Une étude longitudinale de long terme de carrières des époux en Allemagne de l'Ouest. *Revue Française de Sociologie, 39*, 305–351.

Blossfeld, H.-P., Giannelli, G., & Mayer, K. U. (1993). Is there a new service proletariat? The tertiary sector and social inequality in Germany. In G. Esping-Andersen (Ed.), *Changing classes. Stratification and mobility in post-industrial societies* (pp. 109–135). London: Sage.

Blossfeld, H.-P., & Hakim, C. (1997). *Between equalization and marginalization. Part-time working women in Europe and the United States of America.* Oxford: Oxford University Press.

Blossfeld, H.-P., & Hamerle, A. (1989a). Unobserved heterogeneity in hazard rate models: a test and an illustration from a study of career mobility. *Quality and Quantity, 23*, 129–141.

Blossfeld, H.-P., & Hamerle, A. (1989b). Using Cox models to study multiepisode processes. *Sociological Methods and Research, 17*, 432–448.

Blossfeld, H.-P., & Hamerle, A. (1992). Unobserved heterogeneity in event history analysis. *Quality and Quantity, 26*, 157–168.

Blossfeld, H.-P., Hamerle, A., & Mayer, K. U. (1989). *Event history analysis.* Hillsdale, NJ: Lawrence Erlbaum Associates.

Blossfeld, H.-P., & Hofmeister, H. (2006). *Globalization, uncertainty and women's careers: an international comparison.* Cheltenham, UK: Edward Elgar.

Blossfeld, H.-P., & Huinink, J. (1991). Human capital investments or norms of role transition? How women's schooling and career affect the process of family formation. *American Journal of Sociology, 97*, 143–168.

Blossfeld, H.-P., & Jaenichen, U. (1992). Educational expansion and changes in women's entry into marriage and motherhood in the Federal Republic of Germany. *Journal of Marriage and the Family, 54*, 302–315.

Blossfeld, H.-P., Kilpi-Jakonen, E., Vono de Vilhena, D., & Buchholz, S. (Eds.). (2014). *Adult learning in modern societies: Patterns and consequences of participation from a life-course perspective.* Cheltenham, UK, and Northampton, USA: Edward Elgar.

Blossfeld, H.-P., Klijzing, E., Mills, M., & Kurz, K. (2005). *Globalization, uncertainty and youth in society.* London: Routledge.

Blossfeld, H.-P., Klijzing, E., Pohl, K., & Rohwer, G. (1999). Why do cohabiting couples marry? An example of a causal event history approach to interdependent systems. *Quality & Quantity, 33*, 229–242.

Blossfeld, H.-P., Kulic, N., Skopek, J., & Triventi, M. (Eds.) (2017). *Childcare, early education and social inequality. An international perspective.* Cheltenham, UK, and Northampton, MA, USA: Edward Elgar.

Blossfeld, H.-P., Manting, D., & Rohwer, G. (1993). Patterns of change in family formation in the Federal Republic of Germany and the Netherlands: some consequences for solidarity between generations. In H. Becker, & P. L. J. Hermkens (Eds.), *Solidarity between generations. Demographic, economic and social change, and its consequences* (Vol. I, pp. 175–196). Amsterdam: Thesis.

Blossfeld, H.-P., & Mayer, K. U. (1988). Labor market segmentation in the FRG: An empirical study of segmentation theories from a life course perspective. *European Sociological Review, 4*, 123–140.

Blossfeld, H.-P., & Mills, M. (2001). A causal approach to interrelated family events: a cross-national comparison of cohabitation, nonmarital conception, and marriage. *Canadian Journal of Population, 28*, 409–437.

Blossfeld, H.-P., Mills, M., & Bernardi, F. (2006). *Globalization, uncertainty and men's careers: an international comparison.* Cheltenham, UK: Edward Elgar.

Blossfeld, H.-P., & Müller, R. (1997). Sozialstrukturanalyse, Rational Choice Theorie und die Rolle der Zeit. Ein Versuch zur dynamischen Integration zweier Theorieperspektiven. *Soziale Welt, 47*, 382–410.

Blossfeld, H.-P., & Müller, R. (2002/2003). *Assortative partnership selection, division of work in the household, and union separation (I–III).* Three special issues of the *International Journal of Sociology*: Vol. 31, No. 4; Vol. 33, No. 1; Vol. 33, No. 2. Armonk, NY: Sharpe.

Blossfeld, H.-P., & Nuthmann, R. (1990). Transition from youth to adulthood as a cohort process in the FRG. In H. A. Becker (Ed.), *Life histories and generations* (pp. 183–217). Utrecht: ISOR.

Blossfeld, H.-P., & Prein, G. (1998). *Rational choice theory and large-scale data analysis.* Boulder: Westview Press.

Blossfeld, H.-P., & Rohwer, G. (1997a). Causal inference, time and observation plans in the social sciences. *Quality and Quantity, 31*, 361–384.

Blossfeld, H.-P., & Rohwer, G. (1997b). West Germany. In H.-P. Blossfeld (Ed.), *Between equalization and marginalization. Part-time working women in Europe and the United States of America* (pp. 164–190). Oxford: Oxford University Press.

Blossfeld, H.-P., & Rohwer, G. (2002).*Techniques of event history modeling. New approaches to causal analysis.* 2nd ed. Mahwah: Lawrence Erlbaum Associates.

Blossfeld, H.-P., Rohwer, G., & Golsch, K. (2007). *Event History Analysis with Stata.* Mahwah (NJ): Lawrence Erlbaum Associates.

Blossfeld, H.-P., & Roßbach, H.-G. (Eds.). (2019). Education as a Lifelong Process – The German National Educational Panel Study (NEPS). *Zeitschrift für Erziehungswissenschaft, Volume 3*, Wiesbaden: VS Verlag.

Blossfeld, H.-P., & Shavit, Y. (1993). Persisting barriers: changes in educational opportunities in thirteen countries. In Y. Shavit, & H.-P. Blossfeld (Eds.), *Persistent inequality* (pp. 1–23). Boulder: Westview Press.

Blossfeld, H.-P., & Stockmann, R. (1998/99). *Globalization and changes in vocational training systems in developing and advanced industrialized societies,* Vol. I–III. Armonk, NY: Sharpe (Vol. 28, No. 4; Vol 29, No. 1; and Vol 29, No 2 of *International Journal of Sociology*).

Blossfeld, H.-P., & Timm, A. (1997). Das Bildungssystem als Heiratsmarkt. Eine Längsschnittanalyse der Wahl von Heiratspartnern im Lebenslauf. *Kölner Zeitschrift für Soziologie und Sozialpsychologie, 53*, 440–476.

Blossfeld, H.-P., & Timm, A. (2003). *Who marries whom? Educational systems as marriage markets in modern societies.* Dordrecht: Kluwer Academic.

Blossfeld, H.-P., von Maurice, J., Bayer, M., & Skopek, J. (Eds.) (2016c). *Methodological issues of longitudinal surveys.* New York (NY): Springer.

Blossfeld, P.-N., Blossfeld, G. J., & Blossfeld, H.-P., (2016a). Changes in educational inequality in cross-national perspective. In M. J. Shanahan, J. T. Mortimer & M. Kirkpatrick (Eds.) *Handbook of the life course* (pp. 223–247). New York, NY: Springer.

Blumen, I., Kogan, M., & McCarthy, P. J. (1955). *The industrial mobility of labor as a probability process.* Ithaca, NY: Cornell University Press.

Boertien, D., & Mortelmans, D. (2018). Does the relationship between personality and divorce change over time? A cross-country comparison of marriage cohorts. *Acta Sociologica, 61*, 300–316.

Boldrini, M., & Grimaldi, S. (2023). Political careers and stability in office of regional ministers in Italy. *Regional & Federal Studies*, 1-28.

Bollen, K. A. (1989). *Structural equations with latent variables.* New York: Wiley.

Box-Steffensmeier, J. M., & Bradford, S. J. (1997). Time is of the essence: event history models in political science. *American Journal of Political Science, 41*, 1414–1461.

Box-Steffensmeier, J. M., & Bradford, S. J. (2004). *Event history modeling: a guide for social scientists.* Cambridge, MA: Cambridge University Press.

Bras, H., Liefbroer, A. C., & Elzinga, C. (2010). Standardization of pathways to adulthood? An analysis of Dutch cohorts born between 1850 and 1900. *Demography 47*, 1013–1034.

Braun, M., & Müller, W. (1997). Measurement of Education in Comparative Research. *Comparative Social Research, 16*, 163–201.

Breslow, N. (1974). Covariance analysis of censored survival data. *Biometrics, 30*, 89–99.

Brien, M. J., Lillard, L. A., & Waite, L. J. (1999). Interrelated family-building behaviors: cohabitation, marriage, and nonmarital conception. *Demography, 36*, 535–551.

Brückner, E., & Mayer, K. U. (1998). Collecting Life History Data. Experiences from the german Life History Study.In J. Z. Giele and G. H. Elder (Eds.), *Methods of Life Course Research: Qualitative and Quantitative Approaches.* (pp. 152–181). Thousand Oaks, CA: Sage.

Brüderl, J. (1991a). *Bell-shaped duration dependence in social processes. A generalized log-logistic rate model.* University of Bern: mimeo.

Brüderl, J. (1991b). *Mobilitätsprozesse in Betrieben. Dynamische Modelle und empirische Befunde.* Fankfurt and New York: Campus.

Brüderl, J., & Diekmann, A. (1994). *The log-logistic rate model. Two generalizations with an application to demographic data.* München/Bern: mimeo.

Brüderl, J., & Diekmann, A. (1995). The log-logistic rate model. *Sociological Methods & Research, 24*, 158–186.

Brüderl, J., Diekmann, A., & Preisendörfer, P. (1991). Patterns of intragenerational mobility: tournament models, path dependency, and early promotion effects. *Social Science Research, 20*, 197–216.

Brüderl, J., Preisendörfer, P., & Ziegler, R. (1993). Upward mobility in organizations: the effects of hierarchy and opportunity structure. *European Sociological Review, 9*, 173–188.

Brüderl, J., Preisendörfer, P., & Ziegler, R. (1998). *Der Erfolg neugegründeter Betriebe.* Berlin: Duncker & Humblot. 2. Auflage.

Brüderl, J., & Scherer, S. (2004). Methoden zur Analyse von Sequenzdata. *Kölner Zeitschrift für Soziologie und Sozialpsychologie, 44*, 330–347.

Buchmann, M. (1989). *The script of life in modern society*. Chicago: University of Chicago Press.

Buchmann, M., Kriesi, I., & Sacchi, S. (2009). Labour Market, Job Opportunities, and Transitions to Self-Employment: evidence from Switzerland from the Mid 1960s to the Late 1980s. *European Sociological Review, 25*, 569–583.

Bukodi, E., Goldthorpe, J. H., Halpin, B., & Waller, L. (2016). Is education now class destiny? Class histories across three British birth cohorts. *European Sociological Review, 32*, 835–849.

Burt, R. S. (1987). Social contagion and innovation: cohesion versus structural equivalence. *American Journal of Sociology, 92*, 1287–1335.

Burton, M. D., & Beckamn, C. M. (2007). Leaving a Legacy: position Imprints and Successor Turnover in Young Firms. *American Sociological Review, 72*, 239–266.

Butts, C. T., Lomi, A., Snijders, T. A., & Stadtfeld, C. (2023). Relational event models in network science. *Network Science, 11(2)*, 175-183.

Campbell, D. T., & Stanley, J. C. (1963). *Experimental and quasi-experimental designs for research.* Chicago: Rand McNally.

Campbell, R. T., Mutran, E., & Nash Parker, R. (1987). Longitudinal design and longitudinal analysis: a comparison of three approaches. *Research on Aging, 8*, 480–504.

Cantalini, S. (2017). Does education affect the timing or probability of family formation?: an analysis of educational attainment and first union in Italy. *Research in Social Stratification and Mobility, 49*, 1–10.

Carlson, M. J., VanOrman, A. G., & Pilkauskas, N. V. (2013). Examining the Antecedents of U.S. Nonmarital Fatherhood. *Demography, 50* , 1421–1447.

Carroll, G. R. (1983). Dynamic analysis of discrete dependent variables: a didactic essay. *Quality and Quantity, 17*, 425–460.

Carroll, G. R., & Delacroix, J. (1982). Organizational mortality in the newspaper industries of Argentina and Ireland: an ecological approach. *Administrative Science Quarterly, 27*, 169–198.

Carroll, G. R., & Hannan, M. T. (2000). *The demography of corporations and industries.* Princeton, NJ: Princeton University Press.

Carroll, G. R., & Hannan, M. T. (2018). *The demography of corporations and industries* . Princeton, NJ: Princeton University Press.

Carroll, G. R., & Mayer, K. U. (1986). Job-shift patterns in the FRG: the effects of social class, industrial sector, and organizational size. *American Sociological Review, 51*, 323–341.

Carroll, G. R., & Mosakowski, E. (1987). The career dynamics of self-employment. *Administrative Science Quarterly, 32*, 570–589.

Chamberlain, G. (1984). Panel data. In Z. Griliches & M. D. Intriligator (Eds.), *Handbook of econometrics* (Vol. 2, pp. 1247–1317). Amsterdam: North-Holland.

Chaves, M. (1996). Ordaining women: the diffusion of an organizational innovation. *American Journal of Sociology, 101*, 840–873.

Cleves, M., Gould, W. M., & Marchenko, Y. V. (2016). *An Introduction to Survival Analysis Using Stata.* Revised Third Edition, College Station (Texas): Stata Press.

Clogg, C. C., & Arminger, G. (1993). On strategy for methodological analysis. *Sociological Methodology, 23*, 57–74.

Coale, A. (1971). Age patterns of marriage. *Population Studies, 25*, 193–214.

Coleman, J. S. (1964). *Introduction to mathematical sociology.* New York: Free Press.

Coleman, J. S. (1968). The mathematical study of change. In H. M. Blalock & A. Blalock (Eds.), *Methodology in social research* (pp. 428–478). New York: McGraw-Hill.

Coleman, J. S. (1973). Theoretical bases for parameters of stochastic processes. In R. E. A. Mapes (Ed.), Stochastic processes in sociology [monograph]. *The Sociological Review, 19*, 17–28.

Coleman, J. S. (1981). *Longitudinal data analysis*. New York: Basic Books.

Coleman, J. S. (1990). *Foundations of social theory*. Cambridge, MA: Harvard University Press.

Coleman, J. S., & Hao, L. (1989). Linear systems analysis: macrolevel analysis with microlevel data. *Sociological Methodology, 19*, 395–422.

Collier, D., & Messick, R. E. (1975). Prerequisites versus diffusion: testing alternative explanations of social security adoption. *American Political Science Review, 69*, 1299–1315.

Conveney, P., & Highfield, R. (1990). *The arrow of time*. London: Allen.

Corijn, M., & Klijzing, E. (2001). *Transitions to adulthood in Europe: from a matter of standard to a matter of choice.* Dordrecht: Kluwer Academic.

Courgeau, D. (1990). Migration, family, and career: a life course approach. In P. B. Baltes, D. L. Featherman, & R. M. Lerner (Eds.), *Life-span development and behavior* (pp. 219–255). Hillsdale, NJ: Lawrence Erlbaum Associates.

Courgeau, D. (1995). Event history analysis of household formation and dissolution. In E. Van Imhoff, A. Kuijsten, P. Hooimeijer, & L. van Wissen (Eds.), *Household demography and household modeling* (pp. 185–202). New York: Plenum.

Courgeau, D., & Baccaïni, B. (1998). *Multilevel analysis in the social sciences. Population (English Selection), 10*, (pp. 39–70).

Courgeau, D., & Guérin-Pace, F. (1998). Le suivi des itinéraires professionnels des couples par les méthodes de la statistique Textuelle. Lecture des parcours professionnels des couples. In JADT 1998, 4ème journées internationales d'analyse des données Textuelles, (pp. 221–232), Université de Nice-Sophia Antipolis.

Courgeau, D., & Lelièvre, E. (1992). *Event history analysis in demography*. Oxford: Clarendon.

Courgeau, D., & Lelièvre, E. (1997). Changing paradigm in demography. *Population (English Selection), 9*, 1–10.

Courgeau, D., Lelièvre, E., & Wolber, O. (1998). Reconstruire des trajectoires de mobilité résidentielle. Eléments d'une analyse biographique des donnés de l'EDP. *Economie et Statistique, No. 316–317*.

Courgeau, D., & Najim, J. (1996). Interval censored event history analysis. *Population (English Selection), 8*, 191–208.

Covizzi, I. (2008). Does Union Dissolution Lead to Unemployment? A Longitudinal Study of Health and Risk of Unemployment for Women and Men Undergoing Separation. *European Sociological Review, 24*, 347–361.

Cox, D. R. (1958). *Planning of experiments*. New York, NY: Wiley.

Cox, D. R. (1972). Regression models and life-tables. *Journal of the Royal Statistical Society, 34*, 187–220.

Cox, D. R. (1975). Partial likelihood. *Biometrika, 62*, 269–276.

Cox, D. R. (1990). Role of models in statistical analysis. *Statistical Science, 5*, 169?174.

Cox, D. R., & Oakes, D. (1984). *Analysis of survival data*. London: Chapman and Hall.

Cox, D. R., & Snell, E. J. (1968). A general definition of residuals. *Journal of the Royal Statistical Society, B 30*, 248–275.

Cramm, C., Blossfeld, H.-P., & Drobnič, S. (1998). Die Auswirkungen der Doppelbelastung durch Familie und Beruf auf das Krankheitsrisiko von Frauen. *Zeitschrift für Soziologie, 27*, 341–357.

Crouchley, R., & Pickles, A. R. (1987). An illustrative comparison of conditional and marginal methods for analysing longitudinal data with omitted variables. In R. Crouchley (Ed.), *Longitudinal data analysis* (pp. 177–193). Aldershot: Avebury.

Davies, R. B. (1987). The limitations of cross-sectional analysis. In R. Crouchley (Ed.), *Longitudinal data analysis* (pp. 1–15). Aldershot: Avebury.

Davies, R. B., & Crouchley, R. (1985). Control for omitted variables in the analysis of panel and other longitudinal data. *Geographical Analysis, 17*, 1–15.

Davis, J. A. (1978). Studying categorical data over time. *Social Science Research, 7*, 151–179.

Deaton, A., & Cartwright, N. (2016). *Understanding and misunderstanding randomized controlled trails (Working Paper 22595)*. Cambridge, MA: National Bureau of Economic Research.

Defo, K. B. (1998). Fertility response to infant and child mortality in Africa with special reference to Cameroon. In National Research Council (Ed.), *From death to birth: mortality decline and reproductive change* (pp. 254–315). Washington, DC: National Academy Press.

Dex, S. (Ed.), (1991). *Life and work history analyses: Qualitative and quantitative developments*. London: Routledge.

Diamond, I. D., & McDonald, J. W. (1992). Analysis of current-status data. In J. Trussell, R. Hankinson, & J. Tilton (Eds.), *Demographic applications of event-history analysis* (pp. 231–252). Oxford: Clarendon.

Dickerman , B. A., & Hernán, M. A. (2020). Counterfactual prediction is not only for causal inference. *European journal of epidemiology, 35*, 615–617.

Diekmann, A. (1989). Diffusion and survival models for the process of entry into marriage. *Journal of Mathematical Sociology, 14*, 31–44.

Diekmann, A. (1992). The log-logistic distribution as a model for social diffusion processes. *Journal of Scientific & Industrial Research, 51*, 285–290.

Diekmann, A., & Engelhardt, H. (1999). The social inheritence of divorce: effects of parent's family type in postwar Germany. *American Sociological Review, 64*, 783–793.

Diekmann, A., Jungbauer-Gans, M., Krassnig, H., & Lorenz, S. (1996). Social status and aggression: a field study analyzed by survival analysis. *The Journal of Social Psychology, 136*, 761–768.

Diekmann, A., & Mitter, P. (1983). The "sickle hypothesis". *Journal of Mathematical Sociology, 9*, 85–101.

Diekmann, A., & Mitter, P. (Eds.), (1984). *Stochastic modelling of social processes*. New York: Academic Press.

Diekmann, A., & Preisendörfer, P. (1988). Turnover and employment stability in a large West German company. *European Sociological Review, 4*, 233–248.

Diekmann, A., & Weick, S. (Eds.), (1993). *Der Familienzyklus als sozialer Prozeß*. Berlin: Duncker & Humblot.

Dierckx, P. (1975). An algorithm for smoothing, differentiation and integration of experimental data using spline functions. *Journal of Computational and Applied Mathematics, 1*, 165–184.

DiPrete, T. A. (1993). Industrial restructuring and the mobility response. *American Sociological Review, 58*, 74–96.

DiPrete, T. A., & Whitman, T. S. (1988). Gender and promotion in segmented job ladder systems. *American Sociological Review, 53*, 26–40.

Doeringer, P. B. (1967). Determinants of the structure of industrial type internal labor markets. *Industrial Labor Relations Review, 20*, 205–220.

Doeringer, P. B., & Piore, M. J. (1971). *Internal labor markets and manpower analysis*. Lexington, MA: Heath Lexington.

Doksum, J., & Gasko, M. (1990). On a correspondence between models in binary regression analysis and in survival analysis. *International Statistical Review, 58*, 243–252.

Drobnič, S., & Blossfeld, H.-P. (2004). Career patterns over the life course: gender, class, and linked lives. *Research in Social Stratification and Mobility, 21*, 139–164.

Drobnič, S., Blossfeld, H.-P., & Rohwer, G. (1998). Dynamics of women's employment patterns over the family life course: a comparison of the United States and Germany. *Journal of Marriage and the Family, 61*, 133–146.

Drobnič, S., & Wittig, I. (1997). Part-time work in the United States. In H.-P. Blossfeld (Ed.), *Between equalization and marginalization*. Oxford: Oxford University Press.

Dudzevičiūte, G., Česnuityte, V., & Prakapiene, D. (2021). Defence Expenditure?Government Debt Nexus in the Context of Sustainability in Selected Small European Union Countries. *Sustainability, 13(12)*, 6669.

Duncan, O. D. (1966). Methodological issues in the analysis of social mobility. In N. Smelser & S. M. Lipset (Eds.), *Social structure and social mobility* (pp. 51–97). Chicago: Aldine.

Duncan, O. D. (1966). Path analysis: Sociological examples. *American Journal of Sociology, 72*, 1?16.

Eder, D. (1981). Ability grouping as a self-fulfilling prophecy: a micro-analysis of teacher-student interaction. *Sociology of Education, 54*, 151–162.

Eells, E. (1991). *Probabilistic causality*. Cambridge: Cambridge University Press.

Eerola, M. (1994). *Probabilistic causality in longitudinal studies.* New York: Springer-Verlag.

Elbers, C., & Ridder, G. (1982). True and spurious duration dependence: the identifiability of the proportional hazard model. *Review of Economic Studies, 49*, 403–410.

Elder, G. H. (1975). Age differentiation and the life course. *Annual Review of Sociology, 1*, 165–190.

Elder, G. H. (1978). Family history and the life course. In T. K. Hareven (Ed.), *Transitions: the family and the life course in historical perspective* (pp. 17–64). New York: Academic Press.

Elder, G. H. (1987). War mobilization and the life course: a cohort of World War II veterans. *Sociological Forum, 2*, 449–472.

Elzinga, C. H. (2003). Sequence similarity: a non-aligning technique. *Sociological Methods and Research, 32*, 3–29.

Elzinga, C. H. (2010). Complexity of categorical time series. *Sociological Methods and Research, 38*, 463–481.

Elzinga, C. H., Rahmann, S., & Wang, H. (2008). Algorithms for subsequence combinatorics. *Theoretical Computer Science, 409*, 394–404.

Engel, U., & Reinecke, J. (1994). *Panelanalyse*. New York: de Gruyter.

Erikson, R., & Goldthorpe, J. H. (1991). *The constant flux*. Oxford: Clarendon.

Erikson, R., & Jonsson, J. (1996). *Can education be equalized? The Swedish case in comparative perspective*. Boulder, CO: Westview Press.

Erola, J., Härkönen, J., & Dronkers, J. (2012). More Careful or Less Marriageable? Parental Divorce, Spouse Selection and Entry into Marriage. *Social Forces, 90*, 1323–1345.

Esping-Andersen, G. (1990). *The three worlds of welfare capitalism*. Cambridge: Polity Press.

Esping-Andersen, G. (Ed.), (1993). *Changing classes*. London: Sage.

Esping-Andersen, G., Leth-Sørensen, S., & Rohwer, G. (1994). Institutions and occupational class mobility: scaling the skill-barrier in the Danish labor market. *European Sociological Review, 10*, 119–134.

Evertson, M., Grunow, D., & Aisenbrey, S. (2016). Work interruptions and young women's career prospects in Germany, Sweden and the US. *Work, Employment & Society, 30*, 291–308.

Fasang, A. E. (2012). Retirement patterns and income inequality. *Social Forces, 90*, 685–711.

Faulbaum, F., & Bentler, P. M. (1994). Causal modeling: some trends and perspectives. In I. Borg & P. P. Mohler (Eds.), *Trends and perspectives in empirical social research* (pp. 224–249). Berlin: de Gruyter.

Featherman, D. L., Selbee, K. L., & Mayer, K. U. (1989). Social class and the structuring of the life course in Norway and West Germany. In D. I. Kertzer & W. K. Schaie (Eds.), *Age structuring in comparative perspective* (pp. 55–93). Hillsdale, NJ: Lawrence Erlbaum Associates.

Fine, J. P., & Gray, R. J. (1999). A Proportional Hazards Model for the Subdistribution of a Competing Risk. *Journal of the American Statistical Association*, 94, 496–509.

Flinn, C. J., & Heckman, J. J. (1982). Models for the analysis of labor force dynamics. In R. Basmann & G. Rhodes (Eds.), *Advances in econometrics* (pp. 35–95). London: JAI Press.

Fox, J. (1992). Statistical models for nonexperimental data: a comment on Freeman. In J. Popper Shaffer (Ed.), *The role of models in nonexperimental science: two debates* (pp. 63–66). Washington, DC: American Educational Research Association.

Freedman, R. A. (1991). Statistical analysis and shoe leather. *Sociological Methodology, 21*, 291–313.

Freeman, J., Carroll, G. R., & Hannan, M. T. (1983). The liability of newness: age dependence in organizational death rates. *American Sociological Review, 48*, 692–710.

Furr, C. N. R. (2019). Product adaptation during new industry emergence: The role of start-up team preentry experience. *Organization Science, 30(5)*, 1076-1096.

Gabadinho, A., Ritschard, G., Studer, M., & Müller, N. S. (2009). Mining sequence data in R with the TraMineR package: A user's guide for version 1.2. Technical report, University of Geneva.

Galler, H. P., & Pötter, U. (1990). Unobserved heterogeneity in models of unemployment. In K. U. Mayer & N. B. Tuma (Eds.), *Event history analysis in life course research* (pp. 226–240). Madison: University of Wisconsin Press.

Galler, H. P., & Pötter, U. (1992). Zur Robustheit von Schätzmodellen für Ereignisdaten. In R. Hujer, H. Schneider, & W. Zapf (Eds.), *Herausforderungen an den Wohlfahrtsstaat im strukturellen Wandel* (pp. 379–405). Frankfurt: Campus.

Ganzeboom, H. B. G., & Treiman, D. J. (2003). Three internationally standardised measures for comparative research on occupational status. In J. H. P. Hoffmeyer-Zlotnik & C. Wolf (Eds.), *Advances in cross-national comparison. A European working book for demographic and socio-economic variables* (pp. 159–193). New York: Kluwer Academic/Plenum Publishers.

Gardner, W., & Griffin, W. A. (1986). *A structural-causal model for analyzing parallel streams of continuously recorded discrete events.* Unpublished manuscript, University of Washington.

Gesthuizen, M., & Wolbers, M. H. J. (2010). Employment transitions in the Netherlands, 1980–2004: are low educated men subject to structural or cyclical crowding out? *Research in Social Stratification and Mobility, 28*, 437–451.

Giere, R. N. (1999). *Science without laws.* Chicago: University of Chicago Press.

Glenn, N. D. (1977). *Cohort analysis.* Beverly Hills, CA: Sage.

Goldthorpe, J. H. (1987). *Social mobility and class structure in modern Britain.* Oxford: Clarendon.

Goldthorpe, J. H. (1991). The use of history in sociology: reflections on some recent tendencies. *British Journal of Sociology, 42*, 211–230.

Goldthorpe, J. H. (1996). The quantitative analysis of large-scale data-sets and rational action theory: for a sociological alliance. *European Sociological Review, 12*, 109–126.

Goldthorpe, J. H. (2000). *On sociology. Numbers, narratives, and the integration of research and theory.* Oxford: Oxford University Press.

Goldthorpe, J. H. (2001). Causation, statistics, and sociology. *European Sociological Review, 17*, 1?20.

Golsch, K. (2005). *The impact of labour market insecurity on the work and family life of men and women. A comparison of Germany, Great Britain, and Spain.* European University Studies. Frankfurt: Peter Lang.

Goodman, L. A. (1973). Causal analysis of data from panel studies and other kind of surveys. *American Journal of Sociology, 78*, 1135–1191.

Gooley, T. A., Leisenring, W., Crowleym J., & Storer, B.E. (1999). Estimation of failure probabilities in the presence of competing risks: new representations of old estimators. *Statistics in Medicine, 18,* 695–706.

Granger, C. W. J. (1969). Investigating causal relations by econometric models and cross-spectral methods. *Econometrica, 37*, 424?438.

Greve, H. R. (1995). Jumping ship: the diffusion of strategy abandonment. *Administrative Science Quarterly, 49*, 444–473.

Greve, H. R., Strang, D., & Tuma, N. B. (1995). Specification and estimation of heterogeneous diffusion models. *Sociological Methodology, 13*, 377–420.

Grundmann, M. (1992). *Familienstruktur und Lebensverlauf. Historische und gesellschaftliche Bedingungen individueller Entwicklungen.* Frankfurt: Campus.

Guo, G. (1993). Event-history analysis of left-truncated data. In P. Marsden (Ed.), *Sociological Methodology* (Vol. 23, pp. 217–242). San Francisco: Jossey-Bass.

Hachen, D. S. (1988). The competing risks model. *Sociological Methods and Research, 17*, 21–54.

Hägglund, A. E., & Bächmann, A.-C. (2017). Fast lane or down the drain? Does the occupation held prior to unemployment shape the transition back to work? *Research in Social Stratification and Mobility, 49*, 32–46.

Haller, M. (1989). *Klassenstrukturen und Mobilität in fortgeschrittenen Gesellschaften.* Frankfurt: Campus.

Halpin, B. (2013). Sequence analysis. In J. Baxter (Ed.), *Oxford Bibliographies in Sociology.* Oxford University Press, New York.

Halpin, B. (2014). Three narratives of sequence analysis. In P. Blanchard, F. Bühlmann, & J.-A. Gauthier (Eds.), *Advances in Sequence Analysis: theory, Method, Applications.* Berlin: Springer.

Halpin, B. (2016). Cluster analysis stopping rules in Stata. Working Paper WP2016-01, Dept of Sociology, University of Limerick, Ireland.

Halpin, B. (2017). SADI: sequence analysis tools for Stata. *Stata Journal, 17*, 546–572.

Halpin, B., & Chan, T. W. (1998). Class careers as sequences: an optimal matching analysis of work-life histories. *European Sociological Review, 14*, 111–130.

Hamerle, A. (1989). Multiepisode spell regression models for duration data. *Applied Statistics, 38*, 127–138.

Hamerle, A. (1991). On the treatment of interrupted spells and initial conditions in event history analysis. *Sociological Methods & Research, 19*, 388–414.

Hamerle, A., & Tutz, G. (1989). *Diskrete Modelle zur Analyse von Verweildauer und Lebenszeiten*. Frankfurt: Campus.

Han, S.-K., & Moen, P. (1999). Clocking out: temporal patterning of retirement. *American Journal of Sociology, 105*, 191–236.

Han, S.-K., & Moen, P. (2001). Coupled careers: men's and women's pathways through work and marriage in the United States. In H.-P. Blossfeld & S. Drobnic (Eds.), *Careers of Couples in Contemporary Society: from Male Breadwinner to Dual-Earner Families: From Male Breadwinner to Dual-Earner Families*, Oxford. OUP.

Handl, J. (1988). *Berufschancen und Heiratsmuster von Frauen*. Frankfurt: Campus.

Handl, J., Mayer, K. U., & Müller, W. (1977). *Klassenlage und Sozialstruktur*. Frankfurt: Campus.

Hannan, M. T., Carroll, G. R., Dobrev, S. D., & Han, J. (1998). Organizational mortality in European and American automobile industries. Part I: revisiting the effects of age and size. *European Sociological Review, 14*, 279–302.

Hannan, M. T., Carroll, G. R., Dobrev, S. D. Han, J., & Torres, J. C. (1998). Organizational mortality in European and American automobile industries. Part II: Coupled clocks. *European Sociological Review, 14*, 303–313.

Hannan, M. T., & Freeman, J. (1989). *Organizational ecology*. Cambridge, MA: Harvard University Press.

Hannan, M. T., Schömann, K., & Blossfeld, H.-P. (1990). Sex and sector differences in the dynamics of wage growth in the Federal Republic of Germany. *American Sociological Review, 55*, 694–713.

Hannan, M. T., & Tuma, N. B. (1979). Methods for temporal analysis. *Annual Review of Sociology, 5*, 303–328.

Hannan, M. T., & Tuma, N. B. (1990). A reassessment of the effects of income maintenance on marital dissolution in the Seattle-Denver income experiments. *American Journal of Sociology, 95*, 1270–1298.

Hansen, H.-T. (2009). The Dynamics of Social Assistance Recipiency: empirical Evidence from Norway. *European Sociological Review, 25*, 215–231.

Heckman, J. J., & Borjas, G. J. (1980). Does unemployment cause future unemployment? *Economica, 47*, 247–283.

Heckman, J. J., & Singer, B. (1982). The identification problem in econometric models for duration data. In W. Hildenbrand (Ed.), *Advances in econometrics* (pp. 39–77). Cambridge: Cambridge University Press.

Heckman, J. J., & Singer, B. (1984). A method of minimizing the impact of distributional assumptions in econometric models for duration data. *Econometrica, 52*, 271–320.

Heckman, J. J., & Walker, J. R. (1987). Using goodness of fit and other criteria to choose among competing duration models: a case study of Hutterite data. In C. Clogg (Ed.), *Sociological Methodology* (pp. 247–307). San Francisco: Jossey-Bass.

Heckman, J. J., & Willis, R. J. (1977). A beta-logistic model for the analysis of sequential labor force participation by married women. *Journal of Political Economy, 85*, 27–58.

Hedström, P. (1994). Contagious collectivities: on the spatial diffusion of Swedish trade unions, 1890–1940. *American Journal of Sociology, 99*, 1157–1179.

Heinz, W. R. (Ed.), (1991a). *The life course and social change. Status passages and the life course*. Vol. 2. Weinheim: Deutscher Studien Verlag.

Heinz, W. R. (Ed.), (1991b). *Theoretical advances in life course research. Status passages and the life course*, Vol. 1. Weinheim: Deutscher Studien Verlag.

Heinz, W. R. (Ed.), (1992). *Institutions and gate keeping in the life course. Status passages and the life course* (Vol. 3). Weinheim: Deutscher Studien Verlag.

Herle, M., Micali, N., Abdulkadir, M., Loos, R., Bryant-Waugh, R., Bulik, C. M., & De Stavola, B. L. (2020). Identifying typical trajectories in longitudinal data: modelling strategies and interpretations. *European Journal of Epidemiology, 35*, 205–222.

Hernán, M. A. (2018). The c-word: scientific euphemisms do not improve causal inference from observational data. *American Journal of Public Health, 108*, 616–619.

Hernes, G. (1972). The process of entry into first marriage. *American Sociological Review, 37*, 173–182.

Hoem, J. M. (1983). Distortions caused by nonobservation of periods of cohabitation before the latest. *Demography, 20*, 491–506.

Hoem, J. M. (1985). Weighting, misclassification, and other issues in the analysis of survey samples of life histories. In J. J. Heckman & B. Singer (Eds.), *Longitudinal analysis of labor market data* (pp. 249–293). Cambridge: Cambridge University Press.

Hoem, J. M. (1986). The impact of education on modern family-union initiation. *European Journal of Population, 2*, 113–133.

Hoem, J. M. (1989). The issue of weights in panel surveys of individual behavior. In D. Kasprzyk, G. Duncan, G. Kalton, & M. P. Singh (Eds.), *Panel surveys* (pp. 539–565). New York: Wiley.

Hoem, J. M. (1991). To marry, just in case ...: the Swedish widow's pension reform and the peak in marriages December 1989. *Acta Sociologica, 43*, 127–135.

Hoem, J. M., & Rennermalm, B. (1985). Modern family initiation in Sweden: experience of women born between 1936 and 1960. *European Journal of Population, 1*, 81–111.

Hogan, D. P. (1978). The effects of demographic factors, family background, and early job achievement on age at marriage. *Demography, 15*, 139–160.

Hogan, D. P. (1981). *Transitions and social change: the early lives of American men*. New York: Academic Press.

Holland, P. W. (1986). Statistics and causal inference. *Journal of the American Statistical Association, 81*, 945–960.

Holland, P. W. (1988). Causal inference, path analysis, and recursive structural equation models. *Sociological Methodology, 18*, 449?484.

Hser, Y.-I., Yamaguchi, K., Chen, J., & Anglin, M. D. (1995). Effects of intervention on relapse to narcotics addiction: an event-history analysis. *Evaluation Review, 19*, 123–140.

Hsiao, C. (2003). *Analysis of panel data*. Cambridge: Cambridge University Press.

Huinink, J. (1987). Soziale Herkunft, Bildung und das Alter bei der Geburt des ersten Kindes. *Zeitschrift für Soziologie, 16*, 367–384.

Huinink, J. (1989). *Mehrebenenanalyse in den Sozialwissenschaften*. Wiesbaden: DUV.

Huinink, J. (1992). Die Analyse interdependenter Lebensverlaufsprozesse. In H.-J. Andreß, H. Huinink, H. Meinken, D. Rumianek, W. Sodeur, & G. Sturm (Eds.), *Theorie, Daten, Methoden. Neue Modelle und Verfahrensweisen in den Sozialwissenschaften* (pp. 343–367). München: Oldenbourg.

Huinink, J. (1993). *Warum noch Familie? Zur Attraktivität von Partnerschaft und Elternschaft in unserer Gesellschaft*. Habilitationsschrift, Freie Universität Berlin.

Huinink, J. (1995). Education, work, and family patterns of men: the case of West Germany. In H.-P. Blossfeld (Ed.), *The new role of women. Family formation in modern societies* (pp. 247–262). Boulder: Westview Press.

Huinink, J., Mayer, K. U., Diewald, M., Solga, H., Sørensen, A., & Trappe, H. (1995). *Kollektiv und Eigensinn. Lebensverläufe in der DDR und danach*. Berlin: Akademie Verlag.

Hunt, M. (1985). *Profiles of social research. The scientific study of human interactions*. New York: Russell Sage Foundation.

Hutchison, D. (1988a). Event history and survival analysis in the social sciences, part I. *Quality and Quantity, 22*, 203–219.

Hutchison, D. (1988b). Event history and survival analysis in the social sciences, part II (Advanced applications and recent developments). *Quality and Quantity, 22*, 255–278.

Jacobs, S. C. (1995). Changing patterns of sex segregated occupations throughout the life-course. *European Sociological Review, 11*, 157–185.

Jann, B. (2019). HEATPLOT: Stata module to create heat plots and hexagon plots. *Statistical Software Components, Boston College Department of Economics*.

Jayawarna, D., Marlow, S., & Swail, J. (2021). A gendered life course explanation of the exit decision in the context of household dynamics. *Entrepreneurship Theory and Practice, 45(6)*, 1394-1430.

Jenkins, S. P. (2005). Survival Analysis. *Institute for Social and Economic Research, University of Essex, Colchester, UK* (unpublished manuscript).

Jöreskog, K. G., & Sörbom, D. (1993). *LISREL 8. Structural equation modeling with the SIMPLIS command language*. Chicago: Scientific Software International.

Kalbfleisch, J. D., & Prentice, R. L. (1980). *The statistical analysis of failure data*. New York: Wiley.

Kalter, F. (2011). Social Capital and the Dynamics of Temporary Labour Migration from Poland to Germany. *European Sociological Review, 27*, 555–569.

Kaplan, E. L., & Meier, P. (1958). Nonparametric estimation from incomplete observations. *Journal of the American Statistical Association, 53*, 457–481.

Karhula, A., Erola, J., Raab, M., & Fasang, A. (2019). Destination as a process: Sibling similarity in early socioeconomic trajectories. *Advances in Life Course Research, 40*, 85-98.

Kay, F. M. (2023). Revolving Doors: Social Dimensions of Law Firm Culture and Pathways out of Firms. *Law & Social Inquiry, 48(4)*, 1181-1209.

Kelly, J. R., & McGrath, J. E. (1988). *On time and method*. Newbury Park, CA: Sage.

Kenny, D. A. (1979). *Correlation and causality*. New York: Wiley.

Kiefer, N. M. (1988). Economic duration data and hazard functions. *Journal of Economic Literature, 16*, 646–679.

Klijzing, E. (1992). Wedding in the Netherlands: first-union disruption among men and women born between 1928 and 1965. *European Sociological Review, 8*, 53–70.

Klijzing, E. (1993). A method for the simultaneous estimation of parallel processes in the human life course. *Studia Demograficzne, 3*, 111–124.

Klijzing, E., & Corijn, M. (2001). *Transitions to adulthood in Europe*. London: Routledge.

Kneip, T., Bauer, G., & Reinhold, S. (2014). Direct and Indirect Effects of Unilateral Divorce Law on Marital Stability. *Demography, 51* , 2103–2126.

Kohler, H.-P. (2001). *Fertility and social interaction. An economic perspective*. Oxford: Oxford University Press.

Krain, M. (1997). State-sponsored mass murder: The onset and severity of genocides and politicides. *Journal of Conflict Resolution, 41*, 331–360.

Kravdal, Ø., & Rindfuss, R. (2008). Changing Relationships between Education and Fertility: a Study of Women and Men Born 1940 to 1964. *American Sociological Review, 73* , 854–873.

Krempel, L. (1987). *Soziale Interaktionen: Einstellungen, Biographien, Situationen und Beziehungsnetzwerke.* Bochum: Schullwig Verlag.

Kreuter, F., & Kohler, U. (2009). Analyzing contact sequences in call record data. Potential and limitations of sequence indicators for nonresponse adjustments in the European Social Survey. *Journal of Official Statistics, 25*, 203–226.

Kuha, J., & Mills, C. (2018). On Group Comparisons with Logistic Regression Models. *Sociological Methods & Research 20*, 1–28.

Kurz, K. (2000). Soziale Ungleichheit und Wohneigentum. *Zeitschrift für Soziologie, 29*, 28–44.

Kurz, K., & Blossfeld, H.-P. (2004). *Home ownership and social inequality in comparative perspective. Studies in Social Inequality.* Stanford, CA: Stanford University Press.

Laird, N. M. (2020). Longitudinal panel data: an overview of current methodology. *Time Series Models*, 143-176.

Lancaster, T. (1985). Generalized residuals and heterogeneous duration models with applications to the Weibull model. *Journal of Econometrics, 28*, 155–169.

Lancaster, T. (1990). *The econometric analysis of transition data.* Cambridge: Cambridge University Press.

Lauterbach, W. (1994). *Berufsverläufe von Frauen. Erwerbstätigkeit, Unterbrechung und Wiedereintritt.* Frankfurt: Campus.

Lawless, J. F. (1982). *Statistical models and methods for lifetime data.* New York: Wiley.

Lazarsfeld, P. F. (1948). The use of panels in social research. *Proceedings of the American Philosophy of Sociology, 92*, 405–410.

Lazarsfeld, P. F. (1972). Mutual relations over time of two attributes: a review and integration of various approaches. In M. Hammer, K. Salzinger, & S. Sutton (Eds.), *Psychopathology* (pp. 461–80). New York: Wiley.

Lee, N. (2022). Measurement error and its impact on estimates of income dynamics. *Empirical Economics, 63(5)*, 2539–2550.

Leisering, L., & Leibfried, S. (1998). *Time, life & poverty.* Cambridge: Cambridge University Press.

Leisering, L., & Walker, R. (1998). *The dynamics of modern society. Poverty, policy and welfare.* Southampton: Ashford Press.

Lelièvre, E., Bonvalet, C., & Bry, X. (1998). Event history analysis of groups. The findings of an on-going research project. *Special Issue on New Methodological Approaches in the Social Sciences, Population (English Selection), 10*, (pp. 11–38).

Lelièvre, E., & Bringe, A. (1998). *Practical guide to event history analysis using SAS, TDA, STATA.* Methodes et Savoirs, Paris: INED.

Leridon, H. (1989). Cohabitation, marriage, separation: an analysis of life histories of French cohorts from 1968 to 1985. *Population Studies, 44*, 127–144.

Lersch, P. M., & Luijkx, R. (2015). Intergenerational transmission of homeownership in Europe: revisiting the socialisation hypothesis. *Social Science Research, 49*, 327–342.

Lesnard, L. (2008). Off-scheduling within dual-earner couples: an unequal and negative externality for family time. *American Journal of Sociology, 114*, 447–490.

Li, L., & Choe, M. K. (1997). A mixture model for duration data: analysis of second births in China. *Demography, 34*, 189–197.

Liao, T.F., Bolano, D., Brzinsky-Fay, C., Cornwell, B., Fasang, A., Helske, S., Piccarreta, R., Raab, M., Ritschard, G., Struffolino, E., & Studer, M. (2022). Sequence analysis: Its past, present, and future. *Social Science Research, 107*, 102772. .

Liao, T.F. (2021). Using Sequence Analysis to Quantify How Strongly Life Courses Are Linked. *Sociological Science, 8*, 48–72.

Lieberson, S. (1985). *Making it count. The improvement of social research and theory.* Berkeley: University of California Press.

Lieberson, S. (1991). Small n's and big conclusions: an examination of the reasoning in comparative studies based on a small number of cases. *Social Forces, 70*, 307–320.

Liefbroer, A. (1991). The choice between a married or unmarried first union by young adults. A competing risks analysis. *European Journal of Population, 7*, 273–298.

Lillard, L. A. (1993). Simultaneous equations for hazards: marriage duration and fertility timing. *Journal of Econometrics, 56*, 189–217.

Lillard, L. A., Brien, M. J., & Waite, L. J. (1995). Premarital cohabitation and subsequent marital dissolution: a matter of self-selection? *Demography, 32*, 437–457.

Lillard, L. A., & Waite, L. J. (1993). A joint model of marital childbearing and marital disruption. *Demography, 30*, 653–681.

Liu, M. (2013). Migrant Networks and International Migration: testing Weak Ties. *Demography, 50*, 1243–1277.

Lomi, A. (1995). The population and community ecology of organizational founding: Italian co-operative banks, 1936–1969. *European Sociological Review, 11*, 75–98.

Longhofer, W., & Schofer, E. (2010). National and Global Origins of Environmental Association. *American Sociological Review, 75*, 505–533.

Lott, Y., (2020). Is maternal labor market re-entry after childbirth facilitated by mothers? and partners? flextime? *Human Relations, 73(8)*, 1106-1128.

Lu, Y., Liang, Z., & David, M. (2013). Emigration from China in Comparative Perspective. *Social Forces, 92*, 631–658.

Luijkx, R., & Wolbers, M. H. J. (2009). The Effects of Non-Employment in Early Work-Life on Subsequent Employment Chances of Individuals in The Netherlands. *European Sociological Review, 25* , 647–660.

Lyngstad, T. H., & Prskawetz, A. (2010). Do siblings' fertility decisions influence each other? *Demography, 47* , 923–934.

Mach, B. W., Mayer, K. U., & Pohoski, M. (1994). Job changes in the Federal Republic of Germany and Poland: a longitudinal assessment of the impact of welfare-capitalist and state-socialist labour-market segmentation. *European Sociological Review, 10*, 1–28.

Maggs, J. L., Staff, J., Patrick, M. E., & Wray-Lake, L. (2019). Very early drinking: Event history models predicting alcohol use initiation from age 4 to 11 years. *Addictive behaviors, 89*, 121-127.

Magnusson, D., & Bergmann, L. R. (Eds.), (1990). *Data quality in longitudinal research*. Cambridge: Cambridge University Press.

Magnusson, D., Bergmann, L. R., & Törestad, B. (Eds.), (1991). *Problems and methods in longitudinal research: stability and change*. Cambridge: Cambridge University Press.

Mahajan, V., & Peterson, R. A. (1985). *Models of innovation diffusion*. Beverly Hills, CA: Sage.

Manski, C. F. (1993). Identification problems in the social sciences. *Sociological Methodology, 23*, 1–56.

Manting, D. (1994). *Dynamics in marriage and cohabitation. An inter-temporal, life course analysis of first union formation and dissolution*. Amsterdam: Thesis.

Manting, D. (1996). The changing meaning of cohabitation and marriage. *European Sociological Review, 12*, 53–65.

March, J. G., Sproull, L. S., & Tamuz, M. (1991). Learning from samples of one or fewer. *Organization Science, 2*, 1–13.

Marini, M. M. (1978). The transition to adulthood: sex differences in educational attainment and age at marriage. *American Sociological Review, 43*, 483–507.

Marini, M. M. (1984). Women's educational attainment and the timing of entry into parenthood. *American Sociological Review, 49*, 491–511.

Marini, M. M. (1985). Determinants of the timing of adult role entry. *Social Science Research, 14*, 309–350.

Marini, M. M., & Singer, B. (1988). Causality in the social sciences. In C. C. Clogg (Ed.), *Sociological Methodology* (pp. 347–409). San Francisco: Jossey-Bass.

Mason, K. O. et al. (1973). Some methodological issues in cohort analysis of archival data. *American Sociological Review, 38*, 242–258.

Mason, W. M., & Fienberg, S. E. (Eds.), (1985). *Cohort analysis in social research*. New York: Springer.

Mayer, K. U. (1987). Lebenslaufsforschung. In W. Voges (Ed.), *Methoden der Biographie- und Lebenslaufsforschung* (pp. 51–73). Opladen: Leske & Budrich.

Mayer, K. U. (1988). German survivors of World War II: the impact on the life course of the collective experience of birth cohorts. In M. W. Riley (Ed.), *Social structures and human lives. Social change and the life course* (pp. 229–246). Newbury Park, CA: Sage.

Mayer, K. U. (Ed.), (1990). *Lebensverläufe und sozialer Wandel. Sonderheft 31, Kölner Zeitschrift für Soziologie und Sozialpsychologie*. Opladen: Westdeutscher Verlag.

Mayer, K. U. (1991). Life courses in the welfare state. In W. R. Heinz (Ed.), *Theoretical advances in life course research* (Vol. 1, pp. 171–186). Weinheim: Deutscher Studienverlag.

Mayer, K. U. (2015). The German Life History Study: An introduction. *European Sociological Review, 31*, 137–14314.

Mayer, K. U., Allmendinger, J., & Huinink, J. (Eds.), (1991). *Vom Regen in die Traufe: Frauen zwischen Beruf und Familie*. Frankfurt: Campus.

Mayer, K. U., Featherman, D. L., Selbee, L. K., & Colbjørnsen, T. (1989). Class mobility during the working life: a comparison of Germany and Norway. In M. L. Kohn (Ed.), *Cross-national research in sociology* (pp. 218–239). Newbury Park, CA: Sage.

Mayer, U. K., & Baltes, P. B. (1996). *Die Berliner Altersstudie*. Berlin: Academie Verlag.

Mayer, K. U., & Brückner, E. (1989). *Lebensverläufe und Wohlfahrtsentwicklung. Konzeption, Design und Methodik der Erhebung von Lebensverläufen der Geburtsjahrgänge 1929–31, 1939–41, 1949–51. Materialien aus der Bildungsforschung, Nr. 35*. Max-Planck-Institut für Bildungsforschung, Berlin.

Mayer, K. U., & Carroll, G. R. (1987). Jobs and classes: structural constraints on career mobility. *European Sociological Review, 3*, 14–38.

Mayer, K. U., & Huinink, J. (1990). Age, period, and cohort in the study of the life course: a comparison of classical A-P-C-analysis with event history analysis or farewell to Lexis? In D. Magnusson & L. R. Bergmann (Eds.), *Data quality in longitudinal research* (pp. 211–232). Cambridge: Cambridge University Press.

Mayer, K. U., & Müller, W. (1986). The state and the structure of the life course. In A. B. Sørensen, F. Weinert, & L. R. Sherrod (Eds.), *Human development and the life course* (pp. 217–245). Hillsdale, NJ: Lawrence Erlbaum Associates.

Mayer, K. U., & Schöpflin, U. (1989). The state and the life course. *Annual Review of Sociology, 15*, 187–309.

Mayer, K. U., & Schwarz, K. (1989). The process of leaving the parental home. Some German data. In E. Grebenik, C. Höhn, & R. Mackensen (Eds.), *Later phases of the family cycle. Demographic aspects* (pp. 145–163). Oxford: Clarendon.

Mayer, K. U., & Tuma, N. B. (Eds.), (1990). *Event history analysis in life course research.* Madison: University of Wisconsin Press.

McDonough, P., Wort, D., Corna, L. M., McMunn, A., & Sacker, A. (2017). Later-life employment trajectories and health. *Advances in Lifecourse Research, 34*, 22–33.

McGinnity, F. (2004). *Welfare for the unemployed in Britain and Germany. Who benefits?* Cheltenham, UK: Edward Elgar.

McVicar, D., & Anysadike-Danes, M. (2002). Predicting successful and unsuccessful transitions from school to work using sequence methods. *Journal of the Royal Statistical Society (Series A), 165*, 317–334.

Medical Research Council. (1992). *Review of longitudinal studies.* London: mimeo.

Menken, J. (1985). Age and fertility: how late can you wait? *Demography, 22*, 469–483.

Meulemann, H. (1990). Schullaufbahnen, Ausbildungskarrieren und die Folgen im Lebensverlauf. Der Beitrag der Lebensverlaufsforschung zur Bildungssoziologie. In K. U. Mayer (Ed.), *Lebensverläufe und sozialer Wandel* (pp. 89–117). Opladen: Westdeutscher Verlag.

Michael, R. T., & Tuma, N. B. (1985). Entry into marriage and parenthood by young men and women: the influence of family background. *Demography, 22*, 515–543.

Miller, M. L., Ferrer, E., & Ghisletta, P. (2024). The current practice of latent growth curve modeling in the social and behavioral sciences: Observations and recommendations. *International Journal of Behavioral Development, 0*.

Mills, M. (2000a). Providing space for time: the impact of temporality on life course research. *Time & Society, 9*, 91–127.

Mills, M. (2000b). *The transformation of partnerships.* Amsterdam: THELA THESIS Population Studies.

Mills, M. (2011). *Introducing survival and event history analysis.*Thousand Oaks, CA: Sage.

Mills, M., & Trovato, F. (2001). The effect of pregnancy in cohabiting unions on marriage in Canada, the Netherlands, and Latvia. *Statistical Journal of the United Nations ECE, 18*, 103–118.

Mincer, J. (1974). *Schooling, experience, and earnings.* New York: National Bureau of Economic Research.

Mincer, J., & Polachek, S. (1974). Family investments in human capital: earnings of women. *Journal of Political Economy, 82*, 76–108.

Minello, A., & Blossfeld, H.-P. (2017). From parents to children: the impact of mothers? and fathers? educational attainments on those of their sons and daughters in West Germany. *British Journal of Sociology of Education 38 (5)*, 686-704.

Minkoff, D. C. (1997). The sequencing of social movements. *American Sociological Review, 62*, 779–799.

Minkoff, D. (1999). Bending with the wind: strategic change and adaptation by women's and racial minority organizations. *American Journal of Sociology, 104*, 1666–1703.

Momtaz, P. P. (2021). Econometric models of duration data in entrepreneurship with an application to start-ups' time-to-funding by venture capitalists (VCs). *Journal of Applied Statistics, 48(13-15)*, 2673-2694.

Mood, C. (2010). Logistic Regression: why We Cannot Do What We Think We Can Do, and What We Can Do About It. *European Sociological Review 26*, 67–82.

Mood, C. (2013). Social Assistance dynamics in Sweden: duration dependence and heterogeneity. *Social Science Research, 42*, 120–139.

Mooi-Reci, I., & Mills, M. (2012). The Gendered Consequences of Unemployment Insurance Reforms. *Social Forces, 91*, 583–608.

Moreau, T., O'Quigley, J., & Mesbah, M. (1985). A global goodness-of-fit statistic for the proportional hazards model. *Applied Statistics, 34*, 212–218.

Mulder, C. H., & Smits, J. (1999). First-time home-ownership of couples. The effect of inter-generational transmission. *European Sociological Review, 15*, 323–337.

Müller, S., & Schneider, T. (2013). Educational pathways and dropout from higher education in Germany. *Longitudinal and Life Course Studies 4*, 218–241.

Murphy, M. (1991). *The family life cycle.* London: London School of Economics. Unpublished manuscript.

Myers, D. (1997). Racial rioting in the 1960s: an event history analysis of local conditions. *American Sociological Review, 62*, 94–112.

Myers, O., Greenberg, N., Wilson, B., & Sood, A. (2020). Factors Related to Faculty Retention in a School of Medicine: A Time to Event Analysis. *The chronicle of mentoring & coaching, 1(13)*, 334.

Namboodiri, K., & Suchindran, C. M. (1987). *Life table techniques and their applications.* New York: Academic Press.

Nazio, T., & Blossfeld, H.-P. (2003). The diffusion of cohabitation among young women in West Germany, East Germany and Italy. *European Journal of Population. 19*, 47–82.

Needleman, S., & Wunsch, C. (1970). A general method applicable to the search for similarities in the amino acid sequence of two proteins. *Journal of Molecular Biology, 48*, 443–453.

Neels, K., Marynissen, L., & Wood, J. (2024). Economic cycles and entry into parenthood: is the association changing and does it affect macro-level trends? Micro-level hazard and simulation models of Belgian fertility trends, 1960?2010. *European Journal of Population, 40(13)*.

Nelson, W. (1972). Theory and applications of hazard plotting for censored failure data. *Technometrics, 14*, 945–966.

Nieuwenhuis, J., Hooimeijer, P., & Meeus, W. (2015). Neighbourhood effects on educational attainment of adolescents. buffered by personality and educational commitment. *Social Science Research, 50*, 100–109.

Olzak, S. (1992). *The dynamics of ethnic competition and conflict*. Stanford, CA: Stanford University Press.

Olzak, S., & Olivier, J. L. (1998a). Comparative event analysis: black civil rights protest in South Africa and the United States. In D. Rucht, R. Koopmans, & F. Neidhardt (Eds.), *Acts of dissent* (pp. 253-283). Lanham, MD: Rowman & Littlefield.

Olzak, S., & Olivier, J. L. (1998b). Racial conflict and protest in South Africa and the United States. *European Sociological Review, 14*, 255–278.

Olzak, S., & Shanahan, S. (1996). Deprivation and race riots: an extension of Spilerman's analysis. *Social Forces, 74*, 931–961.

Olzak, S., & Shanahan, S. (1999). The effects of immigrant diversity and ethnic competition on collective conflict in urban America: an assessment of two movements of mass migration, 1869–1924 and 1965–1993. *Journal of American Ethnic History, 18*, 40–64.

Olzak, S., Shanahan, S., & McEneaney, E. H. (1996). Poverty, segregation, and race riots, 1960 to 1993. *American Sociological Review, 61*, 590–613.

Olzak, S., Shanahan, S., & West, E. (1994). School desegregation, interracial exposure, and antibusing activity in contemporary urban America. *American Journal of Sociology, 100*, 196–214.

Olzak, S., & West, E. (1991). Ethnic conflict and the rise and fall of ethnic newspapers. *American Sociological Review, 56*, 458–474.

Oppenheimer, V. K. (1988). A theory of marriage timing. *American Journal of Sociology, 94*, 563–591.

Ostermeier, M., & Blossfeld, H.-P. (1998). Wohneigentum und Ehescheidung. Eine Längsschnittanalyse über den Einfluß gekauften und geerbten Wohneigentums auf den Prozeß der Ehescheidung. *Zeitschrift für Bevölkerungsforschung, 23*, 39–54.

Papastefanou, G. (1987). *Familienbildung und Lebensverlauf. Eine empirische Analyse sozialstruktureller Bedingungen der Familiengründung bei den Kohorten 1929–31, 1939–41 und 1949–51*. Studien und Berichte no. 50. Max-Planck-Institut für Bildungsforschung, Berlin.

Peng, H. , Li, J., Song, Y., Yang, R., Ranjan, R., Yu, P. S., & He, L. (2021). Streaming social event detection and evolution discovery in heterogeneous information networks. *ACM Transactions on Knowledge Discovery from Data, 15(5)* , 1-33.

Petersen, T. (1986a). Estimating fully parametric hazard rate models with time- dependent covariates. *Sociological Methods and Research, 14*, 219–246.

Petersen, T. (1986b). Fitting parametric survival models with time-dependent covariates. *Applied Statistics, 35*, 281–288.

Petersen, T. (1988a). Analyzing change over time in a continuous dependent variable: specification and estimation of continuous state space hazard rate models. *Sociological Methodology, 18*, 137–164.

Petersen, T. (1988b). *Incorporating time-dependent covariates in models for analysis of duration data*. CDE working paper.

Petersen, T. (1990). Analyzing event histories. In A. von Eye (Ed.), *Statistical methods in longitudinal research* (Vol. 2, pp. 259–288). New York: Academic Press.

Petersen, T. (1991). The statistical analysis of event histories. *Sociological Methods & Research, 19*, 270–323.

Petersen, T. (1995). Models for interdependent event history data: specification and estimation. *Sociological Methodology, 20*, 317–375.

Pickles, A. R., & Davies, R. B. (1986). Household factors and discrimination in housing consumption. *Regional Science and Urban Planning, 16*, 493–517.

Pickles, A. R., & Davies, R. B. (1989). Inference from cross-sectional and longitudinal data for dynamic behavioural processes. In J. Hauer et al. (Eds.), *Urban dynamics and spatial choice behaviour* (pp. 81–104). New York: Kluwer Academic.

Piore, M. J. (1968). On-the-job training and adjustment to technological change. *Journal of Human Resources, 3*, 435–445.

Plank, S. B., DeLuca, S., & Estacion, A. (2008). High School Dropout and the Role of Career and Technical Education: a Survival Analysis of Surviving High School. *Sociology of Education, 81*, 345–370.

Pollock, G. (2007). Holistic trajectories: a study of combined employment, housing and family careers by using multiple-sequence analysis. *Journal of the Royal Statistical Society (Series A), 170*, 167–183.

Popielarz, P. A., & McPherson, J. M. (1996). On the edge or in between: niche position, niche overlap, and the duration of voluntary association memberships. *American Journal of Sociology, 101*, 698–720.

Popper Shaffer, J. (Ed.), (1992). *The role of models in nonexperimental science: two debates*. Washington, DC: American Educational Research Association.

Pötter, U. (1993). Models for interdependent decisions over time. In J. Janssen & C. H. Skiadas (Eds.), *Applied stochastic models and data analysis* (pp. 767–779). World Scientific Publishers.

Pötter, U., & Blossfeld, H.-P. (2001). Causal inference from series of events. *European Sociological Review, 17*, 21–32.

Preisendörfer, P., & Burgess, Y. (1988). Organizational dynamics and career patterns: effects of organizational expansion and contraction on promotion chances in a large West German company. *European Sociological Review, 4*, 32–45.

Raab, M., Fasang, A. E., Karhula, A., & Erola, J. (2014). Sibling similarity in family formation. *Demography, 51*, 2127–2154.

Rajulton, G. (1992). *Life history analysis: Guidelines for using the program* LIFEHIST *(PC version)*. Discussion paper no. 92-5. Population Studies Centre, London, Ontario: University of Western Ontario.

Ramirez, F. O., Soysal, Y., & Shanahan, S. (1997). The changing logic of political citizenship: cross-national acquisition of women's suffrage rights, 1890 to 1990. *American Sociological Review, 62*, 735–745.

Rasler, K. (1996). Concessions, repression, and political protest in the Iranian revolution. *American Sociological Review, 61*, 132–152.

Reisel, L., & Brekke, I. (2010). Minority Dropout in Higher Education: a Comparison of the United States and Norway Using Competing Risk Event History Analysis. *European Sociological Review, 26*, 691–712.

Rindfuss, R. R., & Hirschman, C. (1984). The timing of family formation: structural and societal factors in the Asian context. *Journal of Marriage and the Family, 55*, 205–214.

Rindfuss, R. R., & John, C. S. (1983). Social determinants of age at first birth. *Journal of Marriage and the Family, 45*, 553–565.

Rodgers, W. L. (1982). Estimable functions of age, period, and cohort effects. *American Sociological Review, 47*, 774–787.

Rogers, E. M. (1983). *Diffusion of innovations*. New York: Free Press.

Rohwer, G. (1995). *Kontingente Lebensverläufe. Soziologische und statistische Aspekte ihrer Beschreibung und Erklärung*. Bremen: University of Bremen.

Rohwer, G. (2006). Verweildauern und Übergangsraten bei mehreren Folgezuständen. In A. Diekmann (Ed.), *Methoden der Sozialforschung* (pp. 348–367). Wiesbaden: VS.

Rohwer, G. (2012). Estimating Effects with Logit Models. *NEPS Working Paper No. 10*. Bamberg: Otto-Friedrich Universitaet, Nationales Bildungspanel.

Rohwer, G. (2016). Treatment and Control Groups in a Dynamic Setting. *International Statistical Review, 84*, 63–78.

Rohwer, G., & Pötter, U. (2007). TDA User's Manual. http://www.stat.ruhr-uni-bochum.de/tda.html.

Roksa, J., & Velez, M. (2012). A Late Start: delayed Entry, Life Course Transitions and Bachelor's Degree Completion. *Social Forces, 90*, 769–794.

Rosenthal, L. (1991). Unemployment incidence following redundancy: the value of longitudinal approaches. In S. Dex (Ed.), *Life and work history analyses: qualitative and quantitative developments* (pp. 187–213). London: Routledge.

Rubin, D. B. (1980). Discussion of ?randomization analysis of experimental data in the Fisher randomization test? by Basu. *Journal of the American Statistical Association, 81*, 961?962.

Ruiter, S., & De Graaf, N. D. (2009). Socio-economic Payoffs of Voluntary Association Involvement: a Dutch Life Course Study.*European Sociological Review, 25*, 425–442.

Sacchi, S., Kriesi, I., & Buchmann, M. (2016). Occupational mobility chains and the role of job opportunities for upward, lateral and downward mobility in Switzerland. *Research in Social Stratification and Mobility, 44*, 10–21.

Sandefur, G. D., & Tuma, N. B. (1987). How data type affects conclusions about individual mobility. *Social Science Research, 16*, 301–328.

Sankoff, D., & Kruskal, J. B. (Eds.), (1983). *Time Warps, String Edits and Macromolecules*. Addison-Wesley, Reading, MA.

Schecter, A., Pilny, A., Leung, A., Poole, M. S., & Contractor, N. (2018). Step by step: Capturing the dynamics of work team process through relational event sequences. *Journal of Organizational Behavior, 39(9)*, 1163-1181.

Schels, B. (2018). Young adults? Risk of long-term benefit receipt and parents? Socioeconomic background. *Acta Sociologica, 61*, 17–33.

Schneider, B., Carnoy, M., Kilpatrick, J., Schmidt, W. H., & Shavelson, R. J. (2007). *Estimating causal effects. Using experimental and observational design.* Washington, DC: American Educational Research Association.

Schömann, K., & Becker, R. (1995). Participation in further education over the life course: a longitudinal study of three birth cohorts in the Federal Republic of Germany. *European Sociological Review, 11*, 187–208.

Settersten, R. A., Dannefer, D., Elder, G. H., Mortimer, J. T., & Kelley, J. A. (2024) Studying social change in human lives: a conversation. *Longitudinal and Life Course Studies, 15(1)* , 5?18.

Settersten, R. A., Elder, G. H., & Pearce, L. D. (2021). *Living on the Edge* . Chicago, IL: The University of Chicago Press.

Sewell, W. H., & Hauser, R. M. (1975). *Education, occupation and earnings*. New York: Academic Press.

Shadish, W. R., Cook, T. D., & Campbell, D. T. (2002). *Experimental and quasi-experimental designs for generalized causal inference.* Boston, MA: Houghton Mifflin.

Shavit, Y., & Blossfeld, H.-P. (Eds.), (1993). *Persistent inequality. Changing educational attainment in thirteen countries.* Boulder, CO: Westview Press.

Shingles, R. D. (1976). Causal inference in cross-lagged panel analysis. *Political Methodology, 3*, 95–133.

Singer, B., & Spilerman, S. (1976a). The representation of social processes by Markov models. *American Journal of Sociology, 82*, 1–54.

Singer, B., & Spilerman, S. (1976b). Some methodological issues in the analysis of longitudinal surveys. *Annals of Economic and Social Measurement, 5*, 447–474.

Singer, J. D., & Willett, J. B. (1991). Modeling the days of our lives: using survival analysis when designing and analyzing longitudinal studies of duration and the timing of events. *Psychological Bulletin, 110*, 268–290.

Singer, J. D., & Willett, J. B. (1993). It's about Time: using Discrete-Time Survival Analysis to Study Duration and the Timing of Events. *Journal of Educational Statistics, 18*, 155–195.

Skopek, J., Triventi, M., & Blossfeld, H.-P. (2022). How do institutional factors shape PhD completion rates? An analysis of long-term changes in a European doctoral program. *Studies in Higher Education, 47*, 2, 318–337.

Smeenk, W. (1998). *Opportunity and marriage.* Utrecht: Thela Thesis.

Smith, S., Maas, I., & van Tubergen, F. (2012). Irreconcilable differences? Ethnic intermarriage and divorce in the Netherlands, 1995–2008,. *Social Science Research, 41*, 1126–1137.

Snyder, L. (1991). Modeling dynamic communication processes with event history analysis. *Communication Research, 18*, 464–486.

Sørensen, A. B. (1977). The structure of inequality and the process of attainment. *American Sociological Review, 42*, 965–978.

Sørensen, A. B. (1979). A model and a metric for the analysis of the intragenerational status attainment process. *American Journal of Sociology, 85*, 361–384.

Sørensen, A. B. (1984). Interpreting time dependency in career processes. In A. Diekmann & P. Mitter (Eds.), *Stochastic modelling of social processes* (pp. 89–122). New York: Academic Press.

Sørensen, A. B. (1986). Theory and methodology in social stratification. In U. Himmelstrand (Ed.), *The sociology of structure and action* (pp. 69–95). New York.

Sørensen, A. B., & Blossfeld, H.-P. (1989). Socioeconomic opportunities in Germany in the post-war period. *Research in Social Stratification and Mobility, 8*, 85–106.

Sørensen, A. B., & Sørensen, A. (1985). An event history analysis of the process of entry into first marriage. *Current Perspectives on Aging and Life Cycle, 2*, 53–71.

Sørensen, A. B., & Tuma, N. B. (1981). Labor market structures and job mobility. *Research in Social Stratification and Mobility, 1*, 67–94.

Soule, S. A. (1997). The student divestment movement in the United States and tactical diffusion: the Shantytown protest. *Social Forces, 75*, 855–882.

Soule, S. A., & Van Dyke, N. (1999). Black church arson in the United States, 1989–1996. *Ethnic and Racial Studies, 22*, 724–742.

Soule, S. A., & Zylan, Y. (1997). Runaway train? The diffusion of state-level reform in ADC/AFDC eligibility requirements, 1950–1967. *American Journal of Sociology, 103*, 733–762.

Spence, A. M. (1973). Job market signaling. *Quarterly Journal of Economics, 87*, 355–374.

Spence, A. M. (1974). *Market signaling*. Cambridge, MA: Harvard University Press.

StataCorp (2017a). *Stata Longitudinal-Data/ Panel-Data Reference Manual. Release 15.*. College Station, TX: StataCorp LLC.

StataCorp (2017b). *Stata Multilevel Mixed-Effects Reference Manual. Release 15.*. College Station, TX: StataCorp LLC.

Steele, F. (2008). Multilevel Models for Longitudinal Data. *Journal of the Royal Statistical Society, Series A*, 5–19.

Strang, D. (1991). Adding social structure to diffusion models. *Sociological Methods & Research, 19*, 324–353.

Strang, D., & Meyer, J. W. (1993). Institutional conditions for diffusion. *Theory and Society, 22*, 487–511.

Strang, D., & Tuma, N. B. (1993). Spatial and temporal heterogeneity in diffusion. *American Journal of Sociology, 99*, 614–639.

Studer, M. (2022). Sequence analysis: Its past, present, and future. *Social Science Research, 107*, 102772.

Studer, M., & Ritschard, G. (2015). What matters in differences between life trajectories: a comparative review of sequence dissimilarity measures. *Journal of the Royal Statistical Society (Series A, Statistics in Society).*

Sudman, S., & Bradburn, N. M. (1986). *Asking questions*. San Francisco: Jossey-Bass.

Surdu, I., Mellahi, K., Glaister, K. W., & Nardella, G. (2018). Why wait? Organizational learning, institutional quality and the speed of foreign market re-entry after initial entry and exit. *Journal of World Business, 53(6)*, 911-929.

Tach, L., & Edin, K. (2013). The Compositional and Institutional Sources of Union Dissolution for Married and Unmarried Parents in the United States. *Demography, 50*, 1789–1818.

Tarone, R. E., & Ware, J. (1977). On distribution-free tests for equality of survival distributions. *Biometrika, 64*, 156–160.

Teachman, J. D. (1983). Analyzing social processes: life tables and proportional hazards models. *Social Science Research, 12*, 263–301.

Thorsen, M. L. (2017). The adolescent family environment and cohabitation across the transition to adulthood. *Social Science Research, 64*, 249–262.

Timm, A., Blossfeld, H.-P., & Müller, R. (1998). Der Einfluß des Bildungssystems auf die Heiratsmuster in Westdeutschland und den USA. Eine vergleichende Längsschnittanalyse der Wahl des ersten Ehepartners im Lebenslauf. *Beiträge zur Arbeitsmarkt- und Berufsforschung* (BeitrAB 215), 129–166.

Treiman, D. J. (1977). *Occupational prestige in comparative perspective*. New York: Academic Press.

Trussell, J., & Richards, T. (1985). Correcting for unmeasured heterogeneity in hazard models using the Heckman-Singer procedure. *Sociological Methodology, 15*, 242–276.

Tuma, N. B. (1980). *Invoking RATE*. Mannheim: ZUMA.

Tuma, N. B. (1985). Effects of labor market structure on job-shift patterns. In J. J. Heckman & B. Singer (Eds.), *Longitudinal analysis of labor market data* (pp. 327–363). Cambridge: Cambridge University Press.

Tuma, N. B., & Hannan, M. T. (1984). *Social dynamics. Models and methods*. New York: Academic Press.

Turkson, A. J., Ayiah-Mensah, F., & Nimoh, V. (2021). Handling censoring and censored data in survival analysis: a standalone systematic literature review. *International journal of mathematics and mathematical sciences, 1*, 9307475.

van Damme, M., & Kalmijn, M. (2014). The dynamic relationships between union dissolution and women's employment: a life-history analysis of 16 countries. *Social Science Research, 48*, 261–278.

van den Hout, A., Chan, M. S., & Matthews, F. (2019). Estimation of life expectancies using continuous-time multi-state models. *Computer methods and programs in biomedicine, 178*, 11-18.

Vaupel, J. A., & Yashin, A. I. (1985). The deviant dynamics of death in heterogeneous populations. *Sociological Methodology, 16*, 179–211.

Vermunt, J. K.(1997). *Log-linear models for event histories*. Newbury Park, CA: Sage.

Visser, M., Gesthuizen, M., Kraaykamp, G., & Wolbers, M. H. J. (2016). Inequality among Older Workers in the Netherlands: a Life Course and Social Stratification Perspective on Early Retirement. *European Sociological Review, 32*, 370–382.

Wagner, M. (1989a). *Räumliche Mobilität im Lebenslauf*. Stuttgart: Enke.

Wagner, M. (1989b). Spatial determinants of social mobility. In J. van Dijk, H. Folmer, & A. M. Herzog (Eds.), *Migration and labor market adjustment* (pp. 241–264). Dordrecht: Springer.

Wagner, M. (1990). Education and migration. In K. U. Mayer & N. B. Tuma (Eds.), *Event history analysis in life course research* (pp. 129–145). Madison: University of Wisconsin Press.

Wagner, M., Schmid, L., & Weiß, B. (2015). Exploring Increasing Divorce Rates in West Germany: can We Explain the Iron Law of Increasing Marriage Instability? *European Sociological Review, 31*, 211–229.

Walder, A. G., & Lu, Q. (2017). The Dynamics of Collapse in an Authoritarian Regime: China in 1967. *American Journal of Sociology, 122*, 1144–1182.

Wang, W., & Zaïane, O. R. (2002). Clustering web sessions by sequence alignment. In *Proceedings, 13th International Workshop on Database and Expert Systems Applications,* (pp. 394–398). IEEE, IEEE.

Weber, M. (1972). *Wirtschaft und Gesellschaft*. Studienausgabe. Tübingen: Mohr.

Wegener, B. (1985). Gibt es Sozialprestige? *Zeitschrift für Soziologie, 14*, 209–235.

Weymann, A. (1995). Modernization, generational relations and the economy of life-time. *International Journal of Sociology and Social Policy, 16*, 37–57.

Weymann, A., & Heinz, W. R. (1996). *Society and biography*. Weinheim: DSV.

Wiggins, R. D., Erzberger, C., Hyde, M., Higgs, P., & Blane, D. (2007). Optimal matching analysis using ideal types to describe the lifecourse: an illustration of how histories of work, partnerships and housing relate to quality of life in early old age. *International Journal of Social Research Methodology, 10*, 259–278.

Willett, J. B., & Singer, J. D. (1991). From whether to when: new methods for studying student dropout and teacher attrition. *Review of Educational Research, 61*, 4.

Wilson, G., & Maume, D. (2014). Men's mobility into management from blue collar and white collar jobs: race differences across the early work-career. *Social Science Research, 46*, 117–129.

Wilson, G., & Roscigno, V. J. (2010). Race and downward mobility from privileged occupations: African American/White dynamics across the early work-career. *Social Science Research, 39*, 67–77.

Wimmer, A., & Feinstein, Y. (2010). The Rise of the Nation-State across the World, 1816 to 2001. *American Sociological Review, 75*, 764–790.

Windzio, M. (2008). Organizational ecology of immigrant employment and organizational buffer zones: who leaves first when the organization gets into trouble? *Social Science Research, 37*, 1171–1187.

Woolley, J.L., & MacGregor, N. (2022). The influence of incubator and accelerator participation on nanotechnology venture success. *Entrepreneurship Theory and Practice, 46(6)*, 1717-1755.

Wu, L. L. (1989). Issues in smoothing empirical hazard rates. *Sociological Methodology, 19*, 127–159.

Wu, L. L. (1990). Simple graphical goodness-of-fit tests for hazard rate models. In K. U. Mayer & N. B. Tuma (Eds.), *Event history analysis in life course research* (pp. 184–199). Madison: University of Wisconsin Press.

Wu, L., & Martinson, B. C. (1993). Family structure and the risk of premarital birth. *American Sociological Review, 58*, 210–232.

Wu, L. L., & Tuma, N. B. (1990). Local hazard models. *Sociological Methodology, 20*, 141–180.

Xie, Y. (2007). Otis Dudley Duncan's legacy: the demographic approach to quantitative reasoning in social science. *Research in Social Stratification and Mobility, 25*, 141–156.

Yamaguchi, K. (1991). *Event history analysis*. Newbury Park, CA: Sage.

Yamaguchi, K. (1994). Some accelerated failure-time regression models derived from diffusion process models: an application to a network diffusion analysis. In P. V. Marsden (Ed.), *Sociological Methodology, 24*, 267–300.

Yamaguchi, K. (1996). Some loglinear fixed-effect latent-trait Markov-chain models: a dynamic analysis of personal efficacy under the influence of divorce / widowhood. *Sociological Methodology, 26*, 39–78.

Yamaguchi, K. (1997). Some cumulative logit models for panel-data analysis: an application to a dynamic analysis of personal efficacy. *Quality and Quantity, 31*, 287–304.

Yamaguchi, K. (1998). Mover-stayer models for analyzing event nonoccurrence and event timing with time-dependent covariates: an application to an analysis of remarriage. *Sociological Methodology, 28*, 327–361.

Yamaguchi, K., & Ferguson,L. (1995). The stopping and spacing of childbirths and their birth-history predictors: rational-choice theory and event-history analysis. *American Sociological Review, 60*, 272–298.

Yamaguchi, K., & Jin, L. (1999). Event history analysis in human developmental research. In R. K. Sibereisen & A. von Eye (Eds.), *Growing up in times of social change* (pp. 319–339). Berlin: Walter de Gruyter.

Yu, J., Xie, Y., & Xie, Y. (2015). Changes in the Determinants of Marriage Entry in Post-Reform Urban China. *Demography, 52*, 1869–1892.

Zielonka, M., & Pelz, S. (2015). *NEPS Technical Report: Implementation of the ISCED-97, CASMIN and Years of Education Classification Schemes in the SUF Starting Cohort 6.* Bamberg: Leibniz Institute for Educational Trajectories (LIfBi).

Zwick, M. (Ed.). (1998). *Einmal arm, immer arm?* Frankfurt: Campus.

About the Authors

Hans-Peter Blossfeld, Prof., Dr. rer. pol. Dr. h. c., has been Emeritus of Excellence at the Graduate Centre Trimberg Research Academy (TRAc) at the University of Bamberg in Germany since April 2020. 2002-2020, he held the Chair of Sociology I at the Faculty of Social Sciences, Economics and Business Administration at the University of Bamberg and was Professor of Sociology at the European University Institute in Florence, Italy. Born in Munich (Germany) in 1954, he studied Sociology, Economics, Social Statistics, and Computer Science at the University of Regensburg, Germany. He received his Doctorate in Economics from the University of Mannheim (Dr. rer. pol., 1984), Germany, and his Habilitation in Sociology (and Venia Legendi, 1987) from the Freie Universität Berlin, Germany. Hans-Peter Blossfeld first worked as a Research Scientist at the University of Mannheim, Germany (1980–1984) and then as a Senior Research Scientist at the Max Planck Institute for Human Development and Education in Berlin, Germany (1984–1992). In 1988–1989, he was a Research Fellow at the Netherlands Institute for the Advanced Study in the Humanities and Social Sciences (NIAS) in Wassenaar, the Netherlands. 1989–1992, he held a Chair in Sociology at the Department of Political and Social Sciences of the European University Institute (EUI) in Florence, Italy. He then moved back to Germany to a Chair for Social Statistics and Sociological Research Methods at the University of Bremen (1992–1998) and a Chair for Theory and Empirical Analysis of Social Structures and Economic Systems at Bielefeld University (1998–2002). 2012–2017, he was on leave from his Chair of Sociology at the University of Bamberg and returned to the EUI to take up a Chair of Sociology at the Department of Political and Social Sciences in Florence, Italy. Over the last 25 years, Blossfeld has directed several large-scale social science projects, including the European Research Council (ERC) funded project "Education as a Lifelong Process – Comparing Educational Trajectories in Modern Societies" (eduLIFE); the National Educational Panel Study (NEPS), funded by the German Federal Ministry of Education and Research, the cross-national comparative project Globalife funded by the Volkswagen Foundation, the TransEurope project funded by the European Science Foundation (ESF) and the Priority Research Programme 1646 funded by the German Research Foundation (DFG).

Blossfeld was a Member of the Steering Committee of the ESF Networks "Quantitative Methods in the Social Sciences 1 and 2" (QMSS1 and QMSS2), and he worked in two DFG Collaborative Research Centers (Sonderforschungsbereiche), namely SFB 3 at the Universities of Frankfurt a. M. and Mannheim, Germany, and SFB 186 at University of Bremen, Germany. Finally, Blossfeld's research has been supported by several grants from the DFG, the ESF, the German Israeli Foundation (GIF), the Humboldt Foundation (AvH), the National Science Foundation (NSF) in the United States, and the European Commission in Brussels (DG5 and DG12). Blossfeld is an elected member of the German National Academy of Sciences Leopoldina in Germany, the Berlin-Brandenburg Academy of Sciences and Humanities (BBAW), the Bavarian Academy of Sciences and Humanities (BAdW) and the European Academy of Sociology. 2004–2009, he was the Chairman of the European Consortium of Sociological Research (ECSR) and 2014–2016, he was President of the Society for Longitudinal and Lifecourse Studies (SLLS). In 2010, Blossfeld was awarded an Honorary Doctorate from the University of Tallinn, Estonia, and in 2012, he received the "Honorary Pin on Ribbon" from the Bulgarian Academy of Sciences in Sofia. 2012–2024, he was IAB Research Fellow of the Institute for Employment Research (IAB) in Nuremberg, Germany, and ROA Research Fellow of the Research Centre for Education and the Labour Market (ROA) at Maastricht University, the Netherlands. He is a Senior Editor of the *Zeitschrift für Erziehungswissenschaft* (ZfE) (Journal for Educational Research). He was a member of the editorial boad of the journal *Studies of Transition States and Societies (STSS)*, the journal *Društvena istraživanja (Journal for General Social Studies)*, the *Journal of Sociology and Social Anthropology* (JSSA), and the Portuguese journal *Revista Sociologia.* 1990–2012, Blossfeld was Editor-in-Chief of the *European Sociological Review* (ESR); 2003–2012, he served as Editor-in-Chief of the *Zeitschrift für Familienforschung* (ZfF) (Journal of Family Research); and 2005–2013, he was Co-Editor of *International Sociology* (IS). Over a period of more than 10 years, he also served on the scientific advisory board of the *Kölner Zeitschrift für Soziologie und Sozialpsychologie* (KZfSS) and the *Zeitschrift für Soziologie* (ZfS). Blossfeld has published 48 books and over 350 articles on life-course research, social inequality, youth, family, educational sociology, labor market research, demography, social stratification and mobility, advanced methods of quantitative social research, and statistical methods for longitudinal data analysis. These have been cited more than 41,000 times (H-Index = 86; I10-Index = 305; see Google Scholar, September 2024).

Götz Rohwer was Professor Emeritus of Methods of Social Research and Statistics at the Ruhr-Universität Bochum. Previously, he was a research fellow at the European University Institute, Florence, and then worked together with Hans-Peter Blossfeld at the University of Bremen. Together with Ulrich Pötter, he is the author of the program TDA (Transition Data Analysis) that is used in Blossfeld and Rohwer (2002). He passed away in March 2021.

Gwendolin J. Blossfeld is Postdoc at the Faculty of Social Sciences, Economics and Business Administration at the University of Bamberg. Her research interests are in the areas of longitudinal analysis methods, life course research, longitudinal data collection, causal analysis, sociology of education, labour market research, social inequality, demography, migration and gender studies. She has published five books and several peer-reviewed journals. In particular, she is the author of the companion book to this volume *Comparison Between Continuous-Time and Discrete-Time Event History Analysis Using Stata* published at Routledge in 2025.

Brendan Halpin, the author of chapter 11, is a associate professor in sociology at the University of Limerick. After a DPhil in sociology at Nuffield College, Oxford, he worked on the British Household Panel Survey at the Institute for Social and Economic Research at the University of Essex, before joining University of Limerick in 2000. His research interests include stratification and social mobility, family formation and dissolution, and labor market processes – primarily from a longitudinal and dynamic perspective. He is also actively interested in methods for analyzing longitudinal data, and he has written Stata packages for multiple imputation of categorical time series such as life-course data and the SADI package for sequence analysis.

Index

FSC
www.fsc.org
MIX
Papier | Fördert gute Waldnutzung
FSC® C083411

Kolibri 360 GmbH
Ferdinand-Jühlke-Straße 7
99095 Erfurt, Deutschland
produktsicherheit@kolibri360.de